Rethinking Theories and Practices of Imaging

Also by Timothy H. Engström

HEALTH CARE REFORM: Ethics and Politics
(*with Wade Robinson*)

Also by Evan Selinger

CHASING TECHNOSCIENCE: Matrix for Materiality
(*with Don Ihde*)

POSTPHENOMENOLOGY: A Critical Companion to Ihde

THE PHILOSOPHY OF EXPERTISE
(*with Robert Crease*)

FIVE QUESTIONS: Philosophy of Technology
(*with Jan Olsen*)

NEW WAVES IN PHILOSOPHY OF TECHNOLOGY
(*with Jan Olsen and Søren Riis*)

Rethinking Theories and Practices of Imaging

Edited by

Timothy H. Engström

and

Evan Selinger
Rochester Institute of Technology, USA

palgrave
macmillan

First published 2009 by
PALGRAVE MACMILLAN

Palgrave Macmillan in the UK is an imprint of Macmillan Publishers Limited, registered in England, company number 785998, of Houndmills, Basingstoke, Hampshire RG21 6XS.

Palgrave Macmillan in the US is a division of St Martin's Press LLC, 175 Fifth Avenue, New York, NY 10010.

Palgrave Macmillan is the global academic imprint of the above companies and has companies and representatives throughout the world.

Palgrave® and Macmillan® are registered trademarks in the United States, the United Kingdom, Europe and other countries.

ISBN-13: 978-0-230-58065-7 hardback

This book is printed on paper suitable for recycling and made from fully managed and sustained forest sources. Logging, pulping and manufacturing processes are expected to conform to the environmental regulations of the country of origin.

A catalogue record for this book is available from the British Library.

A catalog record for this book is available from the Library of Congress.

10 9 8 7 6 5 4 3 2 1
18 17 16 15 14 13 12 11 10 09

Printed and bound in Great Britain by
CPI Antony Rowe, Chippenham and Eastbourne

To

Lara Helena Kuhn and Noreen Selinger

Contents

Figures

Tables

Acknowledgments

We gratefully acknowledge Dean Glenn Kist, Provost Stanley McKenzie, and the College of Liberal Arts Research Fund for their generous support of the project. We wish especially to thank Peggy Noll for her help with the preparation of the final manuscript and Howard Vogl for his invaluable assistance with the digital imaging of the book.

List of Contributors

Chris Burnett is an associate professor in the Department of Art at the University of Toledo, Toledo, OH.

Lisa Cartwright is an associate professor of Communications at the University of California at San Diego, San Diego, CA.

Mildred Cho is the Associate Director of the Stanford Center for Biomedical Ethics, Stanford, CA.

Andy Clark is Professor of Logic and Metaphysics at Edinburgh University, Edinburgh, Scotland.

Alfred Crosby is Professor Emeritus of American Studies, History, and Geography at the University of Texas, Austin, TX.

Raymond De Vries is Associate Professor in the Bioethics Program in the School of Medicine at the University of Michigan, Ann Arbor, MI.

Megan Delehanty is an assistant professor of Philosophy at the University of Calgary, Calgary, Canada.

Timothy H. Engström is a professor of Philosophy at Rochester Institute of Technology, Rochester, NY.

Andrew Feenberg is the Canadian Research Chair in Philosophy of Technology in the School of Communication of Simon Fraser University, Burnaby, Canada.

Brian Goldfarb is an associate professor of Communications at the University of California at San Diego, San Diego, CA.

Patrick Grim is Distinguished Teaching Professor of Philosophy at Stony Brook University, Stony Brook, NY.

Don Ihde is Distinguished Professor of Philosophy at Stony Brook University, Stony Brook, NY.

Judy Illes is Professor of Neurology and Canada Research Chair in Neuroethics National Core for Neuroethics University of British Columbia, Vancouver, Canada.

Thomas Keenan is the Director of the Human Rights Project and Associate Professor of Comparative Literature at Bard College, Annandale-on-Hudson, NY.

Simon Malpas is a lecturer at the Department of English Literature at Edinburgh University, Edinburgh, Scotland.

Alva Noë is an associate professor of Philosophy at the University of California at Berkeley, Berkeley, CA.

Cyril Reade is an assistant professor of Fine Arts at Rochester Institute of Technology, Rochester, NY.

Pamela Schraedley-Desmond is a biostatistician and is on the advisory board of the International Foundation for Research and Education on Depression, Stanford, CA.

Evan Selinger is an associate professor of Philosophy at Rochester Institute of Technology, Rochester, NY.

Vivian Sobchack is a professor of Film and Television Studies at the UCLA School of Theater, Film and Television, Los Angeles, CA.

Introduction

Timothy H. Engström and Evan Selinger

General introduction

This book brings together a range of perspectives and disciplines, theories and practices, for investigating the nature and scope, the consequences and implications of the imaging revolution. The sciences and the humanities, the disciplinary specialist and the lay-citizen alike, are challenged as well as assisted by this revolution, but we are in need of collaborative and cross-disciplinary analysis if we are to understand its full scope and significance. This book is an effort to provide this kind of analysis. It focuses on the impact that theories and practices of imaging have had and are having on inquiry and understanding in a variety of fields. From general concerns about "human nature" and "vision" in relation to imaging technologies to the specific concerns of specific fields of inquiry, this book brings together a range of essays that further our understanding of what the visual is and means in the contemporary.

Has digital imaging ushered in a new range of visual practices and, with them, some implications and consequences that prior technological and visual revolutions could not have propelled? The cumulative force of these new technologies and practices of imaging suggests that the answer is "yes" and that some very large issues are at play: ontological and cognitive, epistemological and aesthetic, political and economic. Not only have new practices emerged, and some new theoretical vocabulary for understanding these practices with them, but these practices and the technologies and artifacts involved have also turned back on us. They have become, in symbiotic and cyborgian ways, part of us, extensions and alterations of who we have been. They thus have challenged some long-standing conceptions of what it is to perceive, to know, and to interact with the "real."

But what kinds of constructions or translations of the "real" does the digital entail? What kinds of cognitive processes and orientations are demanded and/or permitted in relation to digital imaging? How are we situated and/or transformed as embodied social beings in the context of imaging technologies and practices? What forms of reception and/or agency do they induce? What forms of knowing and being, of understanding and intervening in states of affairs, are propelled through contemporary imaging practices and through the commercially and politically shaped contexts in which these practices emerge? In investigating such issues and questions, one will find a range of perspectives employed here–historicist, analytic, pragmatic, phenomenological, empirical, rhetorical, aesthetic, and critical. In their theoretical and practical range, the cumulative point is that the diversity of imaging practices at play in the present suggests that a major transformation is afoot; and this book captures the primary contours of this transformation.

There is little doubt that it is through a sustained investment in modalities of sight and in an accompanying range of technologically mediated forms of seeing that we have hoped most fervently to maintain the presence and availability of the "real," to know and to successfully interact with what we know. The combination of computational and digital development, computer graphic and algorithmic programming, microchip miniaturization and implantation, 3-D display and computational translations of the real, etc., have together fostered a penetration, extension, and transformation of sight that is categorically and quantitatively beyond anything imaginable in their collective absence. The question is: do we have any philosophical or other consensus in terms of which such technological and social trajectories can be evaluated? Visual production is increasingly in the foreground of our cognitive activity. Digitally based visual technologies are increasingly the ones selected for enhancing our own perceptual and cognitive capacities, for representing and organizing what we know. They are the ones with which and on whose terms we interact and, thus, desire to interact. The essays collected here represent an effort to engage comprehensively with the nature of "seeing" in the context of imaging, as these forces have been historically shaped and as they contribute to forming a rapidly emerging future.

Part I Reinventing Sight

In "Reinventing Sight: Theories and Practices of Imaging," Timothy Engström and Evan Selinger establish both an overarching philosophical framework for thinking about sight and a compelling rationale for why a book concerned with theories and practices of imaging is needed. Having taken "the linguistic turn," philosophy has been well prepared to think about the constructive role language plays in realizing any philosophical ambition. Philosophy, they argue, should also make an "imaging turn"; that is, it needs to investigate the role that theories and practices of imaging play in shaping what is meant by "sight" and what we ask of "sight" in organizing our understanding of the contemporary world.

The digital era has not undermined naturalistic appeals to visual perception or to sight's relation to the real. It has, however, provided paradigmatically new means of modeling visual perception, and it has led to new kinds of non-naturalistic, non-analog imaging for inventing worlds of sight that are well beyond the thresholds of naturalistic and realist categories as traditionally understood.

In this first essay of the volume Engström and Selinger provide a sense of what is at stake in this new imaging era, and a sense of how philosophy should respond. In effect, sight has always been of fundamental interest to a range of philosophical endeavors—in ontology (especially regarding accounts of human nature), in epistemology, in aesthetics and the philosophy of art, in social and political theory, and in the philosophy of technology. The questions Engström and Selinger pursue have to do with how these overlapping endeavors are challenged by and might be adjusted as a result of new theories and practices of imaging. In doing so, the authors survey a wide range of theoretical resources—historical and contemporary, analytic and continental—and a wide range of issues—epistemological, aesthetic, and ethical—in order to provide a comprehensive sense of how "sight" has been and should be reconceived in the present.

Part II Perception and Imaging

This section concerns how imaging has itself shaped the kinds of visual perception and reflecting about perception we do. The material, technological, and aesthetic properties of imaging have not been a direct response to or the causal extension of raw perception but, on the contrary, they are forces that have shaped our perceptual capacities and practices themselves. The three contributions to this section develop this general line of analysis from different but related orientations.

In "Visual Apprehension and the European Renaissance," Alfred Crosby situates the current acceleration of modern societies toward ever-greater use of visual forms of representation in the Renaissance. Leonardo da Vinci, he writes, "declared that 'the eye' was the master of navigation, architecture, astronomy, and mathematics, as well as the creator of perspective in painting." This "mastery," however, required inventing a form of "accuracy"—in effect, perspective—by which two-dimensional depiction and representation of three-dimensional space could proceed, a form that emerged in connection with a new "obsession with optics and geometry."

But as rigorous and quantifiable rules for depicting space became available through, for example, the rediscovery of Ptolemy's 1300-year old *Geographia*, the group most quickly and profoundly affected were painters. And Crosby lays out a compelling history—via Alberti and others—of the techniques and technologies that were at play behind such painters; and he explains the enormous impact that such painting has had on the role that visualization and imaging play in the present. He provides a rich array of examples that illustrate that the artists of the Renaissance *avant-garde* were also often architects, engineers, artisans, and mathematicians as well as painters, and that they were thus not simply making compelling images. They were rigorously concerned with space-as-geometry: they were the beginning of a long and sustained trajectory linking imaging and quantification, to capture "realistically" in two-dimensional form the reality of three-dimensional visual experience. But equally important, these invented forms then provided the patterns by which subsequent seeing and "imitation of the real"—the "copying of nature"—could take place.

In explaining this history, Crosby details the ways in which Renaissance "realism" is constructed, not to follow nature but to follow specific and, to some degree, arbitrary conventions of depiction. Their virtue being not the "truth," therefore, of the picturing; rather, Renaissance picture making "may be the only art ever to have led to the creation of a kind of mathematics," in the work of Girard Desargues, Blaise Pascal, and Piero delle Francesca, for example. The moral of Crosby's discussion, particularly of Piero's *Flagellation*, is that Piero's work could be seen as both a symbolic guide to ultimate reality *and* a secular blueprint for cutting the "carpet and wallpaper for the whole scene." His analysis of Renaissance painting provides us, in other words, with an understanding of how science and aesthetics, quantification and imagination were combined to give us an enduring sense of visual confidence, not only that imaging could be exact but that we could create perceptions of reality that kept the "real" within the parameters of our own constructive activity.

In "Experience and Experiment in Art," Alva Noë is also concerned with what an investigation of art can provide for our understanding of perception. Art can furnish us "with the opportunity to have a special kind of reflective experience," one that can contribute to how contemporary philosophy and cognitive science understand perceptual consciousness. Against the background of "the problem of the transparency of experience"—a perennial problem for philosophy, for art, and for cognitive science—Noë fleshes out a conception of experience understood as a mode of interactive engagement with an environment. In effect, vision is not spectation *of* the world but embodied interaction in and with the world. He illustrates this philosophical conception in the context of a discussion of the work of sculptors Richard Serra and Tony Smith.

The "transparency of perceptual experience" is especially difficult to grasp when it comes to sight. That is, when we attempt to make sight itself the object of our reflection, we tend to see through it (so to speak) to the things seen. Thus the philosophy of visual experience has, from Hume and Kant forward, suffered from a conundrum of its own making. While the description and analysis of sight might concentrate on its constitutive parts, the more this becomes our concentration—on blobs (now pixels) of color or on the shapes or forms that contain them or on the purported "inner representations" and "pictures" of the conscious subject looking at objects—the more we falsify our experience. We do not see blobs or pixels or "inner representations" but, rather, the experienced and seen world. In this sense the experience is transparent.

The "puzzle" of the transparency of visual experience, however, is the result of thinking of experiences in terms of "inner pictures and from thinking of reflection on experience as like turning one's gaze inward to those pictures" in the first place. Rather, Noë explains, visual experience is "a temporally extended pattern of exploratory activity." He considers the commonly shared view—that visual consciousness requires the construction of an internal representation—as "an unnecessary shuffle," one that is "phenomenologically wrong-headed." As embodied animals our perceptual experience is a "temporally extended process of exploration of the environment," not a static collection of details stored in a picture in the head. Noë's additional contribution to this line of thought is to propose how the study of some works of art can illustrate this kind of exploration.

What Noë adds to the work of other critics—Rosalyn Kraus, Yves-Alain Bois, Hal Foster—is the effort to coordinate aesthetic analysis of Serra's work in relation to philosophy and cognitive science. That is, our experience of his work illustrates some of the ways in which environmental interaction and visual perception-reflection can be understood. Serra's sculpture does not present an object to be viewed at a glance but an environmental relation to be simultaneously explored and reflected upon. Analogously, Smith's work is an engagement with an environment, but by being *about* geometrical construction, it is less about the psychology of perception than about "universal" and abstract relations, less about what eyes and brains do than about what reflective perceivers do. Noë argues, in effect, that by investigating these different orientations toward two sculptors we are given methods for the study of visual/conscious experience of significance to the philosophy of mind, one that focuses not on a method of introspection but, rather, on acts of "attentiveness to what one does in exploring the world." Certain kinds of art promote this kind of attention and reflection.

In "Phenomenology and Imaging: Incorporating the Material," Don Ihde provides us with a trajectory both of how technologies of sight have developed since the invention of the *camera obscura* and of how they have become active or "constructive counterparts" to our own embodiment in shaping the production of scientific, image-mediated knowledge. In effect, Ihde offers a form of phenomenological analysis for examining the dominant ways in which technological, embodied, and epistemic practices have been interwoven since the Renaissance, but especially regarding their cumulative effect on modern and postmodern, imaging-based science.

Taking his initial cue from Edmund Husserl's effort to overcome René Descartes' legacy—his hope to "replace the notion of knowledge-by-representation" with knowledge of "the things themselves"—Ihde illustrates Descartes' own dependence upon an optical tool of the seventeenth century.

The *camera obscura*, from Descartes forward, became an "epistemology engine," often hidden but implicit, by which modernity modeled the ideal observer: someone able to see both inside the box and observe from a distance the correspondence between the image and the imaged, as if detached and disembodied in the process.

It is Maurice Merleau-Ponty and Martin Heidegger, however, who more than Husserl shape the kind of philosophy of technology that, according to Ihde, will best illuminate how technology becomes incorporated into our own embodiment and more intimately mediates perception, cognition, and the production of knowledge. Contra Heidegger, however, Ihde develops a more pragmatic and empirical (less metaphysical and dystopian) line of inquiry, putting a revised phenomenology to work to investigate the implications of more recent imaging technologies. Ihde explores several examples to demonstrate not simply that science operates by making observations through instruments, but that the instruments must have the capacity to transform data that are beyond the threshold of human sense into the humanly visible. In the absence of this kind of instrumental transformation and mediation, certain kinds of evidence or knowledge would not be available at all. Moreover, because this evidence must be translated into the humanly visible, it is knowledge that is firmly situated within a lifeworld.

By connecting the *camera obscura* (along with other imaging technologies and thinkers about these technologies) to contemporary active-source technologies such as lasers, Ihde provides a succinct history of variations on a theme: imaging technologies have been "co-constructors" of what counts as knowledge and/or the knowable in the first place. His primary point is that our understanding of this knowledge—of its "instrumentally real" nature—is enhanced by our sensitivity to the interrelationship between its materiality and our own embodiment as we engage in forms of inquiry and practice that are fully within the lifeworld.

Part III Epistemology, Identity, and Imaging

This section focuses, first, on how images can be used to advance certain kinds of knowledge claims, and how new imaging practices can enhance our capacity to represent and engage problems that are not traditionally understood in visualized terms. In addition, since not only knowledge claims but human cognition is increasingly interwoven with the work of imaging devices, the essays in this section consider their implications for characterizing the human mind and human nature more generally. Since new technologies and techniques, prosthetic and otherwise, are being used to enhance, transform, and/or correct visual impairments and address perceptual limitations, for example, we are challenged to reconsider how our experiences in this regard relate to conceptions of subjectivity and identity.

Patrick Grim's chapter, "Concrete Images for Abstract Questions: A Philosophical View," begins with the following interrogative: "What are imaging technologies for?" Grim first considers the kinds of images available through cameras, microscopes, and telescopes that allow viewers to amplify or preserve the visible. He then turns to the example of thermal imaging, which through techniques of "false color" allow information regarding something which is not itself visible to be encoded and managed in terms of the visible. Grim argues that similar techniques for making something visible can be applied to abstract, nonphysical information: one can encode abstract properties and relations; one can provide concrete images for dealing with abstract questions.

By appropriating available forms of representation in order to construct concrete images designed to address specific abstract problems, Grim demonstrates that investigators are able successfully to reframe how they examine those problems. Investigations of this type capitalize on the inherent pattern recognition capacity of our visual systems by depicting the items under consideration as visible and enduring presences. These presences reveal complex information at a glance, allow for its capture and retention, facilitate more effective communication among investigators, and can inspire new ways to analyze the information at issue.

The use of concrete images for abstract questions has long been applied to aspects of math and science, but Grim contends that it also is possible to put them to work in philosophy. In this context, Grim presents an overview of research conducted by the *Group for Logic and Formal Semantics*. This interdisciplinary research team specializes in using computational modeling tools to address both traditional and contemporary philosophical problems. On the traditional side, the group has developed new tools and techniques of visual representation to advance our understanding of the phenomena of logical and semantic paradox. On the contemporary side, the group has used spatialized versions of the iterated Prisoner's Dilemma to shed new light on the problem of prejudice reduction—a problem of importance for psychologists and policymakers as well as philosophers. In showing how computational tools can reframe these problems and open them up to a new direction of study, Grim articulates a new approach to philosophical work. In addition to reflection about the social implications of technology, philosophers can take advantage of advances in computational technology in attempting to address long-standing philosophical issues.

Megan Delehanty's chapter, "The Epistemic Status of Functional Brain Images," contributes to our understanding of the processes by which we make sense of and/or misconstrue medical images that are generated by complex mathematical and statistical procedures. For example, rather than depicting brain activity in the form of an analog photographic snapshot that bears an isomorphic resemblance to its referent, new technological approaches to brain imaging, ranging from positron emission tomography (PET) to magnetic resonance imaging (MRI) and functional magnetic resonance imaging (fMRI), center on digital objects. In this context, the causal processes that translate the needed information (which is not available to the naked eye) into a format that physicians, medical technicians, and scientists can find useful includes inputs from a variety of algorithms embedded in software. This translation process is especially complicated, and the "hybrid epistemic objects" at issue—objects that are properly thought of as borrowing from mathematical and photographic conventions, while not being reducible to either—pose a challenge to traditional empiricist assumptions about truth, falsity, and warranted inference. Constructed brain images are "layered": through different formulas for solving different problems, they incorporate non-optical data, accommodate "noise," and synthesize information about brain activity into the form of a stable and idealized object, the "average human brain." Commenting on this dialectic between constraint and revelation, Delehanty notes: "Each layer introduces, to a greater or lesser degree, assumptions about the nature of the data and imposes constraints on how such images can be interpreted. Yet without this extensive process of mathematization, there could be no access to any but extremely partial representations of human brain function."

By opening up the black box of brain imagery, Delehanty's contribution demonstrates that technological resources can eclipse the explanatory and justificatory capacities of traditional philosophy. In this respect, philosophers are challenged to acquire a greater understanding of empirical practice and to generate new concepts that are up to the task of critically addressing shifts in material culture. Additionally, Delehanty's contribution serves as a medical cautionary tale. Patients and physicians alike can benefit from considering her suggestion that the gap between lay and expert understanding of technology can lead to confusion, if not undue deference to visually depicted data, when patients participate in medical practices that revolve around evidence that they interpret through antiquated assumptions about representation.

Andy Clark's chapter, "Technologies to Bond With," is part of a larger philosophical project of trying to understand human nature as "cyborgian". According to Clark, careful study of the evolutionary histories in which human beings use an array of technologies as aids to problem-solving reveals that, at bottom, we are hardwired to be a species of "cognitive opportunists." We regularly render action more efficient by using artifacts to "upgrade" and transform our own processes of thought and communication; and in doing so we, in effect, reconstitute ourselves as "posthuman" subjects. Although in different terminology, the reader has already been primed for Clark's analysis by Grim and Delehanty, where we see our understanding of cognition and subjectivity recast through the advent of "abstract imaging" (Grim) and "hybrid imaging (Delehanty).

It is commonplace to deal with and think about imaging technologies and imaging practices on a case-by-case basis. Researchers concerned primarily with developing (or using) medical technologies, for example, often do not have the opportunity (or need) to study rigorously dimensions of imaging that fall outside of their areas of specialization. Why should a company subsidize the study of non-medical technologies if its goal is to develop specific applications of computer hardware and software that can benefit the medical community, that can improve upon how biological data are computationally translated, represented, and applied in a specific and profitable context? This point is illustrated clearly by the manner in which Kodak organizes its "infoimaging" endeavors. Created by the convergence of imaging science and information technologies, the field of infoimaging is a diverse enterprise, one that is distributed among different imaging groups. While the Health Imaging division devotes itself to "making image capture and diagnosis much less time-consuming," the Entertainment Imaging division concentrates on enhancing the experience of cinema by providing "new options to the film community." A similar trend of imaging specialization structures academic research. For example, a division of labor underlies what film theorists are encouraged to study. Whereas forays into psychoanalysis and the philosophy of mind might be viewed as useful endeavors, departures into digital radiography or MRI seem less relevant.

As a result of this division of labor, the cumulative force and effects of such trends on the reorganization of experience remain unaddressed. Proceeding in this piecemeal way limits each domain-specific enterprise to a partial encounter with imaging culture. Specific interests, goals, concerns, background assumptions, and, of course, technologies, proliferate without adequate focus on the characteristic attitudes, behaviors, and commitments that diverse imaging users and developers share. What makes Clark's account philosophical, then, is that (1) it offers a general account of how seemingly discrete imaging practices overlap and form trajectories (i.e., patterns of development and use that impact communication styles and forms across diverse public and private domains) and (2) it considers how these trajectories affect how we understand ourselves, narrate our identities, pursue knowledge, and justify the beliefs we feel strongest about.

In effect, Clark's attention is not only on concrete examples, it is structured around a synthetic and general analysis of how the mind relates more generally and transformationally to "cognitive technology." This analysis is organized around a range of provocative claims about: what makes tools transparent, how imaging technologies enable us to grapple with otherwise abstract information, the difference between virtual and augmented reality, the structure of knowledge obtained by "hybrid biotechnological systems," and the possibility that as imaging technologies alter our sense of presence and place, so too do they reconfigure who we are and who we become.

Lisa Cartwright's and Brian Goldfarb's, "On the Subject of Neural and Sensory Prostheses," investigates several crucial questions concerning the relation between technology and lived experience. Drawing from Cartwright's own experience with a cervical spine injury and an array of associated literatures, their main focus is on how new conceptions of the nervous system introduced in 1980s and 1990s inspired new understandings of subjectivity and embodiment. In this context, they examine how the use and understanding of neural prosthesis facilitated a reorganization of the body imagined across the discursive realms of biomedical research, medicine, and patient discourse. They frame their inquiry in terms of Gilles Deleuze's and Félix Guattari's notion of "the body without organs."

Given the focus on medical practices and cyborg identities, Cartwright's and Goldfarb's inquiry can be understood as furthering some of the critical conversations found in Delehantey's and Clark's chapters. Of particular importance to the theme of rethinking theories and practices of imaging, however, is their discussion of visual prostheses. As the authors note, "[p]rosthetic devices for replacing or enhancing vision have lagged far behind neural prostheses for hearing." Understanding this trend requires grasping the complexity of visual processes and the difficulties that have limited the effectiveness of technological surrogates. In Cartwright's and Goldfarb's discussion of these issues, they call our attention to a procedure that prosthetically supplements vision by interfacing technology with the visual cortex. What makes this procedure fascinating is that it does not treat the eye as the locus of vision, as does the prospect of retinal implants. Such recasting of vision as a "dispersed" process, the authors argue, will raise a number of pressing phenomenological questions, including how formerly blind subjects who are treated with these prosthetics will come to experience and understand their new visual field. Indeed, now that vision is being understood predominantly as a "brain" issue, it can be expected that many challenges will arise, especially regarding the proposition that the link between organ and function is arbitrary. In effect, it will be contested by prosthetically enhanced subjects whose experience of the world differs significantly from the rest of the non-enhanced populace.

Part IV Art, Aesthetics, and Imaging

This section concerns the ways in which some specific imaging arts reflect on themselves and how they relate across some traditional boundaries of medium. In doing so, the authors engage a very powerful and curious logic that is embedded in the digital: the desire to extend and perfect logics of representation that are associated with pre- and non-digital media, on the one hand, and, on the other hand, the desire to be something different in kind, something whose logic transforms what it comes into contact with. Between these incompatible logics of extension and transformation there emerges a tension, and it is through this tension that the authors in this section explore insights into the contemporary arena of imaging aesthetics.

Vivian Sobchack, in "*Final Fantasy*, or the (Dis)Illusion of Computer Graphic Life," explores precisely this logical tension between what we want of realism, what we want of animation, and why, when computer animation works "too successfully" to imitate the real, it fails as animation. Sobchack offers, however, not simply a compelling explanation of why *Final Fantasy* disappoints. More important, she provides a framework for explaining the internal and incompatible logics that drive much computer-graphic animation and, perhaps more, that confound the computational ambition to simulate algorithmically the organically human.

Final Fantasy "wars with itself—attempting, on the one hand, to *re-create*, through the simulation of cinematic photorealism, 'perfect computer generated human beings ... and, on the other, to *create* and animate an imaginative and irreal (if perceptually realistic) world in—and through—its own CGI imagination.'" In explaining this war, Sobchack draws out several interwoven registers in which these incompatible logics are at work: epistemological, ontological, semiotic, rhetorical, and narrative.

In teasing out the implications of these incompatible logics, Sobchak draws on a range of related examples and a range of theoretical resources—semiotics, rhetorical theory, and film criticism, among others—in order to focus our attention on a persistent and problematic argument embedded within CGI animation. Its progress is predicated upon an "argument of unlimited development" which, paradoxically, "contains the premises for its own rebuttal!" By setting itself up to simulate the human and to match the values of the photoreal, it focuses attention on what it can't do well—simulate the real—while simultaneously failing to do what it purportedly set out to do—provide an *illusion* of life. It gets caught between doing "too much" and "too little." In effect, computer graphic "humans" induce the wrong kind of scrutiny and will thus ultimately be found wanting. In attempting to create the "real," CGI animation "has reached the semiotic and rhetorical limits of its supposedly 'unlimited' development, and its effects will continue to disappoint—serving, indeed, as their own rebuttal to the impossible 'final fantasy' of realizing a 'perfect' simulation of cinematic photorealism."

In "A Scene in the Digital Library: Imaging Literature," Chris Burnett explores the interrelations between imaging technologies and textuality. Beginning with a discussion of Fox Talbot's "A Scene in a Library," Burnett explores how, in the moment photography was extending the possibilities of vision, Talbot put the encounter within a library and speculated on the future of vision in and through the literary: part precedent and part allegory for exploring a range of ways imaging and literature interpenetrate and, in even more unruly and intriguing ways, continue to do so in an era of computation and code. Through a range of examples—from photography, to the X-ray, PET scan, and other imaging technologies—Burnett provides both a genealogy and an interpretation of the literary impulses that linger within the realm of imaging and, equally significant, how transformations of imaging technology interpenetrate and recast the literary.

This is not simply a play of metaphor, argues Burnett. Imaging technologies penetrate and reveal, and also transform and revisualize, literature in materially literal as well as hermeneutically new ways. From revealing and clarifying the Dead Sea scrolls through infrared photography to investigating the work of artists such as Doug Manchee and Peter Greenaway, Burnett explains what can become of the book when made available through a variety of imaging technological projects. In effect, there is a sustained and ongoing interplay among literary meaning, materially imaged literary work, and the forms and patterns of visual production/meaning emerging out of coded and algorithmic activity. The point is to question and illustrate how types of imaging processes can shape strategies of reading, and how literary impulses continue to linger within the imaged.

Code is the most recent "meeting point between visual and textual systems." In this context, Burnett goes in both directions: from imaging artists and their challenges to the literary to poetic strategies for deploying and exploring visualized code, in the work of Alan Sondheim and Christian Bök, for example. Burnett leads us to contemporary "media literature" where identity assignments are abandoned, where the interchange between our capacities both to visualize and to narrate are essential to our experience, where the more we investigate the artifacts of imaging technology (and the experience of extended sight) the more we continue to find literature.

In "William Kentridge: Moving Pictures," Cyril Reade provides an analysis not only of Kentridge's work but of how the visual can continue to demand of us a kind of engagement with conditions of embodiment, history, time, and memory that successfully refuses to be leveled by its purported disappearance within the "postmodern" or within the "virtualization of the real" or because of the circumscription of language. Kentridge is a South African whose animated films—and related drawings, paintings, prints, and photographs—are composed not simply as individual cells, but as moving images that retain ghostly material traces of what was previously seen, that suggest an interplay of present and past: a literal claim on memory, and also a claim about the possibilities and responsibilities of sight and imaging.

Reade investigates Kentridge's work as both an assertion about what methods and formal techniques can serve to convey these claims, and also as images that provide a point of view on history. His images suggest something formally incomplete, contingent, and provisional, suggesting in "their grayness and lack of clarity or sharpness" an engagement with the history of art, with the purposes of art-induced sight, with a post-Holocaust encounter with subjects battling the forces of their own erasure.

By reference to the drawing *Procession of the Dispossessed* (among other works), Reade proposes a precise and allegorically far-reaching interpretation of what contemporary art imaging can provide. It can invert the usual relationship between "text and image" (as described by Roland Barthes and others), between image and commercially induced "social amnesia" (as described by Fredric Jameson and others), between meaning and its disappearance among the "cybernetic simulacra" of the present (as described by Bill Nichols among others), between the body and its dematerialization within the digital. The resurfacing of expressionist style in the work of a South African Jew is, for Reade, more than an identifiable style, "palatable for consumption by the metropolitan art market"; it constitutes a resurgent even if muddling humanism, committed to the work of imaging bodies-in-historical-motion in the shadow of the Holocaust. In effect, Reade offers a sense of how aesthetics and imaging in a digital era can continue to lay claim to our sustained attention as embodied and historically engaged beings.

Part V Ethics, Politics, and Imaging

The chapters in this section address imaging from the perspective of its impact on the ethical and political domains of human experience, and on what counts as meaningful and effective social engagement and agency. From the televisual imaging of humanitarian disasters to the neuroimaging of the brain to the uses of imaging in the educational sphere, imaging technologies inevitably carry a normative force and suggest normative orientations in their wake. However this force is characterized in particular contexts, the essays in this section suggest that there is a more general picture emerging about how imaging technologies can affect our normative orientations toward ourselves, toward others, and toward the ways we understand and act as a result.

In "Publicity and Indifference: Media, Surveillance, "Humanitarian Intervention," Thomas Keenan analyzes the Kosovo War in order to question the degree to which media coverage of human suffering impels viewers to try to alleviate such suffering and/or the degree to which it desensitizes them and, in the process, amplifies the pain of the very victims whose images it captures. Keenan argues that it is not reasonable to expect image and action to exist in a strict cause and effect relation.

In making his case, Keenan takes issue with the postmodern analysis of warfare, at least as it is exemplified by Paul Virilio. Virilio seems to suggest that the practices of global control exemplified by NATO's air campaign are more reprehensible than the actual killings that occurred during the conflict. Keenan finds such a position untenable. While Virilio may express insight into new regimes of power that focus on information and image, Keenan judges his proposal to be laden with hyperbole.

Keenan also criticizes Michael Ignatieff's claim that television serves an "ethical" function by keeping the public informed of conflict around the globe. According to Ignatieff, television enables "nongovernmental private charities and pressure groups ... to mobilize conscience and money on behalf of endangered humans and their habitats." Keenan takes issue with how this position depicts human decision making. It presupposes that once the public sees violence through the proxy of the camera's eye, these viewers will be impelled to righteous action. In addition to begging the question of context and being empirically questionable, such sentiment ignores what has come to be known as the "CNN effect"—the idea that media coverage can function as a weapon of war by influencing the perceptions and judgments of a viewing audience in ideological ways of which the audience might be unaware.

In clarifying why the relation between spectatorship and action is so complex, Keenan highlights two issues that are crucial to any contemporary discussion of the media and politics. The first issue concerns what traditionally has been referred to as "alienation." While it may no longer be fashionable to evoke the discourse of "alienation," it remains important for theorists and citizens alike to understand the type of experience that occurs when people experience violence under the watchful eye of the camera, and find their related expectation that outside help will quickly come is thwarted. Keenan's chapter poses the question of whether this frustration amounts to the addition of a new level of psychological violence. The second issue concerns the possibility that the prospects for genuine political action are being eclipsed by technological practices that stir emotion. At issue here is whether exposure to affectively powerful images of real-time suffering diminishes the distance required for citizens to rationally contemplate which course of political action is best.

In "ELSI Priorities for Brain Imaging," Judy Illes, Raymond De Vries, Mildred Cho, and Pam Schraedley-Desmond critically examine pressing normative-ethical, legal, and social questions concerning the use of neuroimaging tools, especially practices centering on functional MRI (fMRI) technology. As the authors note: "The ability to map how we reason through difficult moral decisions, make conscious and unconscious value judgments, and how we cooperate and compete with others presents a significant challenge to traditional ways of understanding human nature and agency and raises significant and controversial ethical questions." Consulting with experts from across the disciplines who work in both the public and private sectors, as well as with citizen stakeholders, including patient advocate groups, questions are raised about the general implications of framing human experience in biological terms, and the specific consequences that may result from neuroimaging practices which emphasize naturalized accounts of human nature and identity that negate traditional and spiritual perspectives. While definitive answers to these questions are not provided, the contributors accomplish an admirable goal, uncommon in professional philosophy. Rather than confronting the normative issues associated with disasters that have already occurred, this chapter looks to contemporary affairs. The authors construct a framework that allows us to understand and assess a range of ongoing imaging practices, to clarify policy, to enhance deliberative democracy, and to help steer society in a more just direction.

Jean-François Lyotard's philosophy has become intimately associated with postmodern theory. Thus his analysis of technology and culture is often characterized as the quintessential postmodern outlook on these matters. In "The End of the World Picture? Lyotard, Technology, and the Human," Simon Malpas reminds us that in much of the secondary literature, Lyotard is interpreted as categorically rejecting contemporary manifestations of technoscience and the supporting ideas of "modernity" and "reason." Malpas rejects this depiction of Lyotard—and, thereby, postmodernity—as a caricature, contending that Lyotard offers a more nuanced and provocative account of technology in general, and imaging technology in particular.

Using Lyotard's essay "Can Thought go on without a Body?" as a point of orientation, Malpas claims that when Lyotard considers the possibility of life on earth going extinct (due to an explosion of the sun), his goal is not to contrast the "technological" outlook with a phenomenological perspective. The contrast, in other words, is not between the view that human consciousness can only survive the apocalypse through an act of engineering that creates the material conditions for human thought to be transferred to a disembodied medium, and the view that human thought is inherently anchored in embodied (and thereby gendered) perception and action. Instead, Lyotard treats this juxtaposition as a "differend," a case of incommensurable vantage points that cannot be adjudicated via a common standard of judgment. In other words, technologists do not err when they view cyborgian transformation as the only means of long-term survival, and phenomenologists do not err when they view cyborg existence as inhuman. What such a "differend" reveals, Malpas argues, is a fundamental ambiguity in our relations with and attitudes toward technology. Such ambiguity, he insists, has crucial implications for how we should understand the new experiential possibilities offered by imaging technology.

In order to clarify these implications, Malpas briefly explores two of the central views that influence how Lyotard constructs his philosophy. First, Malpas focuses on Martin Heidegger's critique of "technological thinking" as a critique of "humanism." According to this critique, the autonomous subject that humanism posits is undermined by a historicist conception of reality according to which everything that exists, including human beings themselves, has value only as "standing reserve for productive processes." Such a view of reality, Heidegger insists, leaves no room for authentic humanity; it thoroughly devalues our unique capacity to generate meaning by disclosing new worlds. Second, Malpas turns to Jean Baudrillard, a thinker who famously declares that the ubiquity of simulations and the concomitant eclipse of the real define the present period of history. According to Malpas, the combination of these views suggests: "There is no outside of simulation: the Cartesian decision that generates the world picture for the subject has itself been caught up in the picture, and the humanist subject has succumbed to Heidegger's 'supreme danger' by being turned entirely into standing-reserve for the 'combinatory models' of postmodern hyperreality."

As Malpas sees it, Lyotard accepts Heidegger's and Baudrillard's criticisms of the humanist subject, and is sympathetic to how they cast the technological dangers of the present. But unlike these thinkers, Lyotard holds out hope that immanent strategies of resistance can emerge, strategies that allow subjects who are thoroughly mediated by "techno-capitalism" to create new styles of living and new forms of meaning that have more significance than the empty gestures suggested by Baudrillard's notion of superficial performances or Heidegger's conception of appropriated reasoners. Appealing to a utopian sense of the future as a horizon in which the constraints limiting the present imagination are challenged, Malpas interprets Lyotard as claiming that the specific configurations that will make such resistance possible cannot be identified in advance of actual emergent practices of disruption, and also as rejecting the metaphysical and empirical assertions that such outcomes cannot possibly arise. In this context, he identifies imaging technology as an important potential site for this resistance as innovations in imaging continually open new spaces where the "differend" between phenomenological and technological vocabularies might emerge again as opportunities for critique.

In 1981, Andrew Feenberg collaborated with the team that designed the first online educational program. While the enthusiasm for "distance learning" took some time to gain traction, by the late 1990s administrators were smitten and the idea stabilized as a "standard feature of the university." However, not everybody was pleased with this outcome.

In "Word and Image in Online Education," Feenberg assesses the concerns that came to be expressed about distance learning. Feenberg himself attests to having positive experiences teaching philosophy classes online. He even goes so far as to characterize the virtual classrooms in which he cultivated Socratic dialogs as places of "intense intellectual and human interaction." Contrarily, Feenberg notes, many detractors came to depict online learning as an inherently bad thing—as being conducive only to poor pedagogy that saves colleges money at the expense of giving students memorable and rewarding experiences.

As a deflationary counter to the critics who attack online learning for being an artificial medium that necessarily cannot reproduce the repleteness of traditional face-to-face interaction, Feenberg contends that one should directly assess the benefits and the detriments that arise when courses are taught in specific technologically mediated formats. By emphasizing contextual specificity, Feenberg clarifies the problems that can attend courses being offered that over-emphasize visual imagery and undervalue text-based discussions. Indeed, Feenberg argues that low-tech, text-based software can be more beneficial to promoting critical thinking than graphic intensive, cutting-edge products.

Feenberg's intervention is provocative because "the systematic rejection of online education by some faculty has had no effect on the deprofessionalization of higher education." Given this state of affairs, philosophers as well as other educators need to reflect carefully about how they can use technology to promote innovative educational experiences and outcomes. Rethinking the hype associated with tantalizing visual platforms is one positive step in this direction.

Part I Reinventing Sight

1
Reinventing Sight: Theories and Practices of Imaging

Timothy H. Engström and Evan Selinger

It's not long since geometry combined with perspective devices to reshape our habits of picturing, since photography then trumped painting as a pictorial record of the real, since films gave up their histrionic and comic silence to become the new and universal *Gesamtkunstwerk* of social representation, since our bodies and brains gave up their interiors to the penetrating pictures of radiography and the MRI, since textuality lost out to television in the competition for a child's cognitive attention. It is already the case that public action is registered not fundamentally in relation to its location or one's presence in a particular location or in relation to a textual report of that presence; it is seen and surveilled, its movements captured, categorized, and stored by digital means, edited and presented around device-driven conventions of sight. Perhaps social action is now staged in ways that *anticipate* the constructivist nature of its visual representation. Knowing, and being a worthy candidate to become something knowable in the first place, seems increasingly to depend less on underlying, linguistically descriptive practices than on technologies through which the known can be visually identified or constructed and represented. Perhaps the "linguistic turn" has given way to the "imaging turn." Perhaps reinventions of sight are displacing our linguistic categories for capturing and disciplining experience. Or so some are suggesting.

Sight, however technologically mediated, always had its champions as well its philosophical detractors and skeptics. Is there anything new to be said in general, however, about our present circumstances? Has digital imaging ushered in a new range of visual practices and, with them, some implications and consequences that prior technological and visual revolutions could not have propelled? Has there been a "visual turn" whose paradigmatic challenges are, as W. J. T. Mitchell suggests, yet to be fully explored and theoretically contextualized? Our discussion is not so much about vision or visual productions *per se* but about imaging: about capacities to construct "things" to see—and capacities to manipulate and use what's constructed—that, absent the computational underpinnings, would not be possible at all. This is not to say that conventions of digital production and reception don't emerge out of and in relation to previous media or forms of visual production. They do. It is, rather, to say that these previous conventions become transformed as they become digitized, and that the digital embarks on new, additional, and categorically different enterprises than previous media permitted.

Beware, of course, of revolutionary rhetoric, of utopian or dystopian generalizations about what's new or likely to happen because of the new. The cumulative force, however, of these new technologies and practices of imaging suggests that some large issues are at play: ontological and cognitive, epistemological and aesthetic, political and economic. Not only have new practices emerged (and some new theoretical vocabulary for understanding these practices with them), but these practices, and the technologies and artifacts involved, have also turned back on us, have become, in symbiotic and cyborgian ways, part of us, extensions and alterations of who we have been; they have challenged, therefore, some long-standing conceptions of "human nature" itself and the boundaries by which we might have thought we could demarcate the essentially human from the artificially technological (see Clark in this volume).[1]

It might be thought, before the advent of the constructed digital image, that we could maintain confidence in a real and authentic relation between sight and our depictions of sight, and between a given depiction and its truth. Perhaps there is a sense in which constructed and digitized relations put real, authentic, and truthful relations at risk. Not quite. The real and authentic, as captured, for example, through analog technologies such as photography, were themselves conventions, purposeful constructs of the real and authentic all along. Yes, the photograph contains a trace of the real—no need to deny the nature of light-sensitive chemistry—but it is the genres by which the real is framed, the conventions through which we see, interpret, and judge that has counted most. It is a host of related and technologically mediated inferences which permitted images to count as representations of the real and authentic in the first place. The chemistry was essential but not decisive in determining the epistemic and social work images were asked to perform. The retina was an essential means to visual understanding, but it didn't by itself determine that understanding. It is specific devices and their contexts of use—aesthetic, political, and scientific, for example—that have shaped the kinds of inferences that images have been called upon to affirm and the kind of epistemic authority with which these have been affirmed.[2]

Moreover, it is clear that entirely constructed, computationally derived simulations may provide *better* representations of those aspects of the real that concern us. They may capture processes, for example, that the stilled, object-reifying ontological unconscious of the photograph can't capture or represent; they may provide representations which, assuming certain conventions of interpretation and use, give us a better handle on, for example, biological processes, historical processes, and medical interventions; they may give us better and more predictable control of inferential speculations and perceptual extensions of all sorts. Simulating and computationally modeling weather patterns can provide information and inferential guidance that no natural context of sight or actual snapshot of nature is capable of conveying.

The question, then, is not: how do we reconnect seeing to what is real or authentic or natural? But conceding that sight can only be understood as a technologically mediated and culturally trained modality is to say nothing very helpful, either. More specific questions, however, are raised by some paradigmatically new and increasingly ubiquitous practices of imaging: What do digital media of imaging assume vis-à-vis their computational underpinnings? What kinds of constructions or translations of the "real" do these underpinnings entail? What kinds of cognitive processes and orientations are demanded and/or permitted in relation to digital images? How are we situated and/or transformed as embodied social beings in the context of imaging technologies and practices? What forms of reception and/or agency do they induce? What forms of knowing and being, of understanding and intervening in states of affairs, are propelled through contemporary imaging practices and through the commercially shaped contexts in which these practices emerge?

There is little doubt that imaging technologies are having an impact on previously enculturated practices and commitments, that they can exercise a specific kind of intentionality and thus reshape us and our practices in unforeseen ways. We wish to examine some of these ways in what follows. And we hope, thereby, to suggest some contextually relevant trade-offs that these new technologies and practices encourage us to consider, some good and some not so good. In investigating these issues and questions, we propose no single philosophical orientation. One will find a range of perspectives employed here—historicist, pragmatic, phenomenological, empirical, critical, and others left properly unnamed (perhaps they are post-something). In their theoretical and practical range, the cumulative point is that the diversity of imaging practices at play in the present suggests a generally pluralist approach to theoretical questions: we'll take our insights and interventions from wherever we can get them.

An artificial route to the real

From the painted walls of caves to symbolist and surrealist canvases, we have long understood that distortions of the real can improve rather than obstruct our visual understanding. The arts, when they become too real, fail to give us what we want—instructive artifice and form, an awareness of how much the form we give to things matters (see Sobchack and Burnett in this volume). Our forms are what permit them to matter. On the other hand, we've always been realists about the things imaging allows us to do in the world, even when we're not realists about the nature of seeing and imaging itself. Epistemologists, when they have been tempted to assert a realist correspondence between sight and truth—by inventing "propositions" to connect them and "objectivity" to secure the connection—have failed to account for what we do: invent techniques, practices, and conventions of interpretation to achieve a real and desired end. It's not that the real is lost, but that it comes to us under a variety of possible descriptions, through a variety of different media, to serve a variety of interests and uses; and new media can change the kinds of descriptions we thereby seek. It is, however, predominantly through a sustained investment in modalities of sight, and in an accompanying range of technologically mediated forms of seeing, that we have hoped most fervently to maintain the presence and availability of the "real." Hence, our focus on sight, technologies of sight, and metaphors (like "focus") that foreground appeals to sight in explaining and ordering what we know.

We have, of course, become well accustomed to acknowledging the role of the apparatus in assisting sight, in producing representations, and then in sneaking back into our initial conceptions of the seeing and thinking subject. Dürer refers to a number of "perspectival machines," serving a variety of purposes, and already widely distributed in the early sixteenth century. It was a short step to conceptualize the eye and seeing itself mechanistically. Likewise, the *camera obscura* provided the seventeenth century, especially Descartes, with an epistemological analog for the eye and for the knowing subject who purveys, from a (visual) distance, his own epistemic processes (see Ihde in this volume).[3] In the early nineteenth century the *camera lucida* had a similarly important function in helping to ensure that the "correspondence" between seeing and representing would be improved, and so on with photography. The very discourse of "objectivity" has depended upon our trust in the mechanization of the visual.[4] But, as Latour, Mitchell, Manovich, De Landa, and others have pointed out, the traffic between reality and its representations, between what we see and the constructed models we have made to help us structure the seen, has, with the advent of the digital, taken on additional and transformational dimensions. The combination of computational and digital development, computer graphics and algorithmic programming, microchip miniaturization and implantation, 3-D display and computational translations of the real, and so on have together fostered a penetration, extension, and transformation of sight that is categorically and quantitatively beyond anything imaginable in their collective absence.

Aside from many other transformations at play, one metavisual way in which this transformation is important is in relation to the increased automation of seeing itself and, thus, to the related automation of those things, beings, and processes that can be brought into contact with the digital. That is, we are creating machines whose purpose is to recognize other two-dimensional representations themselves—photographs, for example—and relate them back to three-dimensional objects; whose purpose is to acquire and reorder previous visual representations *as* their input. Their purpose is to recognize, furthermore, computational translations or representations of ourselves—in medicine, for example—with which to interact (see Illes and Delhantey in this volume). There are programs whose purpose is to transcribe and re-present the real in computational terms, so as to facilitate an automated and real-time interaction with those other forms of existence—whether bodies or brains—that can be brought into or be translated into such a relation. This is to say that we are no longer talking simply about seeing or imaging *from* the stable perspective of our own independent identity or agency—the conventional conceits of perspectival imaging and early photography—but about digitally mediated and cyborgian relations between ourselves and the digital; and these are relations that foreground vision and visibility in unprecedented ways, that link our cognitive, personal, and embodied social practices to external and commercially produced networks; and these are networks whose purposes and powers are to make things available for representation, distribution, and interaction in ways that do not, therefore, remain within the purview of traditional conceptions of the autonomous, metaphysical, or embodied (rather than computationally abstracted) subject. On the basis of what philosophical and ethical orientations might enthusiasm or caution toward such possibilities be explored?

Caution is suggested because we seem to encounter two very enduring impulses, from Plato's cave to Nazi propaganda to virtual reality projects: what can be brought to the eyes carries with it the implicit conviction that we're having direct experience of the most reliable, confirming, and immediate sort; and that however mediated the experience turns out to be, critical knowledge of how these forms of mediation work tends not to trouble that practical conviction. Vision and presence seem intertwined, no matter what; and being continually reminded that the former doesn't confirm the reality of the latter is of little practical import, so long as the seeing that is mediated is indeed of some practical and institutionally reinforced value: it assists us in the pursuit of some interest or other. Plato expressed a still echoing concern that too much visual immersion and simulation, even when well-intentioned—to reinforce established beliefs—provokes the worry that too much immersion manipulates the contents of one's inner visual memory and potentially impairs one's perception of the real.[5] Yet no other sense modality shares the same privileges and hopes that if something cannot be made available or present to normal perception—whether distant stars or even more distant gods—it can, nevertheless, be represented and made substitutionally or metonymically present to sight in some successful way. And if, on the other normative hand, we're thought to be vulnerable to illusion, deception, or manipulation, sight is again the preferred modality of operation for exploiting these vulnerabilities, too. Sight brings to us what was otherwise absent or distant, not us to it. In effect, it helps us conjure up contexts in which we seem to be able to transcend the limitations of our own bodies: such was both Plato's and Norbert Wiener's curiously shared fantasy, continually perpetuated and reinvented by means of a variety of technologically mediating devices.

Similarly, however important a real and experienced event might be, in a social and communicative context, it is the representation of the event that is of importance, especially when considering how we think about and subsequently act with regard to the event. For what we think about and act upon is the representation, itself shaped by the symbolic context in which it plays. And however faithful the record of an event, it is in the symbolic context of the medium in which it is made available and the narrative company the image keeps that is decisive for subsequent action. Without these contexts, the image is an unintelligible fragment in search of a larger script in which to play a role.[6] Thus, socially speaking, the visibility of an event is impotent, and the distance between an event and a wider public is not overcome, until the event is given a life in representation. And a digitized, distributed and networked representation is of a different kind than has existed before: its "spatial" and temporal simultaneity and ubiquity—its not being hampered by locational or spatial determinants at all—give it a potential force and, at the same time, a potential "nowhereness" (Borgmann's concern) that no previous image or image culture could possess.[7]

It's not, therefore, that we've lost the real: we never had it in any pure sense in the first (or last) place we looked, unassisted by the tools and purposes we have for negotiating the real. But this is not to say that we shouldn't worry. We may not be happy when children get so lost in reading about the "real" that they never go out and play; or so lost in television that other forms of cognitive engagement are never developed; or so skilled at CT scan and radiological reading that the potential and meaningful idiosyncrasies of the person being read recede, in a problematic way, too far into the background. Dreyfus and Borgmann are more worried than Clark about this: they exhibit a greater concern that as we successfully undermine the old and indeed questionable boundaries between the body and its technologically enhanced and internalized appliances, between human nature and technologically assisted alterations of that inheritance, we might be giving up more than old theory. For so long as the desires such appliances serve, and the forms of experience they propel, are shaped by the intentions of those appliances (and the economic forces that market and profit from them), there is reason to be concerned. This philosophical register of concern, however, is less of a traditional ontological sort (what is our nature and doesn't *it* resist such technological reconstructions?); rather, the concern is that investigating imaging and imaging technology is to be obliged to consider normative and political issues, too. We need an understanding of a technology's tacit and latent intentionalities: something we can acquire only by placing it within the larger social, economic, political, and ideological contexts that propel some technologies over others and that endorse some uses over others. That is, we need to be capable not simply of understanding the technologies, but of entertaining a variety of perspectives in terms of which these contextual forces might be described and critically engaged, by considering the ways they (and ourselves in interaction with them) might be employed and redeployed for some purposes rather than others.

Now that we have virtual worlds, temptingly replete but simulated contexts of experience, uses which are themselves induced by the virtual, we can't count on remnants, however abiding—of the real, of proximity, of intimacy, of reference, of indexical trace, of correspondence, etc.—to keep their divergence and difference from the "real" in check. The worry is not whether the virtual can be self-sufficient, self-actualizing, and completely independent of real, material contexts. While dramatic, such hyperbolic questioning is misleading. The question is whether we have any philosophical or other consensus (assuming we'd want one) in terms of which such technological and economic trajectories can be evaluated? In what or whose vocabulary or categories should these trends be described? Are we sufficiently well positioned to consider the purposes to which such technologies get put, to consider the trade-offs they induce, to ensure that the forms of agency we're most in need of assisting are, indeed, assisted? Our critical vocabularies tend to retain the lingering ontological and realist vocabularies of the past—human nature, embodiment, mind and consciousness, reference, trace—vocabularies whose regulative force may be insufficient to guide our engagements with and susceptibilities to the technologies under review. If philosophy has a task, perhaps it is not only, as Hegel suggested, "to capture our time in thought," but also, as Dewey suggested, to wonder actively, critically, and collectively how a better future can be made. As Gadamer, too, suggests, our understanding must be informed by reference to a future we don't yet have but one that informs our analysis of the present. Imaging technologies are surely providing significant tools and suggesting "ways of world making" through which that future is emerging. We hope this volume might contribute to both—to understanding our present and to considering our future.

Some philosophical framing: A genealogical detour

Philosophers themselves have written on just about everything by deploying a wide range (this is sometimes called a "narrow regime") of ocular metaphors, ordered to give the impression that philosophical theory—*theoria*—is itself not writing but seeing, "clearly and distinctly," as Descartes urged. However the writing and theorizing gets done, references to sight and to the implicit authority of things seen have been at issue. Whether sight is considered as a form of seduction into the deceptive world of appearances (as Plato might have worried) or as an immediate and reliable relation to the world itself (as Locke might have hoped), sight has always been a controversial currency for transacting our perceptual and intellectual business. Moreover, both hopes—the transcendentalist hope to outflank the empirical and the positivist hope to empirically purify sight's perceptual claims—have led to a great deal of subsequent skepticism. Either way, it has been argued, we seem continually to seek a complete or total "picture" of things, and appeals to sight and to the metaphorics of sight—clarity, lucidity, focus, bringing things to light, belief in the Enlightenment—have appeared to be required to ground and authorize these general pictures. This is, of course, one way to tell the story of how philosophy and sight have been interwoven.[8]

The philosophy and history of science, technology, and art have gotten in the way, however, of the previous and admittedly simplified plot. There are plenty of philosophical resources that neither essentialize nor privilege sight, either as an immediate ("objective") relation to the world or as a bogy man that needs to be expunged as an inveterate and nasty totalizer. Hume, Smith, and many of their enlightenment friends, for example, put textuality above picturing and, therefore, comparative interpretation above the purported immediacy of vision. It was custom, not perception, which established the context in which sound judgment could be made determinate. The rhetorical tradition was the most widely shared currency of the period. Texts were more widely available than pictures; it was, in fact, primarily through texts that references to pictures themselves became widely distributed. There wasn't yet a public sphere of museums and galleries to see them; there wasn't yet a technology other than the printing press for distributing representations widely; there wasn't yet a commercial sphere dedicated to the production of an imaging culture. "Objectivity," were one to talk in such terms, was the *social* achievement of a historically informed and rhetorically trained interpreter. It was the result of those comparative judgments and practices that survived "the test of time," as Hume put it, not the immediate consequence of direct or brute sensation. Nevertheless it could be argued that a general trend was emerging: to make truth present and objective, one would have to make it transparent to the eyes, ground it in the discourse of sight.

Thus traditional epistemology has been characterized as depending upon ocular metaphors—of mirroring—and then attempting to persuade us to forget that they *were* metaphors: Nietzsche's interpretation. This forgetfulness, coupled with the practical successes and power of modern science and technology, turned much philosophy toward scientism and positivism as ways to realize its ambition—Rorty's interpretation—or toward idealism as a way to transcend the encroachment of positivist naturalism on the ambitions of "mind."[9] In either case, the visual was key to the hopes and linguistic practices of philosophy. The truth was there if we only "copied objectively" enough or "speculated" with sufficient "insight" into the *a priori* and metaphysical structure of things. Both Nietzsche, in one context, and Dewey, in another, harangued the philosophical tradition for continuing to oscillate between two sides of the same metaphysical coin: we have wanted either to mirror reality through the transparent eyes of the body (empiricism) or to elevate ourselves beyond the limitations of the body, speculatively, through the eye of the mind and its internal representations (rationalism). In both cases the currency was ocular-centric: a philosophical economy that puts all its stock in practices and metaphors of vision, in order to ground its language games in something non-linguistic.

If we see this ontological and metaphysical stream of thought in confluence with the evolution of industrial capitalism, the picture might become more ominous. Technological and corporate power are committed to production, to extending the power of the visual, commodifying reality, simulating it, making social life a spectacle (to be consumed) and/or helping to keep that society surveilled and ideologically disciplined (to be controlled). It is this extraordinary growth in the production of spectacle (Debord) and in technologies of surveillance (Foucault), coupled with philosophical discourses and psychosocial practices that privilege the objectifying power of vision—"the gaze" (Lacan)—that have led to much wholesale skepticism of the visual. The concern here is that vision becomes a dominant sense modality of the modern world through the means by which it is produced and disseminated, and these means tend to serve a specific economic structure of profit and domination which is itself not available as an object of sight.[10] Thus to critique this culture there has been a tendency to critique vision itself.

The worry being dramatized above is that the consolidated power of the gaze doesn't admit reciprocity, that once ocular-centric presentational practices are in play, their force is one-way. Vision, if theorized and generalized in this way, is never an epistemically defensible or communicatively democratic modality. It is, rather, a means of seduction, into the unfulfillable desire to see more, to consume more images, to be interested, therefore, only in what can be made visible and commodified through corporately mediated forms of visibility. This, in turn, tends to reinforce a logical (as well as political) assumption that vision is syllogistically bound to cause harm; and the primary harm is the enjoyably infinite simulatability of reality and the simultaneous camouflage this provides for the structural relations that profit from such practices.

This ambivalent relationship to vision has always put avant-garde art and critical theory in a self-referentially paradoxical position. The power of vision—the gaze—"confirms the totalizing structures of society," while the power of art is often considered a refuge—for both melancholy (Adorno) and/or hope (Lyotard)—that might resist these totalizing social forces (see Malpas in this volume). Such paradoxes are generally the consequence of the very tendency to totalize and essentialize ideological forces that, once totalized, consume the very field of possibility which avant-garde art and critical theory require for their own coherence; and one such possibility is visual production.

Although the ideological fear of vision is Marxian in impetus, the philosophical critique of the ontological and epistemological privileging of sight has a largely Heideggerian impetus. Heidegger's concern was that the eye had become *the* normative, metaphysically entrenched, and most instrumentally disastrous form of access to being. In "The Age of the World Picture," Heidegger linked the ontology of vision to the rise of technologies that seek to objectify and dominate the world. Vision positions and fixes things in front of the observing subject, creating a distance—yet assuming a presence—at which we stare, but not to which we "belong"; hence Heidegger's metaphorical shift from the ocular to the aural, to *hören* [hearing] and *Gehören* [belonging] over viewing and dominating.

It is Sartre who has the most frightened conception of the gaze—what Martin Jay calls "ocularphobia."[11] Nothing is more threatening to Sartre's desire for free subjectivity than the power of the gaze of the other, through whom we are forced to see ourselves. For Sartre, the gaze not only structures our social—even our most intimate interpersonal—relations (of being), the modality of the gaze is a form of appropriation endemic to imperialism. The conflation of ontological and political anxiety creates a plot that is considerably less open-ended and dialectical than Merleau-Ponty's, for example. Although the "embodied subject," for Merleau-Ponty, is always an agent being acted upon within the visual field, the interplay—of visibility and invisibility, of light and dark, of cultural production and ontological possibility—remains dialectically open. He is consistently critical of the observer-subject and appreciative of other modalities of sense, but while the visual may provide an inadequate, sometimes deleterious, foundation for being and knowing, the antidote is not to disparage vision altogether.

It is through Lacan that much psychoanalytic theorizing about the role of the gaze in subject-formation comes to visual theory, to film theory and cultural studies, and to add a heightened sense of the role vision plays in securing ideology (especially in the work of Althusser). Lacan's conception reverses the classical priority of idealist and rationalist ontology—where fully constituted subjectivity *deploys* the eyes to see—in favor of considering the subject as someone who is looked at, who comes to self-consciousness by internalizing the gaze of the other. Thus a scopic drive is part of the primary set of forces in the unconscious. It is the acquisitiveness of this scopic drive that, for Althusser, makes it inevitably ideological and, therefore, which makes the approach to vision predominantly negative. (It is in the more pragmatic, more empirical, more phenomenologically inquisitive and less ideologically overdetermined arenas of Science Studies and the Philosophy of Technology that one will find a more richly descriptive and differentiated engagement with technologically mediated practices of seeing—see Ihde in this volume.)

In this context, however, it is Foucault's work that more precisely ties the gaze to the micro-workings of institutions. The eye, as identified by the panoptical arrangement of Bentham's prison, becomes the metaphorical muscle of social control. In the panoptical prison, one eventually and unconsciously internalizes the sense that one could at any time be viewed. The prison becomes metonymically representative of vision: it infiltrates the self in an oppressive way. The eye becomes identified with surveillance and vision becomes analogically and mechanistically attached to the desire, embodied by any social apparatus, to control.[12] The mechanistic, apparatus-orientation of Foucault's discourse ensures that the gaze is always first an issue of power, never a function of reciprocity (relations are not reciprocal in the Panopticon); and it ensures that talk of the eye entails talk of hegemony, of the inevitably dubious course of (scopic) desire. The mechanization of the eye is, on this metaphorical and analogical level, consistent with Debord's thinking as well. Both concede all too many of their metaphors and analogies to the modernist dystopia of industrialization, factory discipline, and mechanics, i.e. to modernist conceptions of and critiques of *techné*.

For Debord, where modern conditions of production rule, there is a massive accumulation of spectacle and a shift from direct experience to a life lived in representation.[13] Although it is not clear what "direct" experience is, or who has ever lived a life that is not within a field of representation, what is decisive is the theoretical connection here between vision and mechanical apparatus, and the opposition that follows from this between authentic and simulated experience, between the lived and the viewed. Such a conceptual structure ensures that a positive appropriation of vision is preempted, and that a more differentiated analysis of the possibilities embedded within contemporary culture for vision will always appear second rate in comparison with the nostalgic tone bestowed upon the (lived) period preceding ours—the period of experience.

For Foucault, "crime" and "transgression" *become* visible objects of knowledge through the bars of the prison; "madness" *becomes* an epistemic concern through the windows of the clinic; the sick body becomes visible and the subject medicalized through the rooms of the hospital. The gaze *constitutes* new subjects and objects of knowledge. The epistemic utility of vision becomes simultaneously a form of power over its objects: from vision to supervision. What is must be visible; what is good—that is, socially sanctioned—must take place within specified (visible) spaces; and agency must be expressed in the context of these already dominant modalities. Is it reasonable to ask, therefore, if or how vision might seek cumulatively different goals or how ocular substrata to epistemology might themselves contribute to representationally and politically more diverse constructs? In this theoretical context, "Le regard" is not, as it was in the work of, say, John Dewey or Iris Murdoch—to suggest two otherwise unrelated allies—a potentially positive or morally good form of attention and concern, one which could permit a kind of surprise and exploratory wonder, where the self confronts contingencies and particularities without a desire for metaphysical consolation, where lucidity is not a form of domination or synonymous with a desire for totality but a form of honesty and respect. Their epistemological and ontological metaphors were not, of course, in search of totality or under the sway of a dystopic conception of mechanics. For Dewey, technologically or instrumentally mediated relations weren't, by themselves, bad or impervious to beneficial intentions. Their value was decidable only in relation to their consequences and to comparative alternatives, and these could be good and bad, depending.

The pragmatic point here is simply that the representations that have accumulated to occupy social life won't be challenged successfully by denigrating the visual altogether with anti-ocular theory. It may be that reference is a function of the production of representations; it doesn't follow, however, that we don't have good pragmatic reasons for trusting some representations over others or for privileging some forms of and sources for our images over others. For the purpose of growing food, a climatic representation of the atmosphere will do better than a street map of Paris. In deciding which simulation to use for which purpose, the notion that they *are* simulations isn't particularly significant. If all we have is representations, as the rhetorical and pragmatist traditions have said for some time, we need to have ways of deciding which ones to use and to trust; we need ways of investigating how new technologies of imaging, in their consequences, extend and/or transform how and what we perceive or need to perceive. The drama of conceiving of everything *as* a representation trades on the degree to which we view this as threatening and disabling, and on an unstated sense that, on a philosophical level, this is viewed as a loss of something we once had—reality or authentic experience or the world itself.

Images are rhetorical in that they are seen in the context of previous habits and codes—loosely rather than computationally understood—of producing, seeing, and interpreting; and these habits are shaped largely by the disciplines and genres within which we locate the objects to be seen and prepare for the kind of experience that is expected from the encounter. However specific and charged the emotional intensity of the image (of death, for Barthes) or ideologically hypnotic the outcome (of cinema for Metz), the grip of imaging on our psyche and social behavior is not identical with these specific genres, nor its potential exhausted by the industrial apparatus through which it becomes available for consideration.

These genres (rather than vision itself) raise, however, specific questions. For it seems difficult (and risks a politically suspicious naiveté) to consider the possibilities of film without considering the implications of the industry as a whole, without seeing the token in terms of the larger type (as Peirce put it) or script (as Latour put it). An analogy may help draw this issue out: Milan Kundera attempts to give to the novel a special and subtle status in the consciousness of the West and to distinguish its art from the political usury to which it and other arts might be put. In his view, the novel has played a central role in shaping the western (comic and ironic) capacity to live with contingency, without totality—political or otherwise.[14] Presumably, these very same capacities can't be applied to cinema. The novel's cognitive operations are not visually dependent or immediate, and so the novel (presumably the good novel) does not lend itself to immediate and mass political utility: it takes longer to consume; the act of consumption is not generally public and spectacular; the literacy it presupposes puts it at a greater distance from the entertainment industry; and perhaps the relentless wash of the moving image on the commandeered eye can be more easily deterred by the agency required of reading.

Rorty, too, depends upon the ongoing currency of the novel to do a kind of meta-philosophical work. The good novel trains our ability to inhabit a perspective that's not our own, to entertain alternative descriptions of what we experience, to appreciate the contingency of such descriptions, and to appreciate the perspectival and revisable character of all vocabularies.[15] Seeing, however, it might be implied, does not entail these mediating cognitive operations or, therefore, open-ended consequences; it is the *descriptions* of the "seen" that more readily admit to these entailments. Images don't come to us with alternatives in their wake, and so they are less likely to suggest within themselves the underdetermining nature of a perspective, or the value of being able to entertain re-descriptions (re-visions).

But is the "communicatively un-reciprocal" screen a generalizable metaphor of one-way visual and ideological force? How do we conceive of some constructive opportunities for vision, opportunities that aren't conceptually frustrated by falling into essentializing antipathies to the visual or to *techné* at the outset? These antipathies seem to slide on rather broad analogical tissues (of the ocular), tissues that are thought to connect together "totalizing desires" (if we grant this for the moment) and Enlightenment philosophical ambitions to "mirror" the real, to connect capitalist economics and appropriating and acquisitive technologies, and to connect mass-communicational politics and ideological manipulation all into one interrelated apparatus. Given such forces, can anything but visual skepticism follow? For even though much anti-ocular theory is a loose collection of overdetermining structural and mechanical metaphors, of grand analogies and suspicious associations, the force of the anti-ocular story is bigger and more compelling than its problematic arguments and elisions.[16]

In response to such essentializing tendencies, both Rorty and Derrida have suggested that the privileging of the ocular is itself written. Philosophy is not seeing; it is *écriture*. And the "presence" such philosophical moves propose will be frustrated, deferred, by the very ocular metaphors that invoke it, precisely because they are metaphors: an ingenious even if materially incomplete explanation of such discursive processes. While ocular-centric discourse might depend upon the slipperiness of its metaphors, the immediacy of appeals to sight and the epistemological functions technologically mediated seeing serves might be more recalcitrant than philosophical deconstruction appreciates. Derrida's point, however, is that traditional philosophical framing devices in aesthetic theory (especially Kantian ones) have hoped to stabilize a boundary between what is "inside" and what is "outside" the work. For Derrida and others, there is no pure image or seeing through images whose boundaries can, with an act of metaphysical discourse, be made decidable.[17] This is not to say, however, that things aren't decidable. For practical purposes and in pragmatic ways, we make them decidable all the time. Rather, it is to say that neither metaphysics (Derrida) nor foundational Philosophy (Rorty's word for the same moves) will be what guarantees or grounds decidability in some *general* sense or adjudicates *degrees* of decidability in particular instances. At the moment a text, in *its* conventional spatial and visual configuration, looks stable, Derrida's text might become graphic, columned, collaged, a postcard. The point here is that visuality and textuality are inseparable and interwoven in ultimately indeterminate ways. While in a Sartrian world the otherness produced by the gaze is always oppressive, acquisitive, and threatening (which is itself dependent upon a conception of subjectivity that is possessive, paranoid, and autonomous), Derrida reads the visual as lacking in total control over its "own" means of operation. It is this double operation of the visual—desiring control while lacking the capacity to enforce it—that is central to much feminist ambivalence toward the visual as well. What is commonly shared in this view is that there is always something outside the frame of visual representation. The very act of framing requires that there be an outside, a position, therefore, from which to question what's inside and to imagine alternative subject-positions in relation to the frame. It is the inevitable act of framing—the act of demarcating an inside from an outside—that, once it is itself made available, permits an alternative view to be proposed.

This is Lyotard's move, too, only in his case the desire for something outside the field of representation is served by the "sublime." The "sublime" becomes emblematic of what is demanded of postmodern imaging. The concept of the sublime posits, since Kant (Longinus had a more rhetorical and less metaphysically ambitious conception), a radically other realm: unrepresentable, awesome, absolute in its resistance to conventions of depiction, and which transcends the merely beautiful. Lyotard adds avant-garde political urgency to this theological remnant, in order to "wage war on totality," and thus confirms perhaps more of the philosophical trappings of metaphysical sublimity than he circumvents.[18]

The generalized philosophical fear that is thereby promoted is that there is no discourse that hasn't depended upon ocular metaphors; no discourse that hasn't found its power in hiding the metaphoricity and latent mechanics of its primary tropes; no cultural form of hegemony that hasn't sought out such philosophy as legitimation of its power; and there is no reason to hope that any representational and/or referential visual practice or visually dependent discourse will but lapse into this dubious hegemonic desire: a metaphysical loop par excellence. The logic of argument moves from suspicion to injunction, from critical interpretation to generalized distrust, from investigating the institutional circumstances of the ocular to essentializing its properties as problematic *tout court*. This is not, of course, the whole story, but telling it this way helps highlight a postwar political ethos that has continued to justify its primary critical tools on the basis of a rhetorical overreach it would otherwise disdain. Some other moves are required if we are to hope that some visual practices and visually dependent narratives might serve alternative social visions, help construct more flexible and beneficial "objects," and help sustain less oppressing forms of intersubjectivity.

Image and language: Underlying structures and unsettled relations

It is worth assuming, for the sake of argument, that although there is a world, what we see it as is largely trained, with the help of discourse; that although there are physiological parameters to vision (and perhaps kinds of visual reception and cognition that acquire shape in advance of language), how we see as social beings is largely trained by cultural practices of seeing and deeply intertwined with language. The world and our eyes for seeing it underdetermine what and how we see. Neither the world nor pictures of it cause vision in any direct and linear sense, although cultural traditions of picturing decisively train vision to operate in specific ways for specific purposes. In effect, not only is vision mediated, but discourse helps shape that mediation. Discourse, too, to avoid tautology, is not self-determining and autonomous; the visual seeps into its operations, its metaphors help consolidate as well as alter its effects. It's not clear, therefore, that there is any theoretical benefit from asserting the autonomy or hegemony of one or the other.

The "linguistic turn" in philosophy has been so ubiquitous, however, that some visual theorists have wanted to resist its imperial ambitions, to carve out an independent sphere for the image that isn't shaped by or contaminated by language. Given the theoretical prominence of semiotic forms of engagement within art history and art criticism, and within film and photo theory, this is understandable. Treating images or pictures in terms of symbol systems and codes, within which they operate and in terms of which they can then be decoded, was itself a metaphorical extension of the linguistic; and it has served many useful purposes in helping us interpret images.[19] It seems reasonable, however, to assume a link between vision and verbalization, to assume that one way in which images can become placeholders for experience, in memory or fantasy, is through their being cross-referenced with verbal orders, but in very elusive and structurally underdetermined ways. It is likely that linguistic forms give a kind of temporal extension and syntactical positioning to the visual that it doesn't otherwise have. The tension, therefore, between the temporalities of language and the spatialities of picturing, between the abstracting tendencies of language and the particularities of vision, needn't be resolved, theoretically or practically. Put more presumptuously, it seems impossible to imagine (or at least to write about imagining) a mental state of affairs that completely segregates vision and language. On the contrary, both vision and verbalization work to assist each other in doing their respective, interdependent things, whatever these turn out to be.

An image is not a statement, in terms of which propositional orientations toward terms like "true" and "false" are likely to be of much help. What provides images with the capacity to operate in proposition-like ways is not a function of an image's "correspondence to reality" or its ever-identical fidelity to code but its contexts of use, its titles and labels. As Gombrich puts it: "Even in scientific illustrations it is the caption which determines the truth of a picture." It is not that the camera itself can or cannot lie, but that the same image can be used in different contexts to affirm very different concerns. What for one constituency is evidence of "the murder of a loved one" is, for another, an image of the "death of an unlovable terrorist." The newspaper decides, the viewer and her constituency contribute, too, by classifying the image and determining the context into which the image is placed and thus framed for the interpretive act. Similarly, physicians don't generally interpret the images they order; they read the description that the EKG interpreter provides, and it is the description that presides over the communication between doctor and patient and that prevails diagnostically in deciding what follows. That the image copies something and is made and reproduced in empirically consistent, methodologically reliable ways, does not determine how it is seen and used.

Computational imaging, however, is paradigmatically different from analog in this methodological regard. Codes are not chemicals: they don't degrade; they're fully determinate and infinitely repeatable, and in this sense they are what analog processes cannot ever become: exempt from the indeterminacies of nature. But interpretively speaking, this is not decisive. Images, like words, however procedurally perfect and repeatable their instances, can become metaphors, can be opened from a variety of perspectives, and tolerate a number of different interpretations and uses. That the digital will convey exponentially more data and maintain its underlying (coded) structures with an infinite sameness does not, by itself, circumvent the discursive contexts in which the coded is put to work.

The debate between copying and constructing, discovery and invention, mimesis and poesis—such dichotomies have been as persistent as they are misleading—is an old one, of course. It is one that Platonists and Sophists formally inaugurated with regard to language some time ago; and this debate has continued to ripple into the present. Similarly, in the visual domain, a case was being made explicit already in the fifteenth century (by Cennino Cennini) for the power of invention over discovery and copying. The visual arts and crafts were, on this view, failing to do all they could do if construed as and constrained by the program of mere copying. Whether one locates one's examples in the acts of Bosch and Dürer, or in later surrealists and cubists, or in more recent digital artists, there is a well-established dynamic at work. Some want our constructivist and representational capacities to remain loyal to "the real" and to constrain their imaginative propensities. Others aim to ensure that our visual productions affirm the very imaginative and productive qualities to which they owe their existence—the invented technologies, forms, and genres through which they become intelligible and instructive artifacts in the first place. The digital era has not left this long-standing debate behind, but it has changed how we understand the compromises each side inevitably makes with the other: inevitable because each side is in as much need of the other as it ever was. Pragmatically speaking, we need representations, whether analog or digital, to do trustworthy and effective work, hence the continuous if misleading talk of "copying the real"; but such representations owe their pragmatic capacity to do the required representational work to our inventive, productive, and interpretive efforts, hence the continuous if misleading appeal to the hyperbolic anxiety that we might be living in "a simulated world without referents."

Underlying structures or codes don't resolve this contest; they don't resolve semantic uncertainty or interpretive debate. When we encounter the foreigner with whom we hope to communicate, we seek a common ground; but raising our voice in English or Esperanto probably won't help, and pointing to a cherished image probably won't be a successful substitute. We may simply fail to understand each other. Moreover, if we have the good fortune to happen upon an underlying and shared linguistic structure—both of our languages are fond, say, of deploying "binary oppositions"—we'll probably not get much further. A shared structure or code doesn't amount to a shared semantic connection or understanding. Similarly, we'll be no better off if we happen upon a picture and some philosophical or scientific assurances that our "visual perceptual consciousness operates in the same way." It is not that there aren't things that underlie other things; it's that these, too, come to us as a result of some kind of description or other, organized to facilitate some purpose or other. Their meaning isn't the direct and causal result of "the operations of perceptual consciousness" or underlying structures—linguistic or mathematical, naturalistically discovered or computationally invented and encoded.

But ambiguity has persisted in the present by thinking of code *as* a new and potentially pernicious, unitary language. Is the digital or computational a new language? And as such, does it threaten to become a new unitary force, excluding memory, place, difference, etc.? The suggestion is misleading (*pace* Chartier): Code is not language; it is not a natural language that works by *virtue* of its capacity to register and convey meaning, memory, place, and difference. Rather, the question is whether what concerns us will be adequately *translatable* into specific coded and visual contexts. It is a question of how code will accommodate and/or transform what matters to us, and do so in the ways we need and want. What tasks will we *wish* to undertake, in part because code has reinforced that particular wish, made certain tasks particularly easy and successful to perform?

Borges and others have imagined, with horror, what would happen if a single language were imposed, if one system of classification and meaning were to follow. But the fantasy is itself the result of a reified and idealist conception of language. For in its use, such a language wouldn't remain very unitary and monomaniacal for long. It is not as if such a code's or language's mere existence would be sufficient for it to determine all subsequent thought, use, and mutation. Rather, the wonderfully absurd fantasy of linguistic oneness trades on the ambiguity between "natural" or "ordinary" language and underlying code, as if having the one code were tantamount to a kind of tyranny over thought and expression. The political register is worth considering; it is where monomaniacal fantasies are all too happily bred. But code is not the culprit; use and the external imposition of use might be, however. The better question is whether there are interests and purposes that lend themselves unhappily to the translation and potential transformation process of encoding, and that thus become available for kinds of utility and appropriation we might wish to resist. The political is rightly raised as a concern, but it is not code all by itself that secures what uses will prevail, or what memories and differences we'll decide to maintain. The initial elision, however, of code and language has an understandable source. As Chartier points out, when books undergo transformation into the digital, they become very different things. They are, in effect, no longer things in some important ways: no longer material, historical, situated objects. The point is not that there are no benefits to the translation and transformation books might undergo, but we need to be aware that a text is not mere code, and that when it is translated into code—i.e., *represented* through code—and made available on a screen, there are trade-offs, not all of them good.[20]

The desire for a universal language through which to formalize and regulate all cognitive and logical operations is a long-standing one: rationalists, analysts, computer scientists, and programmers have revived it from time to time. But formal code and linguistic operations are not the same. From Searle and Davidson to Derrida, Rorty, and others, we've been well instructed to avoid the misleading ambiguity between code and language. Codes don't themselves have a semantic dimension; they don't by themselves simplify, reduce, or literalize languages. Users (and code writers) do this. But it may be that technologies have played a decisive role in keeping the *desire* for universality, for an underlying and all-encompassing code alive. Condorcet's universalizing imagination was stimulated in this way by the printing press itself, as others' have been by the computer. Technologies have regularly been framed within a universalizing and/or reductionist philosophical vocabulary, been put in service of purposes and desires they may very well have contributed to stimulating. But then we're no longer talking about code here but about the dialectical ways in which these desires and technology have interacted.

When it comes to imaging *rather* than language, however, there seems to be a significant difference. While the combination of code and the two dimensional screen might suggest a concern for reductionism, for a tyrannizing universality or an imperial extension of a specific commercially organized culture of imaging, we should be cautious. The explosion of image-based practices and uses that appear to be permitted by (still developing) codes and materials does not, on empirical grounds, endorse this concern. While there may be limits, in some abstract and principled sense, to what codes or screens or simulated and immersive environments can do, it is imaginative or unimaginative users and institutionalized or transgressive practices, and not codes, which establish (and occasionally disestablish) those limits. Just as our capacity for metaphor always eludes the linguistic reductionist, imaging practices and interests elude the would-be reductionist to code. Neither medical diagnosticians nor artists are content to give up their interpretive skills, their hard-won expertise, and their capacities for invention and insight when it comes to seeing what code might reveal or permit.[21]

Reconsidering old orders: From texts to images

We have suggested that technologies of imaging reinforce appeals to sight and that together such appeals require us to consider their collective effects on traditional orders of existence or being, cognition, knowledge, and society. How might technologically mediated sight challenge traditional orders of existence or being—of nature and ourselves—or contribute to our reconceptualization of these orders? How might imaging and its devices reshape or co-constitute cognition? By what means—natural and artificial—is cognition transacted, and in what space or virtual and networked nonspace can it now be transacted? How internally ordered and/or externally directed and device-oriented are cognitive processes becoming? Similarly, how might such technologies mediate and transform previous conceptions of knowledge, and thus contribute to what and how things become represented and described, classified and arranged, and made available for specific tasks? And lastly, how is the social constituted? Through what discursive and/or visual, real and/or virtual and simulated registers is the social transacted, monitored, recorded and reordered, assisted and/or distorted? What, given the ubiquity and force of imaging, now constitutes the public and/or private spheres and the boundaries by which these are thought to be negotiated? What kind of "communicative action"—in both the real and ideal senses Habermas, for example, implores—does a digitally and visually mediated culture enhance and/or transform and/or diminish? In effect, in what ways are these orders in flux as a result of the digital imaging turn? What are the fuller implications of this turn, and what orientation toward them ought to be taken?

Philosophical naturalism seems to provide little help. Imaging technologies aren't grounded in nature; and they aren't merely natural and continuous extensions of previous perceptual processes. They don't perform their tasks as a result of naturalist forces of causation, on the one hand, and, given the plasticity and reorientability of our perceptual habits, their effects on our experience aren't constrained by or made predictable by virtue of our natural endowments for interacting with them. While our experiences of the simulated and constructed are no doubt parasitic upon the (attenuated) indexical traces of natural sight, natural sight doesn't, by itself, determine the degree or range or applicability or meaning of these attenuations. It is not that we are not all naturalists in some metaphysical sense; it's that being a naturalist with regard to sight isn't to be much of anything in particular. Having a retina doesn't, by itself, inform us in a direct and helpful way as to why and when we should prefer a photograph or a painting or a computer simulation to get the job done. It doesn't tell us what jobs are in need of doing.

Ordering cognition through tools

When we consider what constitutes our most highly prized cognitive capacities—to think and reason logically and reflexively, to organize and interpret our perceptual and cognitive activities in a communicatively successful and discursively coherent way, to ensure that these efforts contend meaningfully with previously constituted traditions of thought—we tend to depend upon a schema of cognition that is organized around the priority and centrality of language. It is a schema that is facilitated, no doubt, by technologically mediated forms of textuality—writing and printing—and organized around institutionally sanctioned narrative conventions—whether philosophical, scientific, or legal and political. But linguistic and textual orders do not, of course, hold the field alone. Experientially, the realm of the visual increasingly mediates cognitive development in wealthy societies; visual production is increasingly in the foreground of our cognitive activity; digitally based visual technologies are increasingly the ones selected for enhancing our own perceptual and cognitive capacities, the ones with which and on whose terms we interact and, thus, desire to interact.[22]

It would be a fallacy, though, to suggest that the things and processes that become represented from the textual to the digital and visual—such as material texts themselves—can retain an equivalence. As was mentioned above, the digital text is never a mere translation or copy of the material text. Forms affect meaning; changes in formal and material context influence the perceptual, embodied, social, and semantic dimensions by which we make sense. What New Critics, structuralist, post-structuralist, and deconstructionist critics, and even computational theorists have occasionally shared is the assumption that one can detach linguistic operations from their material contexts, and thus, detach fundamental cognitive operations (as if simply in the head) from the material and technological contexts in which they're embedded (almost all of which are outside the head yet contexts with which what's "inside" must interact to become whatever it is).[23]

About Plato, Dewey wrote:

> Plato's own generation would, I think, have found it difficult to class Plato. Was he an inept visionary or a subtle dialectician? A political reformer or a founder of the new type of literary art? Was he a moral exhorter, or an instructor in an Academy? Was he a theorist upon education, or the inventor of a method of knowledge? We, looking at Plato through the centuries of exposition and interpretation, find no difficulty in placing Plato as a philosopher and in attributing to him a system of thought. We dispute about the nature and content of this system, but we do not doubt it is there. It is the intervening centuries which have furnished Plato with his technique and which have developed and wrought Plato to a system.[24]

One might add that works, including Plato's, are transformed vis-à-vis the technologically mediated forms in which they become available—oral (dialogical) recollection, scroll, serial installment, or bound and hefty book. Absent the intellectual socialization that takes place through the book and the conventions that have "wrought Plato to a system" in some Philosophy curricula, one would not expect web pages to facilitate the same consequences. When one is holding the *Republic* as a book, one feels—is trained to feel—as if one is in fact holding an entire and systematic republic. If approached in the absence of this material whole, through the serial fragments of a screen, one would not have the material encouragement and, similarly, the cognitive orientation toward thinking in terms of a cumulatively systematic and enduring whole. One might very well miss the persuasive affect and enduring political vision that the genre of the book can, by analogy, suggest. (Similarly, of course, there is the pleasure of closing the book and throwing down decisively the whole of it as a historical and political vision.) Moreover, the cognitive capacities and expectations at play may have developed in some measure as a *result* of the material form the Republic now takes—the bound book. In its absence, the question is whether such cognitive habits would emerge or be sustained. The point here is to suggest that a shift in medium—to the digital and screen-based—requires that different modalities of embodied and perceptual and cognitive engagement be acknowledged, and that we are dealing, therefore, with a different (not an equivalent) object or work. In effect, we're not dealing with an object (and, by analogy, with its systematic weight) in the traditional sense at all.[25]

Nor are we able to ground forms of cognition in the mind itself. Neither reading nor seeing occur in the mind. They occur ecologically and relationally, interdependently with the media and contexts through which these activities are learned and subsequently practiced. *Pace* our old friends Descartes and Kant, there is no mental act in the abstract that, after its execution, goes hunting for a medium with which to communicate itself. It is medium-related all the way down, neither mentally internal nor technologically external, but both, dialectically and interactively. Consciousness, as an always already engaged force, tends to works its way toward what is seen, and not delay itself with concern for the qualities of the seeing itself. The benefit of certain kind of art, however (as Noë, Reade, and Burnett in this volume explore), is to guide us in specific ways to consider these qualities as part of what we see. Moreover, it is by connecting recent neurophysiology with these aesthetic opportunities that we might be able to develop a better understanding of what goes into constituting and reconstituting our visual experience and cognition: its temporalities, our agency, its interaction with locational and spatial contexts, the interrelation and traffic between three-dimensional and two-dimensional engagement—both real and imaginary. In effect, specific kinds of art can lead us toward a more attentive and reflective relationship to what goes into visual cognition as an environmentally contextualized practice.

"Active externalism" is one of the ways this interactionism gets explored in the present. As such thinkers as Dennett, Chalmers, and Dretske (and Clark and Noë in this volume), have indicated, "external" artifacts can shape and thus partially constitute our cognitive processes. The very notion of "cognition," therefore, cannot be understood simply as that which occurs "in the head," and we can't presume, therefore, a subject-preserving ontological boundary between the internal and external, between the cognitive and the technological, and, therefore, between who we are and what we make to transform who we are. The traffic goes both ways. Experience, as Dewey put it, is something that is both done and undergone. We make tools and, through them, our cognitive orientations and activities are remade. And one place where these cognitive transformations are most evident is with regard to the production of knowledge and the role imaging technologies increasingly play in its production.

Ordering knowledge through sight: Tools and consequences

For God to be all-knowing, He had to be all-seeing. And He didn't have to be on hand to do all that seeing, or risk becoming diminished by or entangled with what He saw. Sight preserved his distance, his non-contamination by what was seen. Thus seeing was constructed from the first as overcoming distance and transcending the body—the best possible kind of knowing.[26] If He had been all-smelling or all-touching, proximity and engagement rather than sight, surveillance, and distance would have been the epistemic keys. From classical theology to contemporary science, sight has provided the privileged route to total knowledge. It is the mode that even infinite beings prefer. Thus a powerful cluster of related commitments have combined to attach sight to distance and to the principle of non-contamination: theology, science, and aesthetics all, by the eighteenth century and with the help of Kant, thought of distance as a requirement for rational and objective cognition; and sight was the modality that confirmed the possibility of remaining both properly and knowingly related yet distanced from (even disinterested in) the "things in themselves" under consideration. Sight permits both acquaintance and distance at the same time—a move as central to positivist science as it has been to much aesthetics. These assumptions and motives have been curiously reinvigorated by digital technology. Computational enthusiasms, too (Norbert Wiener's, for example) have often been predicated upon the ambition to remove our own bodies, and our embodied relations to perception and cognition, from epistemic processes; and the new and purely computational ontology of the image reinforces this move. For the computational, it is hoped, bypasses the limitations and contaminations inherent in embodied contexts of experience.

Philosophy, science, and technology have generally been oriented toward conceiving of knowledge in relation to a natural order of objects and events. Knowledge is the production of descriptions, and these should correspond to the objects and events described. Similarly, the order of objects and events was communicatively fixed through the production of texts. And the text, too, could be thought of as participating in the order of the object: it takes a material form; this form is relatively enduring; it is dependent upon an identifiable (and authoritative) author—herself an "object" to which we can refer; and the text is stored and accessible in physical space, in an objective order of libraries, universities, and night stands. Digital imaging transforms this order.[27] Computationally constructed representations needn't be of real objects or events at all. And our interventions in the world of otherwise stable and recalcitrant objects can be reimaged and reimagined as a result. Our actions in relation to objects and events can be transacted at a distance and we needn't be in any proximate relation to them in order to see or affect them.[28] Moreover, our representations needn't take textuality—in the linguistic, narratively coherent, and book-producing sense—as their primary form.

The traditional logic of the image—painted or photographed—could be said, however, to participate in the realm of the object. It was an accomplished object or event; it retained a fundamental association with the entire apparatus of authorship (a proximate observer/subject), objecthood, and property; it endured over time (in the case of the photograph, its existence depended upon the sense that it captures a discrete chunk of time) and was thought of as a thing to which one could return without loss of its fundamental integrity, whether in the museum, cathedral, bank lobby, or drawing room—themselves material and spatial confirmations of the relevance of physical object-talk. And the objects and events kept in visual play by means of analog imaging tended not to exceed the threshold of what our eyes could otherwise see. While such epistemological (and ontological) confidence is not without its problems, the logic of the digital confronts us somewhat differently. If associated fundamentally with code, then the traditional cluster of notions—proximate viewer/subject, object production, property, etc.—become mutable, unsettled, and less easily attached to and sustained by the digital and virtually interactive. The hard drive cannot function like the archive or the museum; it cannot, in other words, contribute in the same way to the kind of intelligibility we associate with the authorially produced and materially collected and preserved painting or photograph.[29]

Thus the categories and schemes of genre and classification we developed by reference to previous visual media—often "realist" in undertow: portraiture and perspective in painting, documentation in photography, social narrative in cinema—will not be adequate to the digital, not because these were entirely fixed and stable with regard to the previous media in the first place, and not because they were, on naturalist or realist grounds, direct and unmediated forms of seeing, but because their relative stability was determined in association with forms of materiality that the digital does not require. Of course, painting and photography required time and institutional and mechanical support to become "naturalized" and "transparent" forms of cultural seeing (and in *this* sense, the digital may be no different). Perspectival seeing wasn't accomplished in a day. It was through the theoretical underpinnings of and enthusiasm for geometry that perspectival imaging emerged; it was through devices that one could ensure that viewer, object, and space would be reliably translated and coordinated geometrically, and fixed two-dimensionally; it was through the institutionally sanctioned production and distribution of representational results that the retraining of sight was possible. It was, in other words, not just the theoretical legacy of "realism" but the very materiality of such practices and subsequent representations that helped secure and reinforce the "realist" compliments they received.[30]

Not to get too excited: textual orders of knowledge have themselves depended upon rhetorically educated readers who knew what genre a text belonged to—a scientific treatise is not generally confused with a poem—and who knew what uses, therefore, a text should have—scientific or aesthetic or both. There's no reason to presume that visual or digital production and, therefore, education should ultimately be any different. Images of any use will need to be classified into "genres" that help us pick the useful ones out; and we'll need visually educated interpreters who understand the background conditions—historical and technological—of such production, and who are able to connect habits of sight to the discursive contexts and inferential requirements that go into making knowledge an ordered and intersubjectively consistent affair—a "normal" affair, in Kuhn's sense.

The degree to which the computational and visual turn will involve, as some have suggested, a paradigmatic epistemological mutation—regarding habits of constructing, testing, and confirming epistemic claims, or regarding the granting of scholarly and authorial credit, or conferring the benefits of subsequent intellectual property—remains to be seen.[31] There is no doubt, however, that the virtual and visual modeling and testing that is made available through the combined relations of algorithmic, programmable, and visualizable media are having a profound impact. In a way, the digital construct can be more *generalizably* real than the photograph. For it doesn't participate in the event or the instant; its value, epistemically and ontologically, is in not being constrained by the limitations of a single instant. On the contrary, a digital construct can represent a more general tendency of phenomena to behave in specific and (statistically) repeatable ways. It can achieve much more than a single instant could. In addition, therefore, this changes what we mean by an "adequate representation." If we *were* still tempted to peg the notion of "adequacy" to a fixed and unchanging quality, to the medium's fidelity to the instant, and to its purported ability to guarantee an instant's eternal sameness and repeatability, then the digital will require a change of view. The scientific adequacy and benefit of a digital representation might very well be its transformation of this model; for its benefit can be in its *composite* and *generalizable* reliability, in its being constructed to foreground qualities that an "instant" can't capture, in its being better able to facilitate the observation and manipulation of (computationally available) dimensions that analog images don't possess, given their comparative informational—computational—thinness. In effect, because the digital image *is* data and because it possesses paradigmatically more information than analog forms do, and because digital (data) collection techniques transcend normal and embodied thresholds of sight, they also provide unprecedented opportunities for revising, manipulating, and reconfiguring what wasn't there at all to begin with in a naturalistically analog-sense.

The biggest challenge, therefore, is not likely to be in the more obvious and traditional realm of what constitutes "proof" or "confirmation" and the like, but in the subtler realm of how we understand visually mediated inference and the relationship—now that "correspondence with nature" is even more decisively abandoned—between digital/visual constructions or models and the stuff to which they are said to refer (assuming, of course, that "reference" remains a hallmark, however mutated, of the epistemological enterprise). The issue here is not just one of inference but of translation.

In other words, it seems reasonable to conjecture that the question—of how computational, visual, and constructed models relate to or translate the stuff at issue—is no different in kind from those that have occupied epistemologists all along: Vocabularies, metaphors, mathematics, and models have always operated as middlemen in our efforts to control, predict, and use stuff, as pragmatists and others have long since been arguing. So long as our interests continue to be controlling, predicting, and successfully using stuff, the fact that the middlemen change isn't by itself a new epistemological dilemma.

What may, however, be different are the ways in which these new media relate *us* and reconfigure *us* to the technologies in question, the contributions they make to our choice of problems to engage and experimental designs to consider in the first place. For the technologies will themselves contribute to the selection of those problems to which they can more easily and successfully be applied. They're likely to encourage the collection, selection, ordering, and presenting of "data" that lend themselves most to their own properties and forms of presentation and to serving their primary producers and sponsors. On the other hand, these very same modalities may surprise us in ways that are unanticipated; they may lead to an understanding of phenomena—social and not just natural phenomena—that we might not have thought would lend themselves to such methods;[32] they may promote conjectures and imagined futures that might not have come about otherwise.[33] Thus insofar as our epistemological habits benefit from being disrupted—that is, from reimagined and recontextualized and remediated possibilities—there may be some unanticipated epistemological consequences to follow.[34]

In relatively short order, we moved from the technologically mediated sense—with the help of photography—that the world, action, human affairs could be captured in the discreet and still image, siphoned off from the moving contexts in which the image was taken, to the production of moving images, but of *non*existent worlds. We've moved from the single image as paradigm case—painting and photography—to combinatorial and generated imaging—for which a faithful and transparent relation to the real isn't always or necessarily required or expected. But it is the intervening step that might catch an epistemologist off guard. By increasing our control over a *constructed* (fictional) reality that will never be available to natural sight, we acquire a new level of control over the real. Not because our representations are "correct" or "accurate" in some correspondence-like sense, but because we can now modify the image-as-data without risk of any direct engagement with the real and, thus, apply it successfully later. We can now model complex systems without risking real interventions to test the possible consequences. We can bring representations home from a distance for further work and, after comparing a variety of simulations, intervene in ways that weren't possible otherwise.

To put it somewhat paradoxically, in medicine we now have nonrepresentations, in the realist and analog sense, or fictions of composite data that become direct instruments (in surgery, for example) for real intervention. To make the connection even more intimate, we now have real-time imaging that permits real-time action, contexts in which transmission and development time are absent. In effect, time and time-delay are no longer part of the ontology of the picture. Rather, the digital picture is a means by which we achieve a temporally continuous interaction with the data we ourselves have, in the process and for the sake of medical action, become. Such possibilities don't stop, of course, at medicine.

The medium may not be the message, but it is a form by which a message becomes available; it is a schema whose properties, with the benefit of sufficient use and stable conventions guiding that use, recede into the background. We don't generally get upset at the contribution the medium and some of its favored forms make to our understanding when they remain "transparent"—that is, in the background. When these properties are brought back into the foreground and become themselves the object of analysis, we're better able to see their role in leading us to see in specific ways, better able to see the types of seeing they foster (as several in this volume demonstrate). Gombrich, for example, discussed how illustrated anatomies developed, not as "'transcripts' of things seen but as the work of trained observers who build up a specimen" as much by studying previous conventions and learning a style of illustration as by observation. To gloss Gombrich further, the point is that the style used is task oriented, and its "transparency" is a function of its utility with regard to the task.[35] The properties of the medium are selective; they have their utility precisely because they foreground some features and not others we don't want in the way. We will, therefore, tend to see what is presented, not present what is seen. And where there is a technologically assisted way, there is a will: a will to continue to see in the ways the technological medium makes available and useful. There is no neutral naturalism at work, but technologically organized schemata; and these are required before we undertake to "copy" or "model" the real. This need not prevent us from praising the "accuracy" of the image, so long as we remain aware that the "accuracy" is contingent upon a translation—from stuff to schematically oriented translation of the stuff; from stuff within some natural system or other to *visible* stuff we can now see, with artifactual extrapolation and assistance, for some purpose or other. Thus "accuracy" and "correctness" needn't be abandoned, but they are contingent upon trained viewers and interpreters who implicitly know how to do the viewing. Like any useful map, it's only good—accurate and correct—if its user knows how to translate it back into the space he needs to negotiate, to locate the star or the restaurant or the neuron he's failed to find. Given this context of use, one can certainly talk meaningfully of "accuracy," even "truth". But "truth" in this sense will be a function of the conventions by which the real was translated into the artificial and made the tool of successful description and action.

Ordering existence: Instrumental ontologies

In concert with the epistemological schemas discussed above, our traditional ontological schemas have taken the object as central, and have generally operated on the basis of assumed and necessary boundaries between and among these objects, and between all those objects and us subjects who define all those boundaries. Thus existence gets divvied up between and as belonging to certain kinds—between physical and metaphysical, natural and artificial, and between animate and inanimate forms of existence. It is not that these purported boundaries have not been theoretically and practically challenged on a regular basis, made fuzzy and fluid already; it is that such schemas seem even more obviously inadequate for conceptualizing and managing our digital relations. Our understanding of nature and of ourselves, as well as our interventions into both, are increasingly interwoven with the artificial. The computational has increasingly *become* the schema or provided the vocabulary by which nature and subjects like ourselves become described and interconnected. Existence or Being, whatever that might refer to, is increasingly the reserve upon which and with which new computational and digital schema interact, and this interaction is increasingly organized around forms of visibility. "Being" is what can be seen, given visual representation, made available, distributed, and shared through digital technologies.

Of course, old ontological ambitions might be recontextualized and revived, even enhanced, as they're transformed by new digital means of sight. We might find in the realm of digital/visual technology not more reason for Heideggerian anxiety but, rather, optimistic claims for a new "authenticity," for "existential computing" and "existential technology," as Steve Mann, Ray Kurzweil, and others claim.[36] Once we see ourselves as tools all the way down, we may consider the redesign of the self in terms of what the newest (cyborgian) tools permit. We may hope, for example, to engage a society's commitment to surveillance with our own "sousveillance"—the pleasurable appropriation and internalization of technologies of sight that permit us to look back, record, keep in check, and balance the systems that observe us. We might happily and more deeply embed ourselves within the economies of technology and visuality that prevail, happily consider identity and agency—our "authenticity"—in terms that these technologies themselves provide. Aside from the normative questions that arise when ontology is recast in terms so congenial to the present political economy, our question is, indeed, the role seeing and technologies of sight play in its articulation, in shaping the way in which existential categories will be recontextualized. In the schema suggested above, existential opportunity is understood as being enhanced insofar as one lends one's subjectivity increasingly to the computational, distributed, and networked contexts of commercial gear. Authenticity is understood in visually performative ways: we realize ourselves in the purportedly resistant acts of appropriating, incorporating, and thus looking back at publicly (and privately) ubiquitous systems of observation—we're visually recorded in every commercial space (and increasingly we image ourselves with our cell phones). By uploading one's own images onto "MySpace," for example, we can persuade ourselves that we are using *it*, determining our own form of representation, making ourselves visible in ways that remain under our control. Seeing is not simply believing in some epistemic sense; rather, such seeing is synonymous with being and acting in authentic (observable) ways.

New tools have regularly become the metaphors we literalize into new conceptions of the self or the body: Descartes' clockwork mechanics of the body helped draw it away from the church and into the lab; Galen's hydraulic projections of the pneumatic body helped explain its power; robotic metaphors and fantasies have been abundant since the nineteenth century; and computational constructions of mind and intelligence are now as much a staple in Computer Science and Philosophy of Mind textbooks as they are in contemporary cinema. What may be different about the present application of computational metaphors is their direct and successful application to enhancing sight in particular. Increasingly, what we see as real and the means by which the real is made available is digitally mediated; and these metaphors trade on a traditional techno-fantasy. We have invented something that allows us to see and to overcome the naturalistic limitations of our modes of sight at the same time. More of the world has become available to us than ever before, with apparent immediacy and transparency, and for the sake of knowledge and action that is both intimately near—at the level of neurons and photons—or forever too experientially distant, in an astronomical sense, to be comprehended, except through computation and quantification. Because computation is infinite in principle, we concentrate less on what might not translate well into its operations than on a potentially limitless power whose nature exceeds nature. Unlike the limits of clocks, hydrology, and clumsy and mindless robots, the digital seems to surpass all the conceptions and capacities we have previously attributed to ourselves. And it's good business.

Orders of the image: Forms of sight

Both Kantian-inspired idealists and positivist-inspired empiricists might be said to agree: we operate in the world on the basis of a visual scheme—internally *a priori* and/or empirically and externally caused, as the case may be. But the very idea of a visual scheme in these senses goes against the history of seeing. Habits and conventions of sight also follow the media and technologies by which products of sight are made available as representations, given a symbolic context, and put into social circulation. It is now commonplace to be reminded of the many complex background conditions that have to be met before painting can serve as a symbolic marker of anything real or important. Likewise, the pictorial standards established by specific conventions of painting—medieval, say—were paradigmatically altered with the invention of perspective, as were painterly standards with the advent of the photograph. Not only did photography create new schemata of seeing, but the previously unavailable possibilities for sight that photography enabled also directly affected what painters then saw themselves as doing or being allowed or encouraged to do. In effect, with new technological media of visual production come not only new ways of seeing, and challenges to the visual schemata associated with these older media, but also new ways of using and understanding the older media themselves. Any history of painting, photography, and cinema, and also any history of science, medicine, and history itself, will be a history of "seeing"—a history written in the context of how new technologies and media of sight have transformed their predecessor's operations, how they've changed what we see, the way we see it, and the uses to which these visual products can be put. In effect, modern history can't be written without writing about sight, about the images through which the events of history have been made available to us, and about the social impact such technologically mediated processes of seeing have had. One would not be capable of narrating any recent war, for example, without talking about the images through which that war is "seen" and, equally important, without considering how technologies of visual representation shape how the conflict is waged in the first place: more so in the present than ever before. The traffic goes both ways. Similarly, the history of science can't be written without writing about sight, about the role sight-assisting and sight-transforming technologies play in schematizing, sorting out, and identifying those qualities of the "real" that interest us.

Again, it is not that we know in advance what we need to see and then go looking for a tool to help. The tools shape what we see and think we need to see; they bring new practices and epistemological justifications for those practices into play. It is not as if scientists or artists or historians have prior intentions of what they'd like to see and then pick up a tool to do the job. Their intentions and visual needs are formulated in a context of already-available tools and the conventions of seeing and producing that these tools facilitate. Again, our refrain: there is a kind of "intentionality" that is already associated with the tools themselves and their specific conventions and contexts of use.[37]

In order to dramatize what the digital image is in relation to its predecessors, one might put it, as Manovich does, in a paradoxical form: "the logic of the digital image [is] paradoxical: radically breaking with older modes of visual representation while at the same time reinforcing these modes."[38] That is, the physical modes of production are categorically different, yet the digital also "solidifies (rather than destroys) certain aspects of modern visual culture." It is different at the level of its basic materiality and with regard to the techniques that follow from this materiality, yet at the level of use, application, and reception, much remains the same: we still watch and want to watch certain kinds of movies, see specific kinds of images, and apply them referentially to the real. The digital is applied to serve these ongoing purposes. This isn't really a paradox; it's a question of intellectual concern: if the concern is to describe the material properties, processes, and subsequent technologies of image capture or construction, storage or printing and distribution, then a sense of difference and departure—revolution, perhaps—will prevail.[39] If the concern is cultural function, use, or application—whether epistemic or entertaining—much will remain the same, or at least continuous, because we want it to. There will continue to be things we ask of the digital, precisely because we became invested in asking them of the previous technology. This is not to say that entirely new functions and purpose won't come into play as a result of the digital—they certainly will. This is to say, rather, that neither the language and logic of paradox nor the theoretically motivated language and logic of revolution may adequately capture what's happening. While the persistence of old metaphors of sight—"capturing" or "copying" the real—may obscure what's different about the new, the computational and informational emphasis propelled by the new can obscure the endurance of many of the purposes to which imaging will continue to be put (as both Manovich and Mitchell both explore).

That the boundary of image making and processing is no longer established chemically but computationally (which may appear to transcend all previous conceptions of there being a physical boundary to contend with), this technological difference doesn't, by itself, determine that the purposes to which images will be put and the forms of reception and inference that will be employed for such purposes will change. Some will, some won't. For example, once one concedes that "the real" never was the direct cause or the end result of a technology, but, rather, the achievement of an interpreter, the digital may serve more "realistic" functions than analog imaging ever could. If "photorealism" is the established convention of seeing, the digital may have to be "diluted" to measure up (or down, as the case may be) to its expectations. And, as Manovich suggests, the digital may well be asked to conform to conventions of sight that older technologies established. The point, however, is that with the digital, new expectations, new habits of sight, will emerge. It is not that sight is simply a natural phenomenon looking for technological assistance; it is the result of an interplay between our embodied capacities and artifice, between our eyes and how technologies and cultural practices guide them. In effect, sight has a future that is still in the process of being invented, not a nature yet to be discovered. And the next stage of its future will be the result of both the purposes we ask of it and the technologies that, in turn, contribute to reshaping our capacities and purposes.

While early photographic discourse sought the service of some traditional philosophical binaries that were already part of the intellectual grammar of seeing—presence vs. absence, objective vs. subjective, immediate vs. mediated, inscription vs. interpretation, correspondence vs. construction—the digital makes these even more questionable than they already were. Traditional photography—writing with light—inevitably retained the sense of a required and causal relation between image and referent, between the apparatus and the "direct action of light," between real things and their representations. But the photograph was, at the same time, selective, discrete and atomic, detached temporally and perceptually from the context or ecosystem in which perception takes place.[40] It could only become a representation by virtue of many complex, mediating processes—institutional and semiotic—upon which its representational work depended. Digital imaging, while more (computationally) abstract a process, can, however, become *more* "real and direct," in the sense that it can display ongoing action, rather than temporally discrete, decontextualized, and past action; it can become interwoven in and with the ecosystems or bodies it is "recording"; it can directly mediate and facilitate our action in an environment that is both present and/or absent, immediate and/or distant, too small to engage any other way or too complex to approach without the help of computationally selective modeling. Digital imaging can bypass perspectivally and two-dimensionally fixed viewer/object relations and rotate three-dimensional options without fixing a single or correct position for the viewer, and it can do this by computationally extrapolating from two-dimensional images to three-dimensional perspectives that no real viewer has ever previously enjoyed. It can also represent measurable and remotely sensed depth where two-dimensional images required inference; the digital can, in other words, change previous, perspectivally grounded perceptual habits and expectations. In effect, the older binaries that helped confirm the initial "realism" of perspective or photography, and that accompanied these new procedures and expectations for seeing, are now even more obviously incapable of guiding our sense of what imaging is and does for and to sight.

It doesn't follow, however, as Mitchell and Tomas suggest, that this will "precipitate the cultural dissolution of the photographic eye" or a culture that "has no need for a witness." As has already been argued, the old binaries need not be treated as metaphysically reified and epistemologically naïve bogymen. They'll have some practical currency insofar as they continue to offer some pragmatically useful help for sorting out the images we want from the images we don't want. They'll be an impediment or a quaint remnant of a bygone conceptual era if and when other concepts do a better job at helping us do this sorting. In some cases, the new technology will serve some old and valuable purposes established by some old and venerable interests, ordered and understood via some equally old and venerable concepts—"witnessing the real," for example. In other cases, new post-realist and hyperreal concepts will be needed to accompany what new technologies do—"construct the real," for example, or reorder the real in ways that could not be made available any other way. The point here is that there is no direct, causal link between what technologies do and what concepts or rhetorical strategies we adopt to do the work we ask of concepts when thinking about what these technologies do, or what they might do. Concepts, whatever their initial theoretical origins, aren't required to forever wear the stain of their metaphysical history, or to change by virtue of the technological company they're keeping in the present. These histories are contingent and renegotiable. Thus the meaning such concepts acquire is determined more by the uses to which they're put than by the histories with which they were once associated or the technologies with which they're now associated. In this sense, old binaries—real and unreal, immediate and mediated, transparent and transformed, caused and constructed, etc.—may continue to have a pragmatic and rhetorically successful life long after the metaphysical ambitions that previously propelled them are dumped. The cautionary point here is that one occasionally detects the legacy of a kind of technological determinism in some discussions of the digital, as if its mere existence and difference from previous technologies is sufficient to cause or demand that old concepts be considered passé or inoperative. But the traffic between the technological and the conceptual, like the historical and the conceptual, is neither directly causal nor one-way.

Moreover, whatever the imagery and its purported claims to "objectivity," an image's utility will depend upon the judgments of interpreters. And in the digital realm, this may become even more obviously the case. As Galison suggests, the expert eye is needed to complete the task of the mechanical eye.[41] Thus while digital imaging is on a different informational order than analog, this does not entail proportionally more "accuracy" or "objectivity" and, therefore, less of a role for interpretation. On the contrary, a more finely tuned judgment and interpretive skill may be needed to ensure that the most useful visual information is what's distilled from the image's paradigmatic increase in informational quantity.

Either way, images don't stand alone as uninterpreted icons of the objective. While radiology and mammography, for example, have provided imaged information not available before, they fail conspicuously to provide, on their own, a diagnostic or screening consequence that would live up to our most fervent objectivist and representational hopes. On the contrary, it is the skill of trained interpreters that converts the image into something useful, that gives the image its force, by extracting what's salient; and these skills are neither reducible to nor commensurable with the computationally grounded and formal procedures we depend on for the imaging itself. Images, while automatically taken, are not automatically sorted or understood: sorting and understanding are not algorithmic. Rather, it has become clear that interpreted, holistic (rather than linear or sequential) perception, and trained and unconscious (rather than methodologically and explicitly applied) capacities of pattern recognition are the forces that tend to guide the "seeing" that goes on.[42] With regard to mammography, for example, it's become clear that the image alone provides too little of the right information; and studies have shown that visual dependence and sight-based and mechanistic images of objectivity together can lead us to denigrate other revealing modalities of perception. These studies have shown, for example, that skilled touch can reveal much that doesn't make it into the image.[43] This suggests not only that such skills are not reducible to or able to fit within the paradigm of the "computational apparatus;" it also reminds us that, as Galison writes, "the images of judgment are neither those of truth to nature nor those of mechanical objectivity."[44] We're reminded of what Gombrich demonstrated some time ago with regard to the interplay between science and art: successful depiction is the result of the trained eye, and there is no necessary causal or mechanistic, natural or objective relation that connects the medium (painted or digitized), the object depicted, and the understanding that is extrapolated.

Understanding the social: Visibility, imaging, and agency

Much worry has already been expressed about the transformational effects of social relations when reconfigured along predominantly visual and virtual lines: "the society of the spectacle," "the surveilled society," "the world as simulation," "the absence of communicative reciprocity" associated with visuality, etc. The point is that as the social becomes increasingly presented and represented through the visual, as social activity and interactivity become increasingly transacted through digital and sight-based media, the social order itself is transformed. The order of the social can no longer be predicated upon traditionally realist conventions of proximity and direct exchange, labor, and presence; and the endurance or integrity of these conventions can no longer be affirmed along realist lines of presence, event, journalistic reportage, and/or photo-real depiction. This is not to say that the real disappears but, rather, that it is in the realm of what can make an *appearance*—be seen—that the real will be recognized as real. And as the real becomes increasingly mediated by digital imaging, concerns regarding integrity, reliability, etc. are more vulnerable to and transposed to the technological sphere itself.[45] Thus thinking about the social requires that we consider the ways in which various digital/visual technologies and the practices associated with them interact with it.[46]

Given the speed, immediacy, and saturation of representations, concern might be expressed that there is a risk that all of the potential facts and provocations for action that facts entail might become converted back into "fiction": they will come to us in and through forms of spectation and artifice whose appropriate form of engagement is reception, consumption, and passivity. There is no corresponding and collective agency implicit in or required by our relation to these forms. Thus imaging and correspondingly new forms of visibility suggest new forms of agency in relation to the social.

The visual is, of course, like every other sense or epistemic modality, open to usury. Industries of fashion, cinema, advertising, politics, and war-making all contribute to the formation of visual habits in various problematic ways; and these habits tend to commodify, to derealize gruesome realities, and to shape a compliant and satiated spectator-subject. War and reality TV come to us without visual differentiation. And this has had worrisome consequences regarding our capacity to "see" and respond to some unfortunate circumstances. We need, for example, to see our environment differently if pollution is to be curbed. We can't "see" toxins, however, the way we see and enjoy seeing bodies; we don't "see" temporal processes (especially complex temporal processes of environmental degradation) without strongly argued discursive context. Therefore the visual modality, especially when aligned predominantly with immediate forms of gratification, doesn't make the transition well to new global contexts where a specific kind of visual comprehension is needed and where that comprehension needs to be connected to collective forms of agency to which the initial act of spectation is not related. On the other hand, these very same issues have achieved some alternative and critical representation by virtue of the technologies of simulation now available. We can now remap the globe and simulate critically useful representations that highlight the production and temporally complex transformations of pollutants; we can construct images that revisualize transnational boundaries and nation-states in terms of toxicity, deforestation, energy-consumption, and violence etc.

In medicine, imaging technologies have had a particularly ambivalent record. They have represented a powerful and often gendered confluence of science and cinema, objectifying and depersonalizing the body for the sake of medical control, but not without their successes and benefits (see Cartwright and Goldfarb in this volume). In addition, many of the outcomes of such technologies bypass the stage of object-production altogether. In effect, the Genesee River next to Kodak Park is likely to become cleaner as digital processing and interactivity overtake traditional, chemically based film production and processing.

Baudrillard's concern remains, of course: new media forms induce their own social relations and forms of noncommunication, especially if we assume that such forms continue to project an animus of power, passivity, and receptivity that is characteristic of the vision industry and exacerbated by these new forms. Our purpose here is to suggest, first, that in approaching philosophical issues of vision in connection with new forms of technology, the approach need not be determined by the kind of modernist reflexes that often leave unstated their nostalgia for pre-industrial modes of communication—a sentiment that comes across as nostalgia because these reflexes assume that embodied and pre-industrial communicative modes have been exterminated and that visual communications have become *totally* mediated by the distancing powers of the digital.

Various modes of visual production operate as relational inducements, for sure, but these modes also take shape and acquire functions within various forms of intellectual and social life that are ongoing and not yet dead. The suggestion here is that a pragmatic form of intellectual life is better suited to suggest ways in which vision and critical discourse, vision and possibility, the ocular and the theoretical might be positively aligned. The historically dominant forms of political and economic life that have connected our various and different visual practices probably have a lot to answer for; but the interplay among these modes—whether connected generally by scientism, commercialism, or something else—needs to be imagined through a discourse that also assists us in shaping the kinds of normative affairs we would *like* to see.

While acknowledging the ideological hype that accompanies new technical possibilities and while conceding the powers of appropriation that existing ideologies possess, it seems worth considering how we might like to see a public sphere enhanced through the visual—given the unavoidability and ubiquity of visual modalities. How can communicational reciprocity be enhanced and vision made more viable, after theory has done its thing and made us appropriately suspicious? Assuming there is always a power/knowledge connection, and that this is not *ipso facto* bad, perhaps we can better conceive, in our present technological circumstances, of forms of vision that are admittedly constructions (not discoveries), perspectival (not total), interactively uncoercive (not one-way), contestable and ambivalent (not all-mighty or all-bad). Incredulity and possibility or, as Rorty put it, irony and solidarity, are not necessarily mutually exclusive, even if not necessarily connected either. We can, in other words, maintain a critical perspective theoretically that doesn't undermine our possibilities practically. We need not deny to the visual some of the analogically appropriate capacities often given to critical theory itself: a capacity to be emancipatory, honest, generalizable, and fair; a capacity to be a form of knowledge whose ability to represent a state of affairs is necessary to its reformatory intent. Implied here is an assumption that drives the entire field of cultural studies and pragmatist philosophy. Dominant imagery (like dominant texts) can only be critically understood if there are already transgressive readers or viewers who are not entirely constituted by a given text or gaze or culture of imaging. Critical and transgressive narratives will need transgressive imagery to do their thing just as transgressive imagery will need narratively prepared readers to give it sense and context.

Representation ought to make us nervous, precisely because it is unavoidable and indispensable for every discipline that wishes to be one. It is not that the consequences of representation are always or necessarily bad—pornographic, as Jameson put it—but that there are always some consequences or other for which suspicion is warranted—we are back to ethics and politics again.[47] In this respect, responsibility and representation are inseparable.[48]

There is not much consolation, of course, in a pragmatically toned down philosophical story that acknowledges the complicities of vision in regimes of repression, that remains appreciative of visualized forms of science and art, but which acknowledges the overarching forces of consumerist production in propelling these forms. Of course, as much as one would like to see spaces opened up by technology that would permit greater opportunities for negotiating between just and unjust visions of the world, such technologies are likely to find their primary role in making more of nature available for economic exploitation. But images of destruction and the Internet as a means of distributing such images are contesting this historical trend. Informational imaging permits the regulators and critics to "see" as well; it permits images to be universally distributed to all simultaneously. Although specific representational forms of philosophy, science, cinema, and economics have contributed to the management of the world in some rather dubious ways, the very same visual modalities can serve a variety of interests, and some of them may be what keep us alive. Surveillance can go both ways. Economic and cultural production, as well as organized resistance to deleterious forms of such production, generally depend upon the same underlying modalities of sense-making, world-making, and interpretation. We are not without the visual means to represent the ills of the social narratives we don't like, but it is putting those images into contact with the right narratives that will make the difference.

Notes

1. For further discussion of some of the implications of Clark's work, see our "On Naturally Embodied Cyborgs: Identities, Metaphors, and Models," in *Janus Head*, E. Selinger & T. Engström (2007).
2. Like the interconnected histories of sight, telescopy, and astronomy, the related history of microscopy and germ theory in medicine is an example of the way previously entrenched visual practices can obstruct a new way of seeing and, thereby, impede our capacity to conceive of a new form of existence and agency—germs: a form of existence that became available only by virtue of a new way of seeing—microscopy—that denied the adequacy of the unassisted eye for seeing and recognizing that existence. The ridicule endured by Pasteur is a lesson in how conventionally habituated forms of sight (and blindness) can become "naturalized" to ignore other possibilities.
3. The *camera obscura* was what Don Ihde calls an "epistemology engine" of the seventeenth century.
4. For a detailed discussion of how "objectivity" and mechanization become linked, see Daston and Galison, "Image of Objectivity," *Representations* 40 (1992: 81–128).
5. See Oliver Grau, "History of Telepresence: Automata, Illusion, and Rejecting the Body," in *Robot in the Garden: Telerobotics and Telepistemology in the Age of the Internet*,

ed. K. Goldberg (MIT, 2001: 226–43), for discussion of how this worry was expressed as a result of simulated religious environments in the eighteenth century, too.

6. See John Szarkowski, *The Photographer's Eye*, for a now classic discussion of how photographs came to stand for the "real" (Little Brown, 1980).
7. See Albert Borgmann's *Holding on to Reality: The Nature of Information at the Turn of the Millennium* (Universtiy of Chicago: 1999), and "Information, Nearness, and Farness," in *The Robot in the Garden: Telerobotics and Telepistemology* in the Age of the Internet, ed. K. Goldberg (MIT: 2001).
8. See Martin Jay's *Downcast Eyes: The Denigration of Vision in Twentieth-Century French Thought*, for a thorough account of the ways metaphors of sight have facilitated philosophy's "totalizing" ambitions and, as a consequence of these ambitions, the general ambivalence with which French thought has considered vision (University of California: 1993).
9. See Richard Rorty's *Philosophy and the Mirror of Nature* for a now classic account of the persistence of metaphors of mirroring in shaping the foundationalist ambitions of philosophy, and for a pragmatist's account of why we should abandon this ambition altogether (Princeton: 1980).
10. See Raymond Williams' *The Country and the City* for an account of how conveniently the eighteenth-century aesthetics of landscape painting fit into the rise of capitalism. The working field becomes reified in painting; it is appreciated for its beauty; the labor behind this beauty becomes invisible, and the painting, like the land, together become available as commodity (London: 1985).
11. Jay, op. cit., p. 276.
12. "Our society is . . . one of surveillance. . . . We are neither in the amphitheater, nor on the stage, but in the panoptic machine, invested by its effects of power, which we bring to ourselves since we are part of its mechanism." See Michel Foucault's *Discipline and Punish: The Birth of the Prison* (Penguin, 1979: 217).
13. See Guy Debord's *The Society of the Spectacle* (New York: 1994).
14. See Milan Kundera's, *The Art of the Novel* (London: 1988) and *Testaments Betrayed* (London: 1996).
15. See Richard Rorty's *Contingency, Irony, and Solidarity* (Cambridge: 1989) for the full-blown version of what philosophy has to learn from its literary ally, the novel: how to overcome its Foundationalist ambitions and learn to live with the contingency of its own language games.
16. Jay puts this point superbly: "The antiocularcentric discourse . . . is . . . an often unsystematic, sometimes internally contradictory texture of statements, associations, and metaphors that never fully cohere in a rigorous way. No single figure expresses all of its dimensions and none would be likely to accept them all, even if they were explicitly posed as positive arguments. But as a powerful if at times subliminal context, the discourse . . . has helped shape the attitudes of a wide variety of French intellectuals who share little else in terms of their disciplines, politics, or theoretical self-consciousness. At times, it provides them with a vocabulary to discuss other issues, such as subjectivity, the Enlightenment, and humanism. At others, it seems to determine the way they approach those same issues, as a powerful metaphoric often does, lending arguments an emotional tone and critical energy that would otherwise be inexplicable."
17. See Jacques Derrida's *Truth in Painting* (Chicago: 1987).
18. For a more sustained and pragmatically disagreeable interpretation of Lyotard's conceptions of sublimity see Timothy Engström's "Getting Over the Sublime?: Philosophy, Rhetoric, and Postmodern Evasions" in *Boundary 2* (Duke: 1993).

19. We're treating semiotics here as a theory of interpretation, not primarily as a theory of cognition or a philosophy of mind that reduces all cognitive and/or mental operations to the linguistic.
20. See Roger Chartier, "Languages, Books, and Reading from the Printed Word to the Digital Text, *Critical Inquiry* 31, Autumn 2004.
21. On the question of how expertise and different conceptions of expertise might be understood in this context, see E. Selinger's "Dreyfus on Expertise: The Limits of Phenomenological Analysis" (with Robert Crease), *Continental Philosophy Review* 35 (2002: 245–79) and E. Selinger & R. Crease (eds), *The Philosophy of Expertise* (Columbia: 2006).
22. See Andy Clark's, *Natural Born Cyborgs*, on how and why our inherently "cyborgian" natures are enhanced as a result of what digital technologies make available. We've *always* remade ourselves and enhanced our cognitive capacities, argues Clark, through mergers with technology, whether previously with the help of texts or now with the help of digital devices. And see Steve Mann's "Smart Clothing: The Wearable Computer and WearCam" for even more techno-enthusiasm for the ways digital imaging devices can enhance our cognitive, cyborgian capacities, in *Personal Technologies* 1, no. 1 (1997).
23. One might take Dewey's interactionist and emergent conception of mind and cognition to be the essential philosophical insight needed here. While the empirical details and technologically relevant contexts have changed, we take the prevailing philosophy of mind at play in working out the emerging details of how technology and cognition might be interwoven—"extentionalism," "externalism," etc.—to presuppose this interactionist conception. The benefit of Dewey's model over some recent analytic and phenomenological rivals, however, may be its concern for the role the larger political economy plays in providing and shaping the means available—technological and otherwise—for extending or enhancing (or distorting) our cognitive activities and commitments.
24. See John Dewey, "Ralph Waldo Emerson," *Characters and Events: Popular Essays in Social and Political Philosophy*, J. Ratner (ed.), 1957 (New York: Henry Holt and Company, Orig. 1903).
25. See Feenberg in this volume on some of the implications on cognition and learning when these are contexualized within electronically mediated distance-environments.
26. Risk, and the "intercoporiality" that is presupposed by our capacity to appreciate risk and vulnerability, is at the heart of Merleau-Ponty's phenomenological account of our existence. Dreyfus employs this view in his analysis of what is most importantly absent from visually mediated technologies, especially those whose purpose is to make tele-presence and thus tele-engagement possible. See Hubert L. Dreyfus, "Telepistemology: Descartes's Last Stand," in *The Robot in the Garden: Telerobotics and Telepistemology in the Age of the Internet*, K. Goldberg (ed.) MIT (2001: 48–63).
27. See Roger Chartier, "Languages, Books, and Reading from the Printed Word to the Digital Text," *Critical Inquiry*, Autumn 2004; Roger Darnton, "The New Age of the Book," *New York Review of Books*, 18 March 1999.
28. See K. Goldberg (ed.) *The Robot in the Garden: Telerobotics and Telepistemology in the Age of the Internet* (MIT, 2000) for a range of discussions on the theme of how recent technologies challenge our conceptions of the relations among knowing, doing, and proximity vs. distance.

29. Our general understanding of "culture" has itself been deeply intertwined with the history of *material* collecting, of appropriating, preserving, and placing objects of "other" cultures in our museums. Thus one might say that knowing in the cultural sense, or making a culture intelligible, is linked to this history of collecting. See James Clifford, *The Predicament of Culture: Twentieth-Century Ethnography, Literature, and Art* (Harvard, 1988) for a discussion of the mutual dependence both ethnography and modern art had on the culture of collecting: both Lévi-Strauss and Picasso depended upon bumping into the collected, abstracted, and decontextualized artifacts to be found in New York City "junk shops" for their respective inventions of structuralist anthropology and modern art. See especially Chapter 10: "On Collecting Art and Culture."
30. See Crosby in this volume.
31. See Lawrence Lessig's *The Future of Ideas: The Fate of the Commons in a Connected World* (Vintage: 2001) and *Free Culture: The Nature and Future of Creativity* (Penguin: 2004) for far-reaching discussions of how these issues are being negotiated as a result of digital interactivity. See also A. Ray and E. Graeff "Reviewing the Author-function in the Age of Wikipedia," at http://condu.it.rit.edu/wikitheory/index.php/Umich_conference_paper.
32. See Grim et.al. in this volume on the contribution computational modeling can make to our understanding of prejudice reduction, for example.
33. Such technologies may, for example, propel us to consider forms of subjectivity, citizenship, and embodiment—appliance-facilitated and/or prosthetic—that wouldn't be possible otherwise. See Cartwright and Goldfarb in this volume on some of these considerations.
34. On the notion of "remediation"—the way in which new media depend upon and simultaneously transform what their predecessors make possible—see *Remediation: Understanding New Media*, by J. D. Bolter and R. Grusin, MIT Press, 2000.
35. See E. H. Gombrich, *Art and Illusion: A Study in the Psychology of Pictorial Representation* (6th ed., Phaidon, 2004, p. 83).
36. See Steve Mann, "Existential Technology: Wearable Computing is Not the Real Issue," *Leonardo* 36, no. 1, 2003: 19–25.
37. Modern warfare is an increasingly obvious example not only of the role that technologies of surveillance play but of a practice that is increasingly shaped by anticipatory expectations of what the warfare will look like, how it could or will be represented to a wider public. Technologies of sight are already deeply "embedded" as organizational features of both the making of war and the resistances to the making of war. Political demonstrations themselves are organized not to have a direct impact on the location of their performance; they are performed and recorded simultaneously to be made available as a representation of resistance, as an immediately and universally available sign of resistance. In effect, technologies of sight and communication are already internal to and co-constitute the "event" itself.
38. See Lev Manovich, "The Paradoxes of Digital Photography," in *The Photography Reader*, Liz Wells (ed.) (Routledge, 2003: 242).
39. This is the case in William Mitchell's *The Reconfigured Eye: Visual Truth in the Post-Photographic Era* (MIT: 1992).
40. See David Tomas' "From the Photograph to Postphotographic Practice: Toward a Postoptical Ecology of the Eye," for an ecological account of the optical that is consistent with what's being argued here, in *Electronic Culture: Technology and Visual Representation*, Timothy Druckery (ed.) (Aperture, 1996: 145–53).

41. See Peter Galision's "Judgment against Objectivity," in *Picturing Science, Producing Art*, Caroline A. Jones and Peter Galison (eds) Routledge (1998: 327–59).
42. Galison, op. cit. p. 345.
43. See Malcolm Gladwell's "The Picture Problem: Mammography, Air Power, and the Limits of Looking," *The New Yorker*, December 13, 2004, for discussion of what's needed to supplement, contextualize, and make diagnostically useful a given image, and how regularly we succumb to mechanistic conceptions of objectivity that lead us to denigrate non-mechanistic and non-sight-based skills, especially in the context of medicine.

    In addition, these kinds of skills are often depreciated *because* they're embodied and, moreover, associated with nursing—with women who are often in closer proximity to the bodies of patients than to the machinery of diagnostics and who possess less professional status than the Doctor or technician who is authoritatively wielding the machinery. See the phenomenologically informed work of Patricia Benner, especially *From Novice to Expert: Excellence and Power in Clinical Nursing Practice* (Addison Wesley, 1984) and *Expertise in Nursing Practice: Caring, Clinical Judgment and Ethics* (Springer, 1998).
44. Galison, op. cit. p. 347.
45. After one of the presidential debates in which he participated, Peter Jennings was asked to comment on how it went. He said he couldn't say; he hadn't seen it yet.
46. See Steve Mann, Jason Nolan, and Barry Wellman for an optimistic view of how the personal appropriation and internalization of digital devices could turn the culture of surveillance back on itself, into what they call "souveillance"—citizens congenially performing the "resistant" act of observing and recording the apparatus of surveillance, in "Souveillance: Inventing and Using Wearable Computing Devices for Data Collection in Surveillance Environments," *Surveillance and Society* 1, no. 3 (2003): 331–55.
47. "The visible is essentially pornographic," writes Jameson, in *Signatures of the Visible* (Routledge, 1992: 1).
48. Mitchell puts the hoped-for parallelism this way: "Responsibility, for its part, cannot exist apart from representation. It takes the form of a pledge, warrant, or promise, a representation of how things are or will be. Responsibility is representation, and vice versa." *In Picture Theory: Essays on Verbal and Visual Representation* (Chicago, 1995: 421).

Part II Perception and Imaging

2
Visual Apprehension and the European Renaissance*

Alfred Crosby

The current acceleration of the trend of modern societies toward greater utilization of the visual began in Western Europe almost a millennium ago. Before the fourteenth century pictures always showed God and His aides communicating with humans by speech. Then, soon after 1300, an Anglo-French prayer book showed the Virgin Mary pointing to words in a book. The equivalent today would be a picture of Holy Mary pointing to a computer screen.[1]

Two centuries later Leonardo da Vinci declared that "the eye" was the master of navigation, architecture, astronomy, and mathematics, as well as the creator of perspective and painting. "It advises and corrects," he said, "all human arts."[2] The most renowned product of the Renaissance surge of visualization cited by da Vinci were his own paintings and those of the other master painters of the Italian Renaissance.

Before we approach that artistic eruption we should familiarize ourselves with the way pictures had been made. Let us begin with the "now" of medieval pictures. Today we usually think of pictures as depictions of what existed and was happening in a knife-edge instant. The medieval "now" was closer to the kind described by the American pragmatist William James, that is to say, not a "knife-edge, but a saddle-back, with a certain breadth of its own on which we sit perched, and from which we look in two directions into time."[3] For example, as we pass a cubical building, we perceive it not in a durationless instant in which we can never see more than two walls, but *as we move*, and so we can, sometimes, catch sight of three walls in one "now."

The picture makers of the medieval West not only watched the world from William James's saddle but got down and walked around to get a better look. They were no more reluctant to do so than Shakespeare was, later, to halt action while a protagonist thinks aloud in a soliloquy. If medieval picture makers wanted observers to have a good look at the dishes and food on a table, they tipped the table up like the lid of a trunk—and nothing slid off.

Medieval artists were sure that the rank of their subjects was more important than their actual appearances. Artists often indicated significance by the most obvious means: by size, making the protagonists—Christ, the Virgin Mary, emperors—relatively large and locating them at dead center. Unimportant people and things were small and fitted in along the edges or wherever there was an open patch.

The most distinctive characteristic of medieval art to the modern eye was in the treatment of empty space. For us today, things exist in space like vegetables in an aspic salad. The vegetables may be the chief items of interest, but the aspic is undeniably *there* occupying the area among the items of interest. We do not deny the aspic because it is transparent, and we rarely ignore space even if it is empty.

The Florence painted around 1350 by an unknown artist in the medieval style (Figure 2.1) would not satisfy a twentieth century surveyor, but it is an accurate picture of what the city (a collection of buildings, *not* the nothing between them) looked like to visitors as they passed through its narrow, twisting streets. Medieval space was what it contained.

By 1300 that started to change, which surely was related to the West's new obsession with optics and geometry, Geometry, absent from Dante's *Inferno* and *Purgatorio*, shows up in *Paradiso*, where all is well ordered. In the

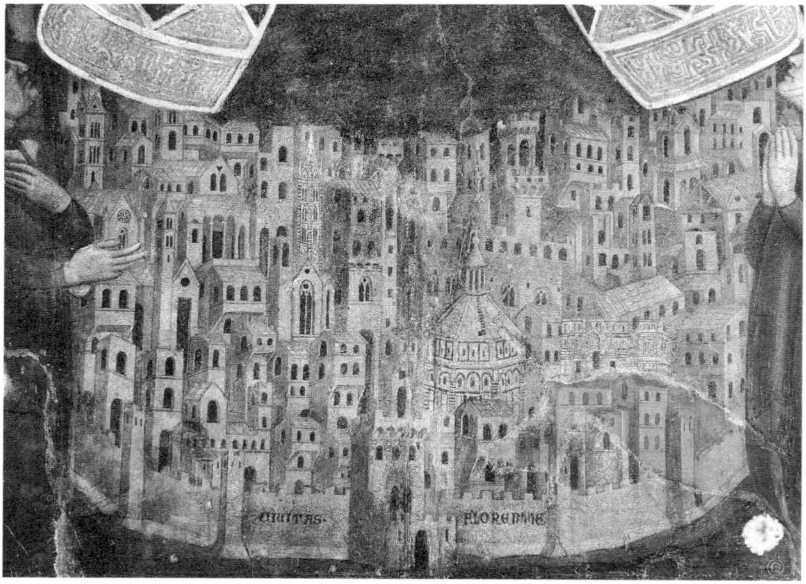

Figure 2.1 Anonymous, panorama of Florence, detail from the Madonna della Misericordia fresco, fourteenth century. Loggia del Begallo, Florence. Courtesy Alinari/Art Resource, New York.

seventeenth canto there is an individual who can see the future "even as earthly minds see that two obtuse angles cannot be contained in a triangle." In the final canto Dante, facing God, the Light Eternal, compares his inability to fathom the relationship between the Deity and humanity to a geometer's inability to square the circle.[4]

"Geometry," Dante wrote elsewhere, "is most white in so far as it is without stain of error, and is most certain in itself, and in its handmaiden who is called perspective."[5] Perspective, then the division of geometry pertaining to light, included within its jurisdiction the making of accurate pictures. By means of pictures, wrote Roger Bacon, "the literal truth might be evident to the eye, and as a consequence the spiritual truth also."[6]

After 1250 the space in Italian paintings began to assert itself; the aspic was starting to stiffen. The Virgin's knee that supported the Christ child began to nudge forward in a shy display of a third dimension. The parallels of walls, ceilings, steps, and moldings of buildings, rooms, and altars announced their relief by easing away from their traditional positioning parallel to the picture plane, and began converging toward some vague area in the back of the picture.[7]

Giotto di Bondone (1267 or 1277–1337) was the first master of the new art. Dante, who may have known him (our most familiar portrait of the poet may be by Giotto), praised him in *The Divine Comedy*.[8] Petrarch called him "the prince of painters." Boccaccio said of him that he had "brought back to light an art which had been buried for centuries beneath the blunders of those who, in their paintings, aimed to bring visual delight to the ignorant rather than intellectual satisfaction to the wise."[9]

Giotto's contemporaries were impressed by the strong organization of his paintings, their combination of high emotion and utter dignity, and the indications of a third dimension (Figure 2.2). To the medieval eye, accustomed to pictures being as flat as blueprints, his seemed deep enough to step into. He positioned buildings and other rectangular structures at an angle to the viewer, with one corner thrusting forward, and walls and edges sweeping back.

Giotto was a genius, but empirically, not scientifically, so. One is usually sure in a Giotto painting which figure is closer to the picture plane than another, but unsure as to how far apart, forward to back, the figures are.[10] He made a brave start at perspective, but his successors made little progress during the rest of the fourteenth century. The problem of "seeing" geometrically was more difficult than we, on this side of their revolution, understand. Taddeo Gaddi, Giotto's student and, in the opinion of some, his successor as the most prominent Italian artist of the century, stuffed *The Presentation of the Virgin* (Figure 2.3) with architecture to indicate the relative position of its many people, but his technique does not fulfill the purpose. If one lived in a world that looked like that, throwing a ball accurately to anybody more than a step or so away would be a matter of pure luck.

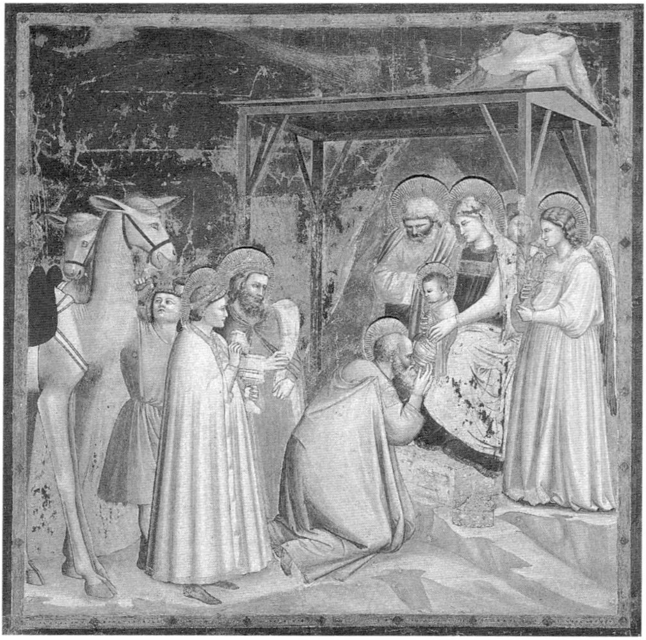

Figure 2.2 Giotto di Bondone, *Adoration of the Magi*, 1306. The Scrovegni Chapel, Padua, Italy. Courtesy Alinari/Art Resource, New York.

Giotto and his school were trying to forge ahead on the basis of no more than artistic instinct. They produced masterpieces—but not geometrically accurate depictions of space. That required something supplemental to artistic genius: theory.

In the fifteenth century the West gained access to original sources of ancient Greek thought. This led to a revival of the Platonic faith that numbers "have the power of leading us toward reality" and that "geometry is knowledge of the eternally existent."[11] The distance between theory and practice vis-à-vis depicting space shrank because now Westerners could take a shortcut through ancient Greece. In 1400 a manuscript of Ptolemy's 1300-year-old *Geographia* showed up in Florence.[12] Ptolemy provided rules for depicting with geometric rigor a curved surface (that of the globe) on a flat surface (a map) via a gridwork (of latitudes and longitudes). The group upon whom, arguably, these rules had the earliest effect was not cartographers but painters.

The identity of the hero or heroes who first actually quantified pictorial art, that is to say, applied Ptolemaic techniques to making naturalistic, two-dimensional representations of three-dimensional scenes as seen by a single observer at a single moment, is not clear. The hero, if there was but one,

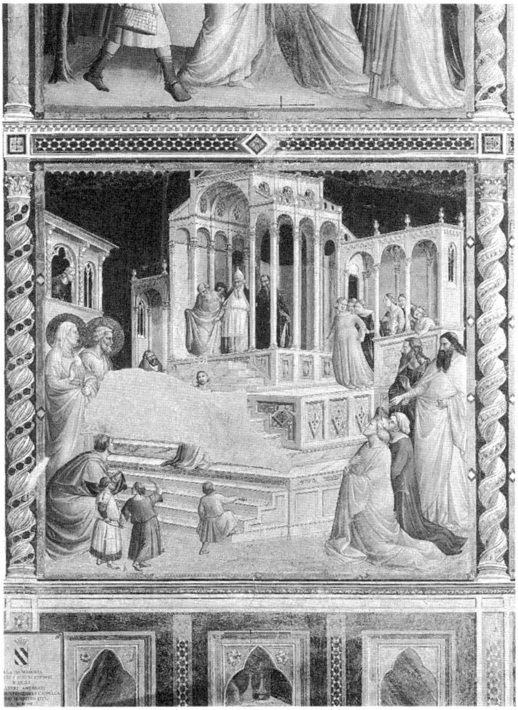

Figure 2.3 Taddeo Gaddi, *The Presentation of the Virgin*, 1332–8. St. Croce, Florence. Courtesy Alinari/Art Resource, New York.

was Filippo Brunelleschi,[13]—clock maker, goldsmith, military engineer, and archaeologist, among other things. He was a fanatical measurer. When he studied the monuments of ancient Rome, he measured and recorded their dimensions as multiples of a basic quantum, not with an unsegmented piece of string or a stick, as was common practice. His ambition was to be so great an architect that his name would be as durable as Giotto's as a painter. This he achieved by designing and directing the construction of the amazing dome of his city's Cathedral of Santa Maria del Fiore.

We can be sure, with that dome as evidence, that Brunelleschi knew enough geometry to understand the problems of perspective. But, like Giotto, he left no autobiography or explanations of his techniques, and the only testimonies to his accomplishment as a perspective painter were written after the fact.[14]

Perhaps the laurel wreath for discovering Renaissance perspective should go to Leon Battista Alberti, who was the first, according to the prominent art historian Michael Kubovy, to present it "as a communicable set of practical procedures that can be used by artists."[15] Alberti was another Renaissance man, a prominent architect, city planner, archaeologist, humanist scholar, natural scientist, cartographer, mathematician, champion of the Italian vernacular, cryptographer, and, like Brunelleschi, inveterate measurer. He offered, if allowed to take precise measurements, to make an exact facsimile in any scale of any statue of whatever quality or size, even as big as Mount Caucasus, even if in two halves in two places, one on the Aegean island of Paros, the other in Lunigiani in northern Italy. He wrote in the 1430s an instructive little book on perspective, a milestone in the history of art.

The Albertian theory of perspective was based on ancient Greek optical theory, interpreted, expanded, and publicized by the Arabs, Grosseteste, Bacon, and others. "Seeing" was a matter of information being acquired by the eye through a cone (or, as it was often called, a pyramid) of light extending out from the eye. An accurate picture was a slice of that cone, vertical to its central axis, made at whatever distance from the eye the picture maker selected. That slice would be identical to what we could produce by sliding a photographic plate across the cone at right angles. Renaissance artists sometimes actually did place a pane of glass across the cone and painted directly on the glass. This would not do for frescoes on walls, but Alberti produced rules that would.

Alberti informed his reader that the first step in producing a picture in proper perspective was to orient the cone or pyramid of the artist's seeing. Its "centric line" would be the shortest possible line from the eye to the center of the scene one wanted to picture. Then, Alberti advised, one should resort to a crude kind of spatial quantification by setting up a veil between oneself and the subject to be painted, a "thin veil, finely woven, dyed whatever color pleases you and with larger threads [marking out] as many parallels as you prefer." (You will remember that Ptolemy's *Geographia*, with gridwork of latitudes and longitudes, was currently de rigueur.) The reality beyond the veil's network should be observed only through the veil, presumably with the head and eye always in exactly the same position. The veil was the picture plane, the slice through the visual cone. One was to paint or draw not what one *knew* to be true about the scene—for instance, with parallel lines always the same distance apart—but strictly what one *saw*. What one saw was parallel lines angling toward each other the farther they extended away from the observer. One could measure how much they converged in appearance by gazing at them through the veil and counting threads. Then one would transfer *that* to a flat surface on which one had carefully drawn lines equivalent to the veil's threads. The veil enabled the painter to quantify not reality, but something more subtle: the *perception* of reality.

Veils and networks proved to be very helpful, but it was hard to "see" only what one actually saw. Some of the early attempts at Renaissance perspective have some very odd features. Structures lean sideways, or do they extend back from the picture plane? One cannot be sure which. (See the strange lean-to at the left side of the building in *The Nativity of the Virgin*, Figure 2.4.) Painters needed, in addition to the veil, geometrical technique.

Alberti supplied that, too. First establish the picture plane, the "window" through which the painter sees what he or she wants to picture. Then draw a person in the foreground, with feet at the bottom of the picture. The head is at the eye level of the artist because it and the artist's head are presumably about the same distance above ground, and it is also at the horizon level because we always see level horizons—oceans, steppes—as level with our eyes. Next divide the height of the person in the foreground into three equal units. These will be the basic units, the quanta, of the picture. Then divide the baseline of the picture into these units. Select a point, the centric point of the visual cone, in the middle of the horizon line. Draw lines from the quanta markers at the base of the picture to this point, which is the

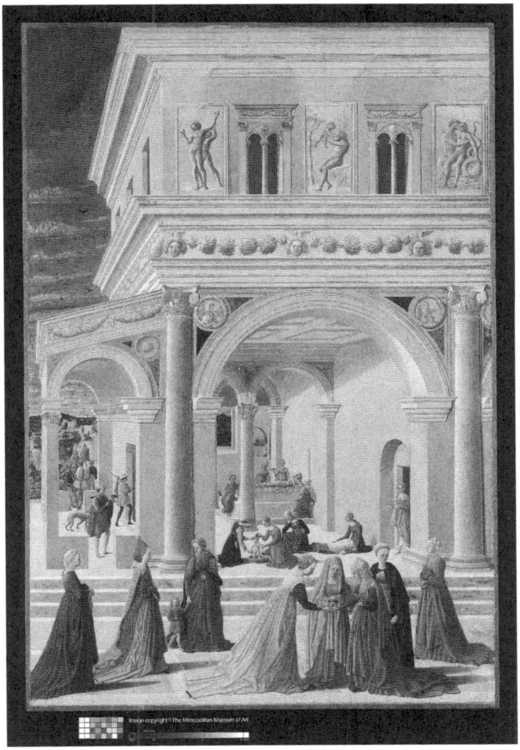

Figure 2.4 Master of the Barberini Panels, *The Birth of the Virgin*, fifteenth century. Courtesy the Metropolitan Museum of Art, Rogers and Gwynne Andrews Funds, 1935 (35.121), New York.

"vanishing point" at which all lines at right angles to the picture plane (orthogonals) meet. (Think of orthogonals as railroad tracks extending straight out from the bottom of the picture, which of course do appear to converge at the horizon.) As these lines converge, so should objects decrease in height and size on the surface of the painting as they recede back from the painter's eye.

Draw horizontal lines across the converging orthogonals. The distances separating the horizontals should diminish at the same rate as the orthogonals converge (this in accordance with one of Alberti's most felicitous inventions, but one too complicated to describe here).[16] Thus we have the chessboard floor typical of so many examples of Renaissance art. (Alberti called the horizontal grid the *pavimento*, the name of the tile floors of houses of his day.)[17] This network, as drawn lines or grooves, can be detected beneath the paint of Masaccio's *Trinity*, painted about 1425, as well as many of the greatest masterpieces of Western art for generations after Alberti. The new perspective, called the *costruzione legittima*, is right out in the open to be seen as tiled floors in scores, perhaps hundreds, of paintings by Leonardo, Raphael, and dozens of lesser artists. These lesser artists, in their innocence, sometimes stood St. John the Baptist on a patch of tiled paving in the wilderness and provided the stable in Bethlehem with similar flooring.[18]

Perspective joined the exalted company of the liberal arts. In 1493 Antonio Pollaiuolo included an allegorical figure of *Prospectiva* as one of the Liberal Arts grouped around the tomb of Pope Sixtus IV.[19]

The sagging tent of medieval space, which had collapsed and billowed in the wind of every influence but Ptolemy's, stiffened and became something to contend with. It had become homogeneous, equal, and preemptory in all its qualities in every vicinity, in all directions, and in all times.[20]

The intellectuals of the Middle Ages respected mathematics in the abstract and tended to veer away from it in practice. Those of the Renaissance respected mathematics, especially geometry, and utilized it extravagantly in practice. Figure 2.5 depicts Albrecht Dürer's print of an artist trying to solve the most intractable problems of foreshortening by glaring through an Albertian veil at a supine nude endwise from the toes up. These and dozens of other examples impress us with the fact that the artists of the Renaissance avant-garde, who were often architects, engineers, artisans, and mathematicians as well as painters, were obsessed with space-as-geometry.

There was also the matter of confidence in contemporary achievement, even in what we can call progress, a sort of faith in short supply among the early medieval intelligentsia and increasingly abundant among the artistic and protoscientific avant-garde in following centuries. Giorgio Vasari, the sixteenth century artist and biographer of artists, praised the painting of his age like an operatic tenor singing of his inamorata. Once upon a time, he said, there had been classical Greek and Roman art, which was very good, and then medieval Western and Byzantine art (with saints "staring as if possessed, with

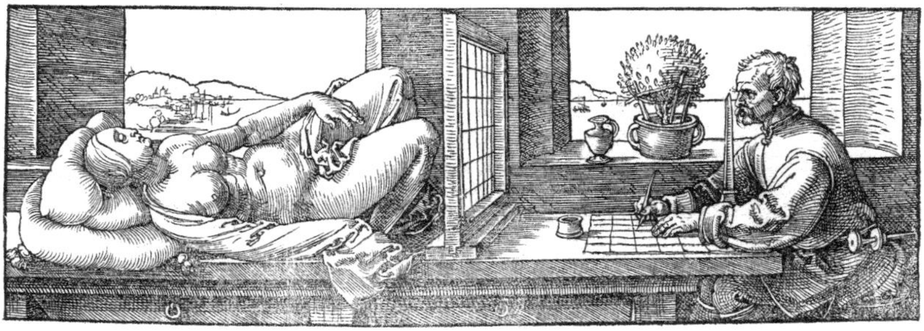

Figure 2.5 Albrecht Dürer, *Draftsman Drawing a Reclining Nude*, 1538. Courtesy the Museum of Fine Arts, Horatio G. Curtis Fund, Boston.

outstretched hands, on the tips of their toes"), which was very bad. Then came Giotto and the rebirth of painting: he and his successors painted in direct imitation of nature. The greatest of them, according to Vasari, was his own contemporary, Michelangelo, who "surpasses not only those whose work can be said to be superior to nature but also the artists of the ancient world."[21]

We pay Renaissance perspective our highest respect: we call it realistic. That, of course, leads to the question, what do we mean by realistic? We do not mean *truly* realistic, because it is rare indeed that we mistake a picture for the real thing. Vasari recorded the story of a painting of a horse by Bramantino so true to life that a real horse kicked it repeatedly, but, as you might expect, Vasari himself had never seen the picture. What we mean by realistic is geometrically accurate; that is to say, you could use a picture constructed according to the principles of *costruzione legittima* as you can use a good map. In contrast, traditional Muslim pictures are exquisitely decorated surfaces without an illusion of real depth; and Chinese landscape paintings, which do offer an impression of great depth, do not have a fixed point of view.[22] Only a boor would not find these beautiful, but you would not want to make your way across even a room, much less a landscape, carrying a tray of full glasses with these pictures as your only guide.

In order to paint pictures that would be realistic by Western Renaissance standards, the practitioners of *costruzione legittima* were obliged to make choices as arbitrary as any made by Islamic or Chinese artists. To cite a few examples, Westerners depicted scenes, they thought, as if viewed in a single instant by a single eye. Most of us have two, which produces stereoscopic vision, but no matter. The eye can focus in a single instant on only the center of a scene, but no matter again. Giotto, Alberti, and company drew and painted scenes as they appeared to be at a single instant—and then took the time to move up and down, back and forth, in order to focus on their several parts.[23] Helpful, useful, justifiable, but in its way as arbitrary.

The Renaissance perspective masters chose to obey the laws of optical perspective as they apply to parallel lines stretching out in front of an observer, which appear to converge, but to ignore the fact that parallel lines that stretch out laterally appear to converge, as well. For the artist to draw these as he or she actually sees them would be to draw parallel lines that converge toward *two* different vanishing points, left and right. That would mean these straight lines would have to appear to *bend*. The only twentieth century artists who consistently obey this optical truth are, oddly enough, comic book artists seeking exaggerated effects.

After the quattrocento the stream of creativity that originated with Giotto, Brunelleschi, and Alberti split and ran off in two directions. One led to more art, eventually to the tortured perspectives of the sixteenth century mannerist painters. The other led to more mathematics: projective geometry, invented by Girard Desargues (1593–1662), advanced by Blaise Pascal (1623–62), and today one of the principal branches of mathematics. Renaissance picture making may be the only art ever to have led to the creation of a kind of mathematics.[24]

Picture making veered close to mathematics, even melded with it, in the fifteenth century. Piero della Francesca, who was born about the time when *costruzione legittima* was invented, and who died the year that Columbus set sail for what turned out to be America, provides evidence in support of that statement. Of Renaissance painters none was a greater master of mathematics, and of Renaissance mathematicians none was a greater painter.[25]

Piero wrote three treatises pertaining to arithmetic, geometry, and painting. The simplest of them instructed merchants and artisans in the use of the counting board and in commercial procedures. For example, this is how to gauge the volume of a barrel:

> There is a barrel, each of its ends being 2 bracci in diameter; the diameter at its bung is 2 and 1/4 bracci and halfway between bung and end it is 2 and 2/9 bracci. The barrel is 2 bracci long. What is its cubic measure?

The answer, obtained after a flurry of calculation, is 7 and 23600/54432 bracci,[26] a calculation and answer that show how deeply at least some of the Renaissance Neoplatonists were acquainted with practical quantification (and also illustrate how much Renaissance mathematicians needed decimals!).

Piero's other two books, among the most important scientific texts of the fifteenth century, were technical treatises on the making of pictures and geometry. Though a master of the subtle use of colors, he ignored color in *De prospectiva pingendi*, a work refining the Albertian principles of picture making. Color was secondary, geometry primary. He devoted the third and last of his major writings to the five regular bodies of geometry. These, the tetrahedron, cube, octahedron, icosahedron, and dodecahedron, had fascinated Plato and would obsess Johann Kepler a century later.[27]

Piero's devotion to Neoplatonism, mathematics, and his art is nowhere more apparent than in his enigmatic masterpiece, *Flagellation* (Figure 2.6). Its Albertian vanishing point is rigidly certain, but where is the center of interest? Is it the three men to the right in the foreground who stand together but seem to be ignoring one another? Or is it the party of men in the background centering on Christ (Christ in a background?), who is being whipped in a scene as bereft of direct emotional expression as a still life of a bowl of fruit?

The *Flagellation* is full of symbols of a Platonized and personal Christianity, and we do not and probably never will understand most of them, but (and this is the painting's special significance for us) they are almost entirely quantitative and geometrical. In their meanings, whatever they may be, they urge the viewer toward mysticism. In the nature of their language, they hustle the viewer toward a mathematical perception of reality.

Painter-mathematicians of the quattrocento painted with a picture unit, a quantum, in mind. Alberti favored dividing the height of a human figure drawn in the extreme foreground into thirds, and used that third as his quantum.[28] The quantum Piero chose for *Flagellation* seems to have been the distance on the surface of the painting between the floor and the point at which the level glance of the painter strikes the wall at the Albertian vanishing point behind the man with the whip. Most of the floor of the

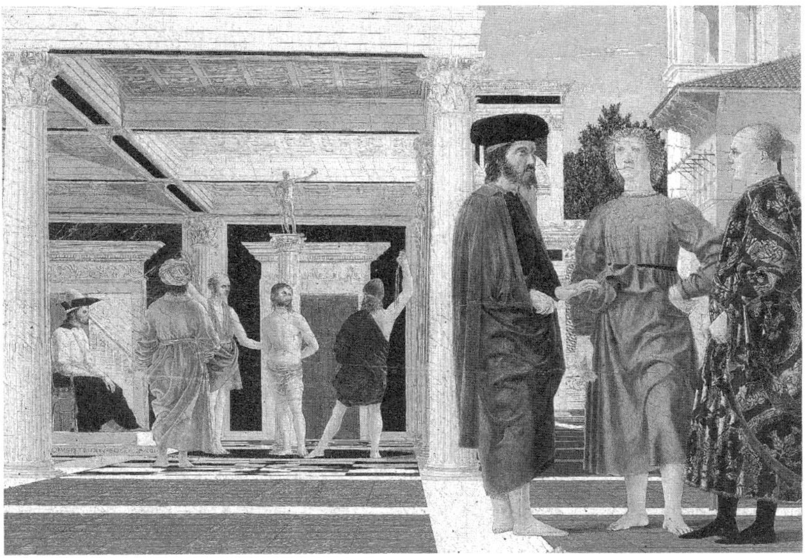

Figure 2.6 Piero della Francesca, *Flagellation of Christ*, 1450s. Galleria Nazionale delle Marche, Urbino, Italy. Courtesy Alinari/Art Resource, New York.

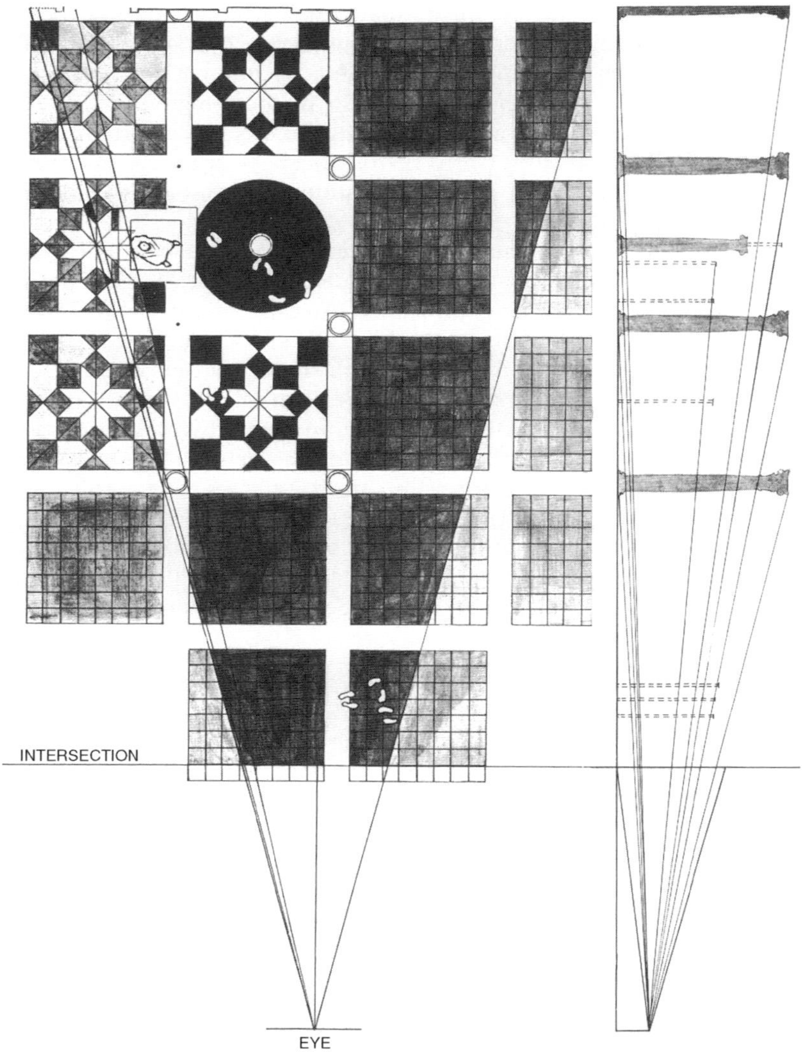

Figure 2.7 "Reconstruction of Plan and Elevation of Piero della Francesca's *Flagellation of Christ.*" R. Wittkower and B. A. R. Carter, "Perspective of Piero della Francesca's 'Flagellation,'" Journal of Warburg and Courtauld Institutes, 16 (July–Dec. 1953), plate 44

visible area is taken up by large squares of brown tiles, each square eight tiles in width and eight in depth. Each of the tiles in the extreme foreground measures two quanta by two, and, therefore, each large brown square measures 16 by 16 of the quanta. The square at the center of which Jesus stands is composed of differently colored tiles in a complex geometrical pattern, but the whole square appears also to be 16 by 16 quanta. The distance between the centers of the two columns close to the picture plane is 19 quanta. It is 38 quanta, twice 19, from the group in the front to the closest of the figures in the group in the rear, the turbaned figure seen from the back. From him to Christ is another 19 quanta. Christ's column, including the statue on top, is 19 quanta high. The distance from the painter's eye to the plane of the picture, which can be calculated geometrically, is 31e and a half quanta; Christ's column is 63 quanta, twice times 31 and a half, behind the picture plane. All the main features of the painting—the foreground group, the nearest column, the turbaned figure, the man with the whip—are at distances from the observing eye that can be expressed in multiples of the quantum times the forever mystical π. And on and on and on into the mazes of mystical mathematics.[29]

If you were a Neoplatonist Christian, you could consult Piero's *Flagellation* as a guide to the ultimate reality. If you were a crass secularist, you could confidently use it to buy and cut carpet and wallpaper for the whole scene[30] (Figure 2.7). The age of exact imaging not merely of reality—that would be easy—but of the perception of reality had dawned.

Notes

*This chapter first appeared in a longer version as Chapter 9, of my book, *The Measure of Reality: Quantification and Western Society, 1250–1600* (Cambridge: Cambridge University Press, 1997).

1. Crosby, *Measure of Reality*, 136.
2. *The Literary Works of Leonardo da Vinci*, trans. and ed., Jean P. Richter (London: Phaidon, 1970), 1: 112, 177.
3. *The Human Experience of Time: The Development of Its Philosophical Meaning*, ed. Charles M. Sherover (New York: New York University Press, 1975), 371.
4. Dante Alighieri, *The Divine Comedy: Paradiso*, trans. Charles S. Singleton (Princeton, NJ: Princeton University Press, 1975), 146–7, 186–7, 376–9.
5. *Dante's Convivio*, trans. William W. Jackson (Oxford: Clarendon Press, 1909), 111.
6. *The Opus Major of Roger Bacon*, trans. Robert B. Burker (New York: Russell & Russell, 1962), I, 238–42.
7. White, *Art and Architecture in Italy*. (New Haven: Yale University Press, 1993), 143–224.
8. Giovanni Boccaccio, *The Decameron*, trans. G. H. McWilliam (Harmondsworth: Penguin Books, 1972), 494; Dante, *Paradiso*, canto 11, lines 94–6; Giorgio Vasari, *Lives of the Artists*, trans. George Bull (Harmondsworth: Penguin Books, 1965), 68; Thomas C. Chubb, *Dante and His World* (Boston: Little, Brown, 1966), 505–7;

Patrick Boyde, Dante *Philomythes and Philosopher: Man in the Cosmos* (Cambridge University Press, 1981), 350.

9. Chubb, *Dante and His World*, 505–7; Boccaccio, *The Decameron*, 493–5; Theodor E. Mommsen, *Medieval and Renaissance Studies*, ed. Eugene F. Rice, Jr. (Westport, CT: Greenwood Press, 1966), 212.
10. Edgerton, *Renaissance Rediscovery of Linear Perspective* (New York: Basic Books, 1975), 97.
11. *The Republic of Plato*, trans. Francis M. Cornford (New York: Oxford University Press, 1945), 241, 244.
12. Edgerton, *Renaissance Rediscovery of Linear Perspective* (New York: Basic Books, 1975), 93–7.
13. Martin Kemp, *The Science of Art: Optical Themes in Western Art from Brunelleschi to Seurat* (New Haven, CT: Yale University Press, 1990), 9, 12–14.
14. Vasari, *Lives of the Artists*, 135–6; Antonio di Tuccio Manetti, *The Life of Brunelleschi*, trans. Catherine Enggass (University Park: Pennsylvania State University Press, 1970), 42–6; Edgerton, *Renaissance Rediscovery of Linear Perspective*, 143–52; Lawrence Wright, *Perspective in Perspective* (London: Routledge & Kegan Paul, 1983), 55–9; Eugenio Battisti, *Filippo Brunelleschi: The Complete Work*, trans. Robert E. Wolf (New York: Rizzolli, 1981), 102–11; Michael Kubovy, *The Psychology of Perspective in Renaissance Art* (Cambridge University Press, 1986), 32–9.
15. Ibid., 32–8.
16. Alberti, *On Painting* (London: Routledge & Paul, 1956), 43–56.
17. Edgerton, *Heritage of Giotto's Geometry* (Ithaca: Cornell University Press, 1991), 156; Edgerton, *Renaissance Rediscovery of Linear Perspective* (New York: Basic Books, 1975), 45.
18. Wright, *Perspective in Perspective* (London: Routledge & K. Paul, 1983), 82.
19. Edgerton, *Renaissance Rediscovery of Linear Perspective* (New York: Basic Books, 1975), 91–2.
20. William M. Ivins, Jr., *On the Rationalization of Sight* (New York: Da Capo Press, 1973), 7–10, and Samuel Y. Edgerton, Jr., "The Art of Renaissance Picture-Making and the Great Western Age of Discovery," in *Essays Presented to Myron P. Gilmore*, eds Sergio Bertelli and Gloria Romalus (Florence: La Nuora Italia, 1978), 2: 144; Edgerton, *Heritage of Giotto's Geometry*, 107.
21. Vasari, *Lives of the Artists*, 36–8, 45–7, 89, 93, 253–4.
22. Wright, *Perspective in Perspective*, 305; Edgerton, "The Art of Renaissance Picture-Making," 2: 135; Yi-Fu Tuan, "Space, Time, Place: A Humanistic Frame," in *Making Sense of Time*, eds Tommy Carlstein, Don Parkes, and Nigel Thrift (New York: Wiley, 1978), 7–16.
23. Wright, *Perspective in Perspective*, 1–32; *Dante's Convivio*, 98; Graham Nerlich, *The Shape of Space* (Cambridge University Press, 1976), 63–4.
24. Morris Kline, *Mathematics for the Nonmathematician* (New York: Dover, 1985), 232–41.
25. Vasari, *Lives of the Artists*, 191; E. Emmett Taylor, *No Royal Road: Luca Pacioli and His Times* (Chapel Hill: University of North Carolina Press, 1942), 191; Kenneth Clark, *Piero della Francesca* (London: Phaidon, 1969), 70; Marilyn A. Lavin, Piero della Francesca (London: Allen Lane, 1972), 12.
26. Michael Baxandall, *Painting and Experience in Fifteenth-Century Italy*, 2nd ed. (Oxford: Oxford University Press, 1988), 86.
27. Clark, Piero, 70–4; Arthur Koestler, *The Sleepwalkers: A History of Man's Changing Vision of the Universe* (Harmondsworth: Penguin Books, 1964), 251–4.

28. Edgerton, *Renaissance Rediscovery of Linear Perspective*, 42–3, 195.
29. R. Wittkower and B. A. R. Carter, "Perspective of Piero della Francesca's 'Flagellation,' "*Journal of Warburg and Courtauld Institutes*, 16 (July–Dec. 1953), 293–302. For further quantitative analysis of this painting, see Kemp, *Science of Art*, 30–2. See also Marilyn A. Lavin, Piero *della Francesca*: "*The Flagellation*" (London: Allen Lane, 1972).
30. Wittkower and Carter, "Perspective of Piero della Francesca: "The Flagellation," Plate 44.

3
Experience and Experiment in Art*

Alva Noë

> *The danger that lies in describing things as more simple than they really are is today often very overestimated. This danger does actually exist in the highest degree for the phenomenological investigation of sense impressions. These are always taken to be much simpler than they really are.*
>
> Ludwig Wittgenstein (1964, p. 281)[1]

> *To be an artist is not a matter of making paintings or objects at all. What we are really dealing with is our state of consciousness and the shape of our perception.*
>
> Robert Irwin (1972)[2]

A significant impediment to the study of perceptual consciousness is our dependence on simplistic ideas about what experience is like. This is a point that has been made by Wittgenstein, and by philosophers working in the Phenomenological Tradition, such as Husserl and Merleau-Ponty. Importantly, it is an observation that has been brought to the fore in recent discussions of consciousness among philosophers and cognitive scientists who have come to feel the need for a more rigorous phenomenology of experience. (See, for example, the papers collected in Varela & Shear 1999 and also Pessoa, Thompson & Noë 1998; Noë, Pessoa & Thompson 2000).

The central thought of this chapter is that art can make a needed contribution to the study of perceptual consciousness. The work of some artists can teach us about perceptual consciousness by furnishing us with the opportunity to have a special kind of reflective experience. In this way, art can be a tool for phenomenological investigation.

The chapter has three sections. First, I present what I call the problem of the transparency of experience. This is a problem for philosophy, for art and for cognitive science. Second, I present an alternative conception of experience as a mode of interactive engagement with the environment. Finally, against the background of this conception, I discuss, briefly, the work of the sculptors Richard Serra and Tony Smith.[3]

3.1 The transparency of perceptual experience

Part of what makes perceptual consciousness so difficult to grasp, whether in science or art, is its transparency. We can illustrate this by means of a fable. Once upon a time there were artists who sought, in their art, to depict reality. For these "realists," art was a way of investigating the world. Then someone noticed that knowledge of the world can be an obstacle to its successful depiction. What matters is not how the world is, but how it presents itself to us in experience. "Free the eye from the dominion of the understanding" became the slogan of the new "experientialist" artists who thus came to overthrow realism. But before long experientialism, in its turn, was thrown into crisis. For to capture in a picture how the world presents itself to us in experience—to make a picture of how things truly appear—is just to make a picture of that which is experienced, of that which appears, namely the world. The subject matter of art-making, then, is not experience itself, but the experienced world, and so art must direct itself to the world. Experientialism in this way collapses back into realism and the struggle begins all over again. An oscillation ensues between realism and experientialism.

This fable is meant to get at the following idea. When we try to make perceptual experience itself the object of our reflection, we tend to see through it (so to speak) to the objects of experience. We encounter *what* is seen, not the qualities of the seeing itself.

This idea was noticed by Grice, who wrote: "if we were asked to pay close attention, on a given occasion, to our seeing or feeling as distinct from what was being seen or felt, we should not know how to proceed; and the attempt to describe the differences between seeing and feeling seems to dissolve into a description of what we see and what we feel" (1962, p. 144).

This is a familiar theme in philosophy. An accurate description of visual experience will confine itself to those objects whose presence is guaranteed by the experience itself, e.g. blobs of color. Or so at least Hume (1740/1978), and philosophers working in his tradition, such as Price (1940) and Ayer (1973), have argued. When we talk of what we see (e.g. deer grazing on a lawn), we "go beyond" what is strictly given to us in experience.

Kant (1787/1929) attacked this idea of Hume's and insisted that we falsify experience when we attempt to describe it in these supposedly neutral terms (as noticed by Strawson 1979). I am not *more* faithful to my experience of the deer, but less, when I try to describe it in terms of my experience of brownish blobs on a green background. To be faithful to experience, I must talk about the way the experience purports to represent the world (Strawson 1979). To describe experience *is* to describe the experienced world. And so experience is, in this sense, transparent.

The transparency of experience, it should be clear, poses a problem for any attempt to make perceptual experience itself the object of investigation in the way that has interested philosophers. But it is important to recognize that this problem of transparency arises no less for the empirical (psychological, neuroscientific) study of consciousness.

Consider Ernst Mach's famous drawing (Figure 3.1) of the visual field (Mach 1886/1959). The drawing is meant to serve as a representation not of the room, but of the visual experience itself, capturing on paper the distinctive character of visual awareness (as the artist sits with fixed monocular gaze looking out on his room). Notice that the drawing indicates the presence of the nose, mustache, belly, body at the periphery, that it represents the field as in sharp focus and uniform detail from the center out to the periphery, where things go indistinct.

Mach's drawing fails (as noticed by Wittgenstein).[4] The picture does not capture what the visual field is like. It is true that the visual field is indistinct at the periphery. But the character of this indistinctness is not captured by the fade-to-white. That is, it doesn't seem to us as if the visual field fades to white at the edges. Consider also that the visual field is not in this way uniformly detailed and in sharp focus from the center out to the periphery. You only see sharply that which you fixate, but you don't fixate such an expanse all at once.

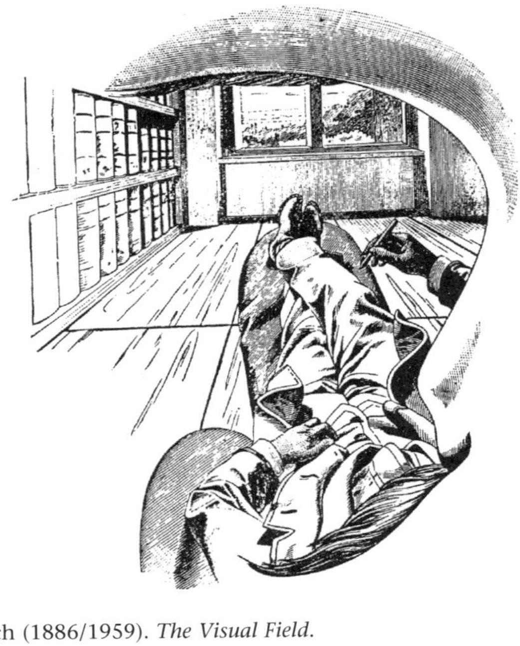

Figure 3.1 Mach (1886/1959). *The Visual Field.*

Mach, it seems, misdescribes what it is like to see. At best, he succeeds in depicting his room. We see the familiar oscillation at work. He tries to direct himself to his experience, but, like the experientialists of the fable, he directs himself instead to the world. Experience, when thought of along these lines, is too transparent to capture in thought or on paper.

3.2 Experience as a temporally extended pattern of exploratory activity

The puzzle of the transparency of experience results from thinking of experiences as like inner pictures and from thinking of reflection on experience as like turning one's gaze inward to those pictures. But this is a false characterization of experience. In experience we are aware not of inner pictures, but of the things around us in the environment.

To defend this claim, consider Mach's drawing again. We noticed that the drawing represents the visual field as too sharp throughout, as too defined, and that the characteristic indistinctness of the visual field at the periphery is not captured by the fade-to-white. We can summarize the picture's misrepresentation of experience as follows: the picture gives expression to the idea that when you see, you have all the environmental detail *in your consciousness at once*. Let's call this the "details in the head" conception of perceptual (or visual) experience.

It is worth noticing that the "details in the head" conception is, in fact, shared by many visual and perceptual theorists. Indeed, visual scientists have long tended to think of their central task as that of trying to determine how the brain gives rise to richly detailed picture-like experiences of this sort, on the basis of the paltry information about the environment projected onto the retina.[5]

Consider an example from touch that illustrates this established problematic.[6] Suppose you hold a bottle in your hands with your eyes shut. You feel it. You have the feeling of the presence of the whole bottle even though you only make finger-to-bottle contact at a few points. The standard account of this phenomenon proposes that the brain takes the little information it receives (at the isolated points of contact) and uses it to build up an internal model of the bottle (one capable of supporting the experience).

But consider: this positing of a process of construction of an internal representation may be an unnecessary shuffle. For the bottle is right there, in your hands, to be probed as occasion arises. Why should the brain build models of the environment if the environment is present and so can serve as its own model, as an external but accessible repository for information (as has been argued by Brooks 1991 and O'Regan 1992)?

Recently good deal of empirical support has been found for this "world as its own model" view (e.g. work on change blindness (reviewed by Simons 2000), inattentional blindness (Mack & Rock 1998), animate vision

(Ballard 1991) and embodied cognition (Brooks 1991). But anecdotal evidence will suffice to make us suspicious of the need for an internal model. Consider that we have all encountered the following familiar stunt. You are dining with a friend. You give a start and exclaim, "Hey isn't that so and so?" Your friend turns around. While she looks away you take a french fry. When she turns back, you shake your head and say that you must have been mistaken. She didn't see you take the fry. She doesn't notice that a fry is missing. Why does she fail to notice that the fry is missing? A number of candidate explanations suggest themselves. One such explanation is as follows: in examining her food she never built up a detailed internal model against which to compare her impression of the now altered pile of fries.

But there is a deeper problem with the "details in the head" conception of experience. It is, I believe, phenomenologically wrong-headed. (For more on this, see Noë et al. 2000).

Consider the bottle again. It is quite true that you have a sense of the whole bottle as present in your grasp. But surely it does not seem to you as if, all at once, when you hold the bottle, you in fact make contact with each and every part of its surface! No. It seems to you, rather, as if the bottle is there in your hands and as if you have access to it by movements. You are, in fact, as a perceiver, master of the sensorimotor skills required to exploit such access. The basis, then, of the feeling of perceptual presence of the bottle is just this skill-based confidence that you can acquire the information at will by probing the world (O'Regan 1992; O'Regan and Noë, 2001).

And so for vision. When you see, you take yourself to be aware of a densely detailed world, to be sure. But you do not take yourself to have all that detail in consciousness all at once. The "details in the head" model falsifies experience. Rather, you take yourself to have access to the detail by the flick of the eye or the turn of the head. The seeing, the experiencing of all that detail, is not something static, but something temporally extended and active.

The upshot of this discussion is that perceptual experience, in whatever sensory modality, is a temporally extended process of exploration of the environment on the part of an embodied animal. This is the key that unlocks the puzzle of transparency, and so the problem of phenomenology. If perceptual experience is in fact a temporally extended process, then to investigate experience we need to turn our gaze not inward, but rather to the activity itself in which this temporally extended process consists, to the things we do as we explore the world.

3.3 Toward an art of experience

I have proposed that to investigate visual experience—that is, to do visual phenomenology—we must investigate the temporally extended pattern of exploratory activity in which seeing consists. I now propose that to study some works of art is to undertake precisely this sort of investigation.

The study of such works of art can serve as a model of how to study experience and can also reveal how art can be, in the sense of Irwin's quote given at the outset, not only concerned with the making of objects, but more significantly with the investigation of perceptual consciousness.

I will focus on Richard Serra, whose work very perfectly exemplifies the ideas I have in mind. However, there are and have been numerous other artists whose projects are *experiential* in the sense I have in mind (e.g. Smithson, Irwin, Turell, to name only three).[7] What I shall argue is that Serra's work (and also the work of these other artists) enables us to catch ourselves in the act of perceiving and can allow us thus to catch hold of the fact that experience is not a passive interior state, but a mode of active engagement with the world. In this way, Serra brings to rest the troubling oscillation between experientialism and realism.[8]

Before turning to the work itself, let me be very clear about what I am doing and about what I am not doing. I am not arguing that the conception of experience I advocate fixes or solves any aesthetic problems. I argue, rather, that—against the background of this conception of experience—we can appreciate how Serra's work (and the work of others) helps clarify certain theoretical problems about consciousness. Now it might in fact be the case that the analysis here offered does in fact help to solve certain puzzles of an aesthetic nature about Serra's and Smith's work. If so, this remains secondary to the main purpose.

A final preliminary: I do not claim that the interpretation of Serra and Smith here offered is particularly original. Since developing these ideas I have learned that others—Rosalyn Kraus, Yves-Alain Bois, Hal Foster—have anticipated me in this or that respect. I am not, however, aware of anyone who has coordinated these various points in relation to philosophy and cognitive science in just the way I have attempted to do.

I turn now to the large-scale sculptures in metal that have been central to Serra's work since the 1970s. A striking example is *Spin Out: For Bob Smithson*, which consists of three large plates of steel (each ten feet by forty feet by one and a half inches), each embedded into the earth along the side of a hill in a roughly circular array, as illustrated in Figure 3.2.[9]

Let us note four characteristic features of Serra's sculptures. First, many of them are, as I will put it, *environmental*. That is, they are *intrinsically* site-specific. What I mean by this is that the sculptures are not merely enhanced by their locations, but are made for their locations and are meant to become part of the environment. Some pieces (such as, for example, works in the recent *Torqued Ellipse* series) lack site-specificity, in so far as they can be moved from one site to another. But it would be a mistake to conclude from this that these works are not environmental. Crucially, these pieces are able to transform a location and so produce a new environment.

Second, like Borges' map of the world which is built on the same scale as the world itself, most of Serra's sculptures are *lacking in perspicuity*, that is, you cannot take one in at a glance. This has to do not only with their scale, but also with their complexity; to be understood they need to be explored.[10]

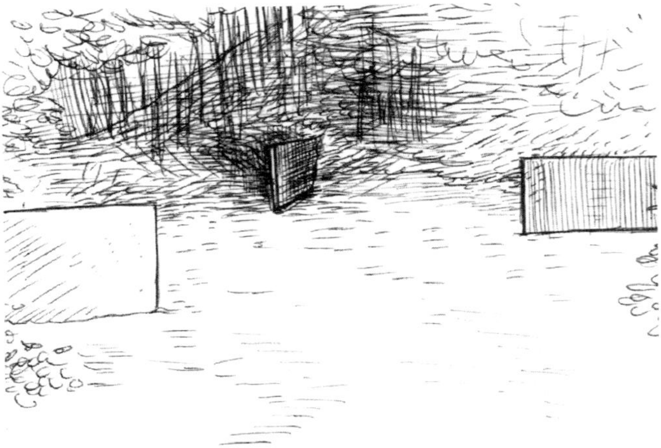

Figure 3.2 Richard Serra *Spin Out: For Bob Smithson* 1972–3. Illustration by Miriam Dym.

Third, a typical Serra sculpture, thanks to its scale, its surprising curves and apparent tilt, is overwhelming and disorienting, sometimes even frightening, almost always intimidating. It demands a reaction. Fourth, and as a consequence of the first three points, the pieces are, as I shall put it, particulars. By this I mean that they are unique, concrete entities. To encounter a Serra sculpture is to get to know an individual. The significance of this point will emerge in a moment.

With these features in mind—environmental character or site-specificity, lack of perspicuity, overwhelming scale and particularity—we can start to lay out the method of phenomenological investigation deployed by the artist in these works. When one first encounters a piece, such as *Running Arcs (For John Cage)* (Figure 3.3), one is liable to be struck by the scale and the visually inscrutable orientations and distributions of weight. One not only notices these qualities but is disturbed by them. One puzzles: what stops these giants from falling over? To wander around or through a piece such as this can cause a loss of balance. In this way the works make us reflect on how we *feel*, perceptually, in their presence. And they direct our attention to the complexity of our experience, a complexity we easily overlook. The loss of balance, for example, introduces us to what are strictly non-visual (e.g. vestibular, kinesthetic) components of our 'visual' experience.

If we press on and explore a piece by walking into or around it, we come to feel at ease in or near it. In this way, we gain a kind of practical knowledge. We gain knowledge about the spaces they occupy as places, or niches or environments. The process of exploring the piece is a process of exploring

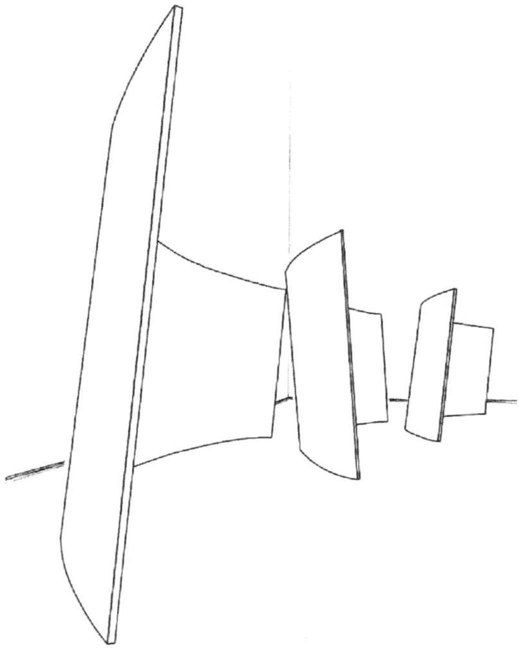

Figure 3.3 Richard Serra *Running Arcs (For John Cage)* 1992. Illustration by Miriam Dym.

the place. It is likewise a process by which we come to understand how experience can be, in this way, a form of openness to the environment. In light of the foregoing discussion of perceptual experience as a mode of active exploration of the world, it should be clear that the process of exploring the art work (and thus the environment in which it is situated) is at once a process of exploring one's experience of the world. And the knowledge one thus attains is knowledge of the character of one's experience.[11]

In this way, Serra's work brings the oscillation between experientialism and realism to a halt. Perceptual experience is transparent to the world precisely because experience is an activity of engagement with the world. To attend to the exploration of the world is thus to attend to the quality of experience. We can think of Serra's work, and that of other experientialist artists, as providing opportunities for first-person phenomenological investigation.

Now there is of course a sense in which one could say of *any* object (and certainly any work of art) that it provides its viewer with an opportunity to reflect on what is like to perceive it. What I am calling *experientialist art* is art that finds its home in this self-reflective moment. While just about anything can be made the occasion of such a self-reflective act, it is important

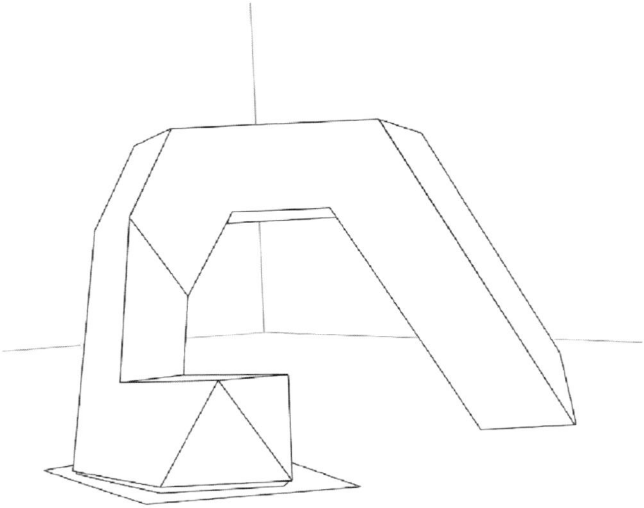

Figure 3.4 Tony Smith *Cigarette* 1961. Illustration by Miriam Dym.

to notice that not all forms of art and not all artists undertake their activity in this vicinity. We can deepen our grip on what distinguishes Serra's project *as experientialist*, by contrasting his works with those of a superficially similar, and no less important artist, the sculptor Tony Smith (Figure 3.4).

There are similarities of scale and material in the work of Smith and Serra—each makes metal sculptures suitable for outdoor installation at least twice the height of a normal person. Smith's work, however, in contrast with Serra's, is utterly nonpsychological. Where Serra's works are experimental, Smith's works are *geometrical* or *mathematical*. Smith's sculptures may be compared with what mathematicians call constructive existence proofs. A constructive existence proof demonstrates the existence of a mathematical entity by actually producing a relevant instance. Smith's constructions realize certain ways of combining shapes (e.g. tetrahedrons) in order to fill out space and so demonstrate that it is possible thus to fill out space. The space we learn about when we understand a Smith sculpture is not the space we experience, but it is, rather, the objective, abstract space of mathematics.

It is this mathematical character of Smith's work that explains another important feature that places it in direct opposition to that of Serra. Serra's pieces are, as I've emphasized, particulars whose effectiveness depends on their environmental character, their scale and their complexity. They confront a person the way a steep incline on the way home from work confronts a person. In contrast, Smith's pieces are *representative* or universal. A crucial fact about a geometrical construction is that it instantiates in the particular what are in fact wholly universal relations. The particular triangle on which the geometer builds his or her construction stands for all triangles of the relevant kind (equilateral, say). Its particularity does not matter. It is irrelevant to the significance of the demonstration whether it is written in chalk on the board or in pencil on paper, or whether it is drawn to one scale or another. (This is a Kantian point. See Kant 1787/1929, pp. 576–93.) In precisely this sense, Smith's sculptures have a kind of *immateriality*. That is, their material is immaterial. They are multiply realizable, lending themselves to reproduction on different scales and out of different materials.[12] They lack site-specificity.[13]

Serra once wrote that he is not interested in sculpture "which is solely defined by its internal relationships" (Serra 1973). Smith said in an interview: "I don't make sculpture, I speculate in form" (Gossen 1981, cited in Storr 1998). These comments bring out the difference between their projects. Smith is concerned precisely with the sculpture of internal relationships, with geometry, with form. Serra's is a sculpture of consciousness.

3.4 Conclusion

My aim in this chapter has been to propose that we can think of the work of certain artists (but not all) as providing methods for the study of experience. Serra's works, I have suggested, constitute experiments in a kind of phenomenological psychology. A Serra sculpture is an object for us to experience which functions to draw our attention to, what we do when we experience it and to how things are with us perceptually. In doing so, I have suggested, the work enables us to appreciate that experience is a mode of direct contact with and exploration of the world. In this way, the works enable us to understand both how experience can be transparent and why its transparency is no obstacle to scientific, philosophical and artistic investigation of experience. Experience is transparent to the world because it is just a mode of active engagement with the world. A phenomenological study of experience is not an exercise in introspection, it is an act of attentiveness to what one does in exploring the world. To reflect on the character of experience, one must direct one's attention to the temporally extended, fully embodied, environmentally situated activity of exploration of the environment. Experiential art enables us to do this.

Acknowledgments

This chapter is based on talks presented at the California College of Arts and Crafts in July 1998 and October 1999, at the Getty Research Center in May 1999, and at the meetings of the College Art Association in New York in February 2000. I am grateful to audiences at those meetings for helpful criticism, especially to Patricia Churchland, Tom Crow, David Freedberg,

Ellaine Scarry and Barbara Stafford. In addition, several people have read earlier versions of this chapter and have helped me to improve it. I would like to thank Miriam Dym, Daniel Guevara, David Hoy, Alexander Nagel, Hans Noë, Lawrence Rinder, Alexi Worth and Erika Belsey Worth. I am very grateful to Miriam Dym for her illustrations of the art works.

Notes

*The editors and author are grateful for permission to reprint this chapter which first appeared in the *Journal of Consciousness Studies*.

1. My translation of the German original.
2. This passage is cited in the brochure accompanying Irwin's 1998 exhibition at the Dia Center for the Arts entitled *Part I: Prologue: x183, Part II: Excursus: Homage to the Square3*, April 12, 1998–June 13, 1999. Elsewhere Irwin writes: "The act of art has turned to a direct examination of our perceptual processes" (Irwin 1972b). This is also cited in the same brochure.
3. Images of works of Serra and Smith may be viewed at various sites on the Internet. Interested readers may want to visit <http://www.artcyclopedia.com/artists/serra_richard.html> and <http://www.artcyclopedia.com/artists/smith_tony html>.
4. Wittgenstein remarked that Mach's drawings are one of the clearest examples of confusion between physical and phenomenological modes of description. Referring to the inadequacy of Mach's representation of the blurredness of the visual field by means of a different kind of blurredness in the drawing, Wittgenstein remarked: "No, one cannot make a visible picture of the visual field" (1964, p. 267). See Thompson, Noë & Pessoa 1999 for more on this. A point very similar to Wittgenstein's is made by Dennett 1991, p. 55, who writes: "One can no more paint a realistic picture of visual phenomenology than of justice or melody or happiness".
5. Recently this consensus has begun to come undone. A group of new thinkers have come to endorse what I have called the new skepticism about visual experience (Noë, 2001). Traditional skepticism about experience questions whether we can know, on the basis of experience, that things are as we experience them as being. The new skepticism, in contrast, questions whether we even have the experience we think we have. Whereas traditional visual science wondered how we can have such rich visual impressions on the basis of such impoverished retinal input, the new visual science that takes its start from the new skepticism targets a rather different question: why does it seem to us as if we see so much when, in fact, we see so little! The leading spokesperson for the new skepticism is Daniel Dennett. See Noë (in press) forthcoming for extended discussion of this point.
6. This often used example is due to Donald MacKay (1962, 1967, 1973).
7. My use of the term "experiential art" is similar to the use made by Rinder and Lakoff (1999) of the term "consciousness art."
8. Some readers might be struck by the fact that I now turn to sculpture after developing the to-be-criticized conception of experience with reference to a *drawing* of Mach's. Mach's drawing, I think, illustrates the problematic conception of experience that is my real object. It is not my intent, however, to suggest that there is an intrinsic connection between drawing as a medium and the problematic conception. For reasons that will, I hope, become clear, some sculpture provides an apt method for investigating experience thought of as a form of activity. But it is no part of my purpose to suggest that picture making cannot

also perform this role. Cézanne is an excellent example of a painter who is, in my sense, an experiential artist. Merleau-Ponty (1948/1964) has made much the same point when he observed that Cézanne's is an art directed to the phenomenal which nevertheless avoids the "ready-made alternatives" that preoccupied impressionism (e.g. "sensation versus judgment," "the painter who sees against the painter who thinks"). Merleau-Ponty emphasizes that although Cézanne remains faithful to the phenomenal, there is a clear sense in which he "returns to the object." He does so in a way, however, that allows him to discover a "lived perspective" that is at once experiential and also directed to the world. Cézanne's pictures represent the environment *as experienced*. In this way, he is able to discharge the tension between experientialism and realism.

9. In an interview, Serra described the piece as follows: "The plates were laid out at twelve, four and eight o'clock in an elliptical valley, and the space in between forms an isosceles triangle, 152 feet on the long side, seventy-eight and seventy-eight feet on the legs. Each plate is ten feet high by forty feet long by 1 1/2 inches thick hot-rolled steel sunk into the incline at an equal elevation" (1973, p. 16).
10. The aim of philosophy, Wittgenstein held, is the attainment of a "perspicuous overview" (übersichtliche Darstellung) of a region of grammar or conceptual space. To gain such a perspicuous representation is to gain understanding and to see things aright. My use of the term "perspicuity" is meant to invoke this idea. Serra's sculptures provide us opportunities to understand the environment we live in by exploring bits of it in order to attain a perspicuous overview where at first there is none.
11. The dynamic character of the appreciation of the work of art is underscored by Serra in his discussions of the pieces. Consider further his remarks about *Spin Out*: "First you see the plates as parallel; when you walk left, they move right. As you walk into them, they open up, and there's a certain kind of centrifugal push into the side of the hill. In fact the people at the Kröller-Müller wanted to call the piece 'Centrifugaal' in Dutch. They talked a lot about vorticism. And then when you walk above it, there's another path which connects the two sides of the valley… . There's a ridge which encircles the whole space at about 150 feet. When you walk on the ridge, there's a contraction and the space becomes elliptically compartmentalized, which you can't see as you walk through it, and it's a different way of understanding your relation to the place: you're overhead looking down" (1973, p. 16).
12. The same cannot be said of Serra's work. Small-scale maquettes of the *Torqued Ellipses* utterly lack the aura of their full-scale counterparts.
13. It is interesting to note that if this reading of Smith is right, then Fried (1967/1998) was mistaken to select Smith as an example (indeed, as *the* example) of "theatrical art." Smith's works are self-standing wholes; they are concerned with the exploration of objective space (as it were) and not with provoking a reaction in a viewer. In contrast, I have been emphasizing ways in which Serra's work *is* theatrical. Serra's pieces perform their phenomenological function only as completed by the presence of a spectator. Indeed, Bois (1978, p. 52) has remarked that "all of Serra's oeuvre is an implicit reply to Michael Fried's text" (cited in Taylor 1997).

Works cited

Ayer, A. J. *The Central Questions of Philosophy*. New York: Holt, Rinehart & Winston, 1973.

Ballard, D. H. "Animate Vision." *Artificial Intelligence* 48 (1991): 57–86.
Bois, Y-A. "A Picturesque Stroll around *Clara-Clara*," In E-G. Güse (ed.) *Richard Serra*. New York: Rizzoli, 1978.
Brooks, R. "Intelligence without Representation," *Artificial Intelligence* 47 (1991): 139–60.
Dennett, D. C. *Consciousness Explained*. Boston: Little, Brown & Co, 1991.
Fried, M. "Art and Objecthood," in M. Fried (ed.) *Art and Objecthood*. Chicago: University of Chicago Press, 1967/1998.
Gossen, E. C. "Tony Smith: 1912–1980," *Art in America* 69 April 11, 1981.
Grice, H. P. "Some Remarks about the Senses," in R. J. Butler (ed.) *Analytical Philosophy*. Oxford: Basil Blackwell, 1962.
Hume, D. *A Treatise of Human Nature: Analytical Index by L. A. Selby-Bigge, Second Edition*. Oxford: Oxford University Press, 1740/1978.
Irwin, R. "The State of the Real. Part I. A conversation with Jan Butterfield," *Arts* 46, no. 10. (1972a) (June).
Irwin, R. "Reshaping the Shape of Things. Part 2," *Arts* 47, no. 1 (1972b) (September).
Kant, I. *Critique of Pure Reason*. Translated by Norman Kemp Smith. London: Macmillan, 1787/1929.
Mach, E. *The Analysis of Sensation*. Translated by C. M. Williams. New York: Dover, 1886/1959.
Mack, A. & Rock, I. *Inattentional Blindness*. Cambridge, MA: The MIT Press, 1998.
Merleau-Ponty, M. *Sense and Non-Sense*. Translated by H. L. Dreyfus and P. A. Dreyfus. Chicago: Northwestern University Press, 1948/1964.
Noë, A., Pessoa, L. & Thompson, E. "Beyond the Grand Illusion: What Change Blindness Really Teaches us about Vision," *Visual Cognition*.7, No. 1/2/3 (2000): 93–106.
Noë, A. "Experience and the Active Mind," *Synthese*, 129, 1 (2001): 41–61.
O'Regan, K. J. "Solving the 'Real' Mysteries of Visual Perception: The World as an Outside Memory," *Canadian Journal of Psychology* 46, 3 (1992): 461–88.
O'Regan, J. K. & Noë, A. "A Sensorimotor Account of Vision and Visual Consciousness," *Behavioral and Brain Sciences*, 24, 5 (2001): 939–1031.
Pessoa, L., Thompson, E. & Noë, A. "Finding Out about Filling-In: A Guide to Perceptual Completion for Visual Science and the Philosophy of Perception," *Behavioral and Brain Sciences* 21, 6: (1998) 723–802.
Price, H. H. *Hume's Theory of the External World*. Oxford: The Clarendon Press, 1940.
Rinder, L. & Lakoff, G. "Consciousness Art: Attending to the Quality of Experience," in L. Rinder (ed.) *Searchlight: Consciousness at the Millennium*. New York: Thames & Hudson, 1999.
Serra, R. (1973) "Document: Spin Out '72–'73: Interview by Liza Bear," *Arts Magazine*, April, in R. Serra's *Writings Interviews* (Chicago and London: University of Chicago Press, 1997).
Simons, D. "Current Approaches to Change Blindness," *Visual Cognition* 7 (1/2/3) (2000): 1–15.
Storr, R. "A Man of Parts," in R. Storr (ed.) *Tony Smith: Architect, Painter, Sculptor*. New York: The Museum of Modern Art, 1998.
Strawson, P. F. "Perception and its Objects," in G. A. MacDonald (ed) *Perception and Identity: Essays Presented to A. J. Ayer with his Replies*. Ithaca: Cornell University Press, 1979.

Taylor, M. C. "Learning Curves," in *Richard Serra: Torqued Ellipses*. New York: Dia Center for the Arts, 1997.
Varela, F. & Shear, J. "The View from Within: First-Person Approaches to the Study of Consciousness," *Journal of Consciousness Studies* (Special Issue), 6, No. 2/3 (1999).
Wittgenstein, L. *Philosophische Bemerkungen*. Oxford: Basil Blackwell, 1964.

4

Phenomenology and Imaging: Incorporating the Material

Don Ihde

Edmund Husserl's phenomenology begins in the early twentieth century. Similarly, the first philosophers to begin to make technology thematic (with one exception) also began in the early twentieth century.[1] I shall take as an opening task for this chapter, a showing of how phenomenology and philosophy of technology came to relate to each other.

Were one to believe early intepretations of Husserl's phenomenology, it would not seem to be a very promising candidate for philosophizing about technologies, which necessarily must deal with materiality. Husserl's early thought has usually been interpreted as a "philosophy of consciousness," a variation upon the philosophy of mind, with a method usually termed "subjectivist," and its goal to describe "appearances" or phenomena. And much of Husserl's early work and the language he used left itself open to this interpretation. In my opinion, this was because the philosophical framework against which Husserl was reacting, that of early modernity, more specifically the kind of epistemology associated primarily with Rene Descartes. Cartesian epistemology is a theory of knowledge production which is laid out in terms of an "ego" or "subject," which must infer what the "external" world is like based upon "ideas," or "sensations," or "representations" which are directly present to the subject, while "external reality" is objective and elsewhere. Edmund Husserl, very unfortunately to my mind, adapted this very language, particularly in his *Cartesian Meditations* (1931), although he did so in order to overturn or negate virtually every Cartesian point: Husserl wanted to overcome the subject–object distinction; to replace subjectivity with intersubjectivity; and to replace the notion of knowledge-by-representation with knowledge of "the things themselves" which he called *phenomena*. Careful interpreters have, of course, noted this as well. But I want to give a unique twist to this problem and show that from its very inception Descartes had employed a *technology* to model epistemology. That technology was a very popular optical tool of the time, the seventeenth century, the *camera obscura*.

Essentially, the *camera obscura,* literally "dark room," is a room or box enclosed, with a small opening—with or without a lens—with a blank wall or screen at the opposite side from the opening, upon which is cast [inverted and in two dimensions] an *image* or representation of some lighted object or light source itself from outside the box. The optical effect was known in antiquity, but was first fully described in a technical way by Al Hazen in 1037. By Renaissance times, *camera obscurae* were being used by artists to produce pictures with high verisimilitude and by the seventeenth century they were also being used in optical and scientific experiments. Here is Descartes from *La Dioptrique* (1637):

> If a room is quite shut up apart from a single hole, and a glass lens is put in front of the hole, and behind that, some distance away, a white cloth, then the light coming from external objects forms images on the cloth. Now it is said that this room represents the eye: the hole, the pupil; the lens, the crystalline humor [etc.].[2]

In *La Dioptrique* Descartes simply repeats what was already a tradition making the *camera* an analog of the eye. But, once one notes the features of his subsequent epistemology, one can discern that Descartes took a second, new step: the *camera* is to the eye *as the eye is to the subject [ego].*

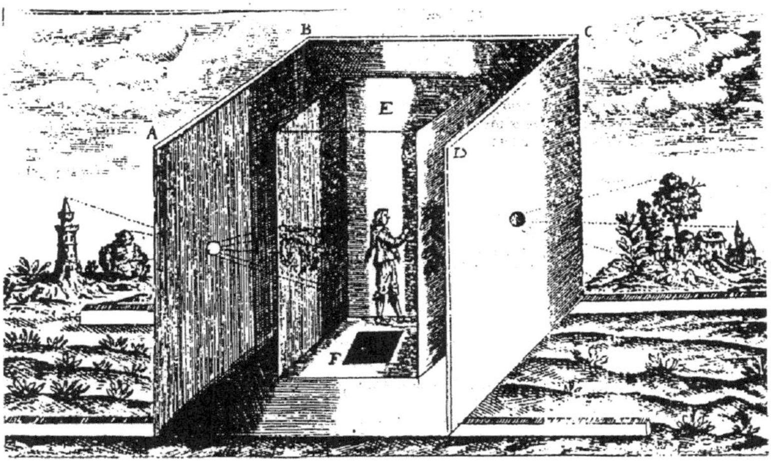

The large camera obscura, constructed in Rome by Athanasius Kircher in 1646, is shown with the top and front cut away. It was a small portable room that could easily be carried by an artist to the scene that he wished to depict: he would climb inside through a trap door. In this engraving he is tracing, from behind, an image cast on transparent paper that hangs opposite one of the lenses. George Eastman House, Rochester, New York

Figure 4.1 Camera Obscura.

My animation of Cartesian epistemology above shows all the important features of early modern thought, based upon the "epistemology engine" of the *camera obscura*: (a) the "subject" is the Einstein homunculus in the box, (b) the "image" is any and all "representations" which are directly presented to the mind, but these are caused by (c) the causal origins, "objects" in "external" [to the box] reality. The relation (b) to (c) must be inferred via the "geometrical" method. The reliability of this correspondence (b) to (c) is, however, only guaranteed if there is an ideal observer (d) who is able to see at one and the same time, both inside and outside the box and thus see the correspondence—(d) is, of course, Descartes' God, but I claim it is actually Descartes who sees both inside and outside the *camera* device which he is using. In short, the *camera*, later hidden or implicit, is the secret model for early modern epistemology.

I do not think Husserl knew the *camera* lay behind Descartes' model for knowledge, but by following the language of that model, he was left in an unfortunate position. But, in retrospect, one can say that Descartes' taste for the "mechanical" is clearly present from this beginning because it was a mechanical (optical) device which he used to model knowledge itself.

If Descartes showed by his use of the *camera obscura*, his predilection toward technologies, the same cannot be said for Husserl who rarely used technological examples. I shall here leave the Husserl of the *Cartesian Meditations*, and move to his later work where he developed the notion of a *lifeworld* and gave hints about the possible roles of technologies. His late work, *The Crisis in European Philosophy and Transcendental Phenomenology* (1936), developed a highly embodied and perceptualist phenomenology. The sciences, he claimed, particularly those driven by mathematization and idealization, had "forgotten" the bodily-perceptual being of humans in a cultural–historical lifeworld. In analyzing this crisis Husserl made two moves which vaguely anticipate a possible role for a philosophy of technology.

The first move was a passing recognition of a "technology" which effectively moved humanity to a different level of lifeworld activity, inscription technologies or *writing*. Husserl claimed, "[t]he important function of written, documenting linguistic expression is that it makes communications possible without immediate or mediate personal address; it is, so to speak, communication become virtual. Through this the communalization of man is lifted to a new level."[3] As usual, this example remains sparse and abstract as do most of Husserl's examples, but it indicates that a technology can change the very way in which linguistic communication can occur and thus produce a change in the lifeworld itself.

A second move was made in "The Origin of Geometry" wherein Husserl claims that geometry arises out of praxical activity, an activity within *material culture*:

> The empirical art of measuring and its empirically, practically objectivizing function, through a change from the practical to the theoretical interest,

> was idealized and thus turned into the pure geometrical way of thinking. The art of measuring thus became the trailblazer of the ultimately universal geometry and its "world" of pure limit-shapes.[4]

To locate an abstract science by noting its practices within material culture was an insight which 50 years later could be considered "normal" in today's versions of "science studies," but was unusual this early in the twentieth century.

Yet both these foresights, if we term them that, are recognizable only retrospectively, and if we do so, we should also be equally aware that Husserl missed his greatest opportunity to situate science practice in a bodily-perceptual lifeworld. Husserl's Galileo remained the "mathematizing" Galileo, a Galileo *without a telescope*, which, had Husserl followed his lifeworld praxis notion and applied it to Galileo's telescopic praxis, would have placed Galileo at a juncture which raised the level of humankind's perception through a technology. Husserl did not *do* philosophy of technology.

Maurice Merleau-Ponty, whose own phenomenological work was responding to the later Husserl of the lifeworld period, perhaps came a little closer to an appreciation of a "phenomenology of technics." His rich and suggestive examples include those of an embodiment extended through material artifacts, for example, the lady with a feathered hat:

> A woman may, without any calculation, keep a safe distance between the feather in her hat and things which might break it off. She feels where the feather is just as we feel where our hand is. If I am in the habit of driving a car, I enter a narrow opening and see that I can "get through" without comparing the width of the opening with that of the wings, just as I go through a doorway without checking the width of the doorway against that of my body.[5]

This immediate sense of bodily ability extended through an artifact, I later termed an 'embodiment relation' of humans-plus-technologies.[6] Merleau-Ponty also noted that in this phenomenon, the technology becomes, as it were, incorporated into one's bodily sense:

> The blind man's stick has ceased to be an object for him and is no longer perceived for itself; its point has become the area of sensitivity, extending the scope and active radius of touch and providing a parallel to sight. In the exploration of things, the length of the stick does not enter expressly as a middle term: the blind man is rather aware of it through the position of objects than of the position of objects through it.[7]

Phenomenologically, this human-technology relation is generalizable—it is through this perceptual-bodily incorporation that scientific and musical instruments are also "embodied," and through which even the kinesthetic and tactile becomes a "distance sense." Clearly there are implications here for a philosophy of technology.

But most historians of phenomenology would agree that it was Martin Heidegger who first developed a technology incorporating phenomenology and thus a recognizable "philosophy of technology."[8] Already in *Being and Time* (1927), chronologically preceding both Husserl's and Merleau-Ponty's examples just cited, he undertakes his famous "tool analysis." In an extended analytic description, Heidegger shows how a hammer is "embodied" by withdrawing in use; how it belongs, not as an object, but as an equipmental element in a system of assignments or context, and how it becomes a mere object only when broken or missing. From this he draws an *ontological* conclusion: human praxical activity is "[t]he kind of dealing which is closest to us is … not a bare perceptual cognition, but rather that kind of concern which manipulates things and puts them to use; and this has its own kind of 'knowledge.'"[9] The hammer is not first an object-in-itself, but equipment within a use-context which includes a whole set of material entities (the nails, the shoes to be soled, the workshop, etc.). This emphasis on the priority of praxis remains throughout this early philosophy of technology.

Later, Heidegger was to develop farther his philosophy of technology, particularly in his "Question concerning Technology" (1954). By the mid-twentieth century, he had made "Technology" a virtual metaphysics, a way of seeing the entirety of the natural world as a resource well for human uses and purposes—and he saw this as a major danger coming from modernity itself.[10] But this kind of philosophy of technology, by its very elevation to a metaphysics, becomes so "high altitude" that its usefulness as a phenomenological analysis must be questioned. Indeed, in his last 1949 Bremen Lectures, he equates modern agriculture with its transformation of foods as equivalent to the holocaust with its transformation of humans as part of the same technological metaphysics!

I now leave these early-to-mid twentieth century developments within phenomenology and philosophy of technology, to jump to the present and its juncture with the twenty-first century. Hans Achterhuis, himself a very well-known Dutch philosopher of technology, undertook a project to update philosophy of technology with a book, *Van stoommachine tot cyborg—denken over techniek in de niew wereld* (1997) [From steam engine to cyborg—thinking technology in the new world]. In a new edition, updated with analysis of work done between 1997 and 2000, it was translated into English as *American Philosophy of Technology: The Empirical Turn* (2001). What Acterhuis, with five of his colleagues did, was to analyze the work of six contemporary philosophers of technology: Albert Borgmann, Hubert Dreyfus, Andrew Feenberg, Donna Haraway, Don Ihde, and Langdon Winner. Achterhuis argued that these twenty-first century philosophers of technology, unlike their largely European and American forefathers, were not dominantly dystopian concerning technologies; were more "empirical" in that they dealt with specific analyses of types of technologies; and were more pragmatic than first generation philosophers. He could have pointed out, additionally, that four of the six had significant backgrounds in Heideggerian philosophy of technology although some have become highly critical of it, and that phenomenology remains operative in different degrees among different backgrounds.[11] Examples of a pragmatic "empirical" turn include Borgmann's recent concern with information technologies; Dreyfus's with the Internet; Feenberg's with technologies and multiculturalism; and mine with imaging technologies. In short, I take this as a certain trajectory coming out of earlier twentieth century phenomenology and the philosophy of technology.

With these historical relations now established, I now turn to some more concrete examples of how phenomenology relates to issues within the philosophy of technology. It is important to note that the epistemological context today is considerably different than it was in Husserl's time, now a century ago. And, one could equally observe that today's technologies are also considerably different than they were at mid-century when Heidegger did his work on technology. My task, then, is to pick examples which will be contemporary and show how a phenomenology relates to them; while at the same time bringing the style of phenomenology out of its early modern setting and into a contemporary pragmatic and praxical one.

I shall take as my first example a piece of contemporary scientific knowledge which came into being *after* the philosophical work of all three of my previously cited classical phenomenologists ceased. Here is the piece of knowledge: roughly at the center of the Crab Nebula there is a rapidly spinning pulsar from whose axes there are being emitted two jets of X-ray radiation which are shooting out from the pulsar at near light speeds. Admittedly, this sounds quite esoteric and strange, and it is a phenomenon not known until late in the twentieth century. But, on the other hand, the evidence for this phenomenon is now well known; has been repeatedly displayed; and can be pulled up on the Internet from NASA imaging at any time. It is from the Chandra X-ray source, an orbiting imaging device launched in 2000—thus we are now in the twenty-first century. Figure 4.2 shows the Chandra X-ray source image of the pulsar and its radiation jets.

I begin with some common sense and naturalistic observations about this phenomenon. First, what could we know about the Crab Nebula were we limited to the plenary, full body presence *without* technologies? The answer has to be: not much. We could visually locate the nebula, but even with the keenest eyesight, it would be at most a tiny blur of light in the nighttime sky. Secondly, what could we know of it through a science of "pure mathematics?" Again: not much more. Yes, we might be able to do some geometric measurements regarding relative location, but without any instruments, the ultimate yield would be little. This is, of course, not the way contemporary science operates. Instead—and I argue it has operated this way at least since Galileo [and earlier!]—science operates by

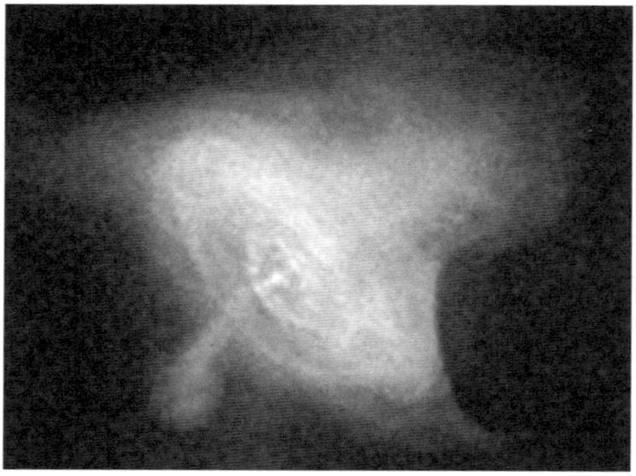

Figure 4.2 The Crab Nebula in X-Rays.
Source: Chandra X-Ray Observatory, *NASA*.

making observations through instruments. But, interestingly, in one essential sense, astronomy from Galileo *until the mid-twentieth century* actually made its observations within the approximate limits of Galileo's telescope, that is, within the limits of optical or emissions limited to optical "light." The *technologies* of observation within astronomy began to change only since the mid-twentieth century, first with the discovery of radio astronomy, related to radar and radio waves, and new radio-telescopes led eventually led to the discovery of the universe's background radiation (invisible to optical telescopes) and to "radio sources" not always associated with light sources. Radio sources come with wave frequencies much longer than optical light waves, thus radio technologies "opened" a source of celestial emissions previously unknown. Since then, new instruments have been invented and constructed which "open" observation in astronomy to a very wide range of frequencies, from gamma frequencies to radio frequencies. Scientifically one result is the recognition that previous astronomy had been limited to a very narrow range of phenomena. Returning now to my example of the pulsar and radiation jets at the center of the Crab, the observation technology which makes this phenomenon visible is the Chandra X-ray source which images emissions within the "slice" of radiation we call "X-ray" radiation.

This common sense and naturalistic description, however, hides a host of philosophical problems, problems which I now want to elucidate by turning to *phenomenology* and *philosophy of technology*. A phenomenological framework is one which utilizes an ontology of "intentionality." It must analyze the dynamic interrelationships between *embodied humans*, in their fully bodied actions, within the context of a concrete environment or *world*, within which they focus upon and interact with chosen phenomena. In the past I have usually diagramed this as:

Human—interaction—World

The "human" includes the sense of fully embodied and also fully social; the World includes the full range of possibilities and the dashes are interactive, with human experiencing a world, but the world acting upon the human.

Where, then, technologies? In my use of phenomenology, technologies can and do often occupy positions within and along the interactive range of relations within, to, and with an environment. Were I to return to Merleau-Ponty's feather, one could note:

Woman—feather—doorway

Here the feather is "incorporated" into her now extended sense of embodiment and she experiences herself within the limits of body + feather fitting within the doorway. This is a "technology" in embodiment position in my language. A full set of descriptions may be found in *Technology and the Lifeworld* (1990). Here, however, I want to relate phenomenology to my pulsar example.

Although highly complicated, my pulsar example remains one of a (or multiple) human observer(s) bodily related to a phenomenon, the pulsar, *through* a complicated set of technologies. An echo of Husserl's notion of plenary perception, an embodied human relating to a project, could go two directions here: on the one hand, were the observer to be limited to face-to-face or direct bodily relations to the Crab and its pulsar, we would be left as indicated above, with a poor level of knowledge. But, if the technologies—Chandra X-ray source with its capacities to transform data into images and vice versa (and one might add the whole set of operations which go with this instrument)—are now taken into the experience, then our knowledge rises to a new level somewhat parallel to that noted in writing technologies by Husserl. Two further interesting phenomenological points can be made here: first, in the case of the pulsar, *no scientific knowledge of its shape, structure, behavior, etc., is possible at all except through the mediation of the technology*. Once a phenomenon is beyond face-to-face range, by distance, size, or structure, only *instrumentally mediated* evidence is possible. This is the case whether we are taking account of genetic marking, of galaxies and black holes, or any macro- or micro-phenomenon lying outside our ordinary bodily range. All such science, I claim, is *instrumentally real*.[12]

There is a second point as well, to become evidence, the evidence must "*take account of human embodiment*." My pulsar example shows this quite dramatically—without instrumental mediation, we could not "see" the pulsar structure, shape or activity at all, but, secondly, because all this activity *lies outside the capacities of our perception entirely, to be 'seen' it must be translated into the humanly visible*. This is what the Chandra X-ray imaging process does—it transforms the data it has gathered, into a shaped, visible, but "false colored" image which we can "see" as a normal visual gestalt, pulsar plus jets. Thus the technologies, instruments, make the phenomenon available, but by means of transforming it into an image (or data) perceivable by embodied (or interpreting) humans. The instrument, positioned within intentional relativity, must both detect the emissions and translate them into humanly accessible form. Were I to revert to older terminology, this sounds an awful lot like the active model of *genetic phenomenology* Husserl favored in his last period, but it is a technologically embodied phenomenology.

Through this 'constructed' imaging process, then, the new levels of knowledge made possible technologically, also remain within a new shaping of a *lifeworld*. Put in another way, science *cannot forget the lifeworld* because it must make its evidence into that which is bodily perceivable and interpretable. Or, in another way of putting it, the "crisis" within Husserl's *Crisis*, in part lies with the erroneous notion of science he elaborated. Science is much more than a mathematization process—it was primarily taken as that indeed, for the early-twentieth century philosophers of science and for the logical positivists of Husserl's time, and apparently Husserl borrowed too much of this view of science from these sources. Had he followed his own better insights by paying attention to lifeworld practices, not only the measuring practices of the Egyptians, but those of even the Galileans, a different notion of science might have emerged. I am suggesting that a revised phenomenology, more attuned to the material and material cultures, re-situates science within a lifeworld.

With this in mind, I want now to return to my opening technological example, the *camera obscura*, with a very different purpose in mind. Descartes used the *camera* as a model for the mind. I shall regard it as a suggestive imaging device, rejecting it as a metaphor for mind but accepting it as an instrument for knowledge production. We, like Descartes and/or his God-trick perspective, can easily see both from the inside and outside the *camera* and thus discern *both* the similarities and differences between the image and the imaged. The *camera* is not perfectly isomorphic—rather, it transforms the imaged into image. It has never been a perfect "correspondence" machine. In its first version, the image is *inverted* [at least if there is no inverting lens to "correct" the inversion] and *converts* three-dimensionality to two-dimensionality. But, this has certain advantages: Renaissance painters used the advantage of two-dimensionality to do drawings; the *camera* "automatically" produced "Renaissance perspective" on its *tabula rasa*. Alberti, even da Vinci, practiced this type of art production. In short, the *camera* became one of many kinds of technologically mediated vision. Dürer has depicted many other of the devices used in the fifteenth and sixteenth centuries, such as drawing frames with crossed wires, sighting devices, and the like. Each "aided" vision to help make geometrically accurate drawings. This was seeing via or mediated by instruments. (In actuality, this practice simply transferred to art practice what was a much more ancient type of "measuring perception" such as used in ancient astronomy. If Stonehenge,

MesoAmerican spiral sun graphs, and the like were solstice measurement devices, then these, too, produced the same measuring perceptions for a different practice.) Both scientific knowledge and artistic production, then, can develop practices using mediational devices.

The *camera obscura*, however, historically turned out to be a most productive imaging invention. I have claimed, if we follow its trajectories, that there is a sense in which the *camera obscura* "invents" much of modern science. What I am going to show will offend careful historians because I shall skip the struggles, ambiguities, failures which belong to all histories of development, but instead shall highlight results which show that a long history of science finds fulfillment in a simple set of variations upon the original, antique *camera obscura*. I shall follow this progeny and show some of the variations:

- The imaging machine which is closest to the antique *camera*—although its development violates the actual chronology of such—is the sixteenth century *photographic* camera. Earlier antique *cameras* often were literal with respect to their name, dark *rooms*, into which the observer could enter. But by the late seventeenth to nineteenth centuries, box cameras with viewing tubes or openings to view the interior screen were common. And it was this style of mini-camera which became the body of the photographic camera. Using silver salts, sensitive to light, several attempts were made to "fix" images. Joseph Niépce partially succeeded by 1826. Of course the long time exposures needed restricted the images to objects which were immobile—thus architecture was an early subject. Niépce called his projects *heliographs*. Louis Daguerre bought the process, improved it, and published it by 1839, and what was first a daguerreotype later became "photography." But, just as with the antique *camera*, the image produced *transformed* what was imaged. In addition to the inversion and 3-d to 2-d transformation of the antique *camera*, the photographic camera transformed the perception of time. For example, the chemical "fixing" produced a "still." It "froze" time and thus early metaphors about photographs often included the sense of "death-like" although also the realism or isomorphism was simultaneously "life-like." Photography was used by science almost immediately—the gibbous moon was photographed by 1840, and spectrographic photography followed in a few more years.
- But, if we return more strictly to the chronological trajectory from the *camera obscura* to its second radical reinvention, one then turns to a variation introduced by Isaac Newton roughly two centuries after the painters we noted using the antique *camera*. Newton's new instrumental variation simply replaced the optical lens in the camera aperture with a
- prism. [He speaks of placing a prism next to a hole in his "window shuts" or shades to produce the rainbow spectrum.] What we have here is

a very small variation on the antique *camera*. Newton retains the circular aperture but places a prism instead of a lens at the opening and this produces the "rainbow spectrum." From the results, Newton derives his theory of light, arguing that "white" or composite light consists of color bands which the prism separates. His *Opticks* was published in 1704, the dawn of the eighteenth century. This is clearly imaging, but it is equally clear that it is no longer isomorphic imaging—the rainbow on the *tabula rasa* is not "like" its source, the sun. Rather, I shall call it *non-isomorphic*, and note that it is much more like a coded "writing" which needs to be interpreted. But this will take a step or two more to see why. Newton's color spectrum had "fuzzy" blendings of colors, one into the other. But in spite of this, he recognized that the colors had what today we would recognize as different frequencies which did not focus at the same spot, and he ingeniously used this newly produced knowledge to develop the *reflecting telescope* which uses a parabolic shaped mirror to refocus all colors at the same focal spot.

- Scientists never ceased to be fascinated by their optical toys and from Newton on they continued to play with both variations of the camera, isomorphic and non-isomorphic. A breakthrough occurred with another variation on the camera. Chemists had been using prisms to produce spectra since Newton, but in the nineteenth century William Wollaston tried yet another variant for the camera. In 1802 he decided to try a slit instead of the usual circular aperture, and this change immediately produced clear color distinctions compared to Newton's fuzzy, overlapping spectrum. Moreover, dark lines were present between some of the primary colors. In short, the late modern spectrometer had been invented. Cutting the story short, within a few more decades the now *spectroscope* users discovered that each element cast a different spectrum, a *chemical signature*, as it were. The most common setup was a light source (usually a candle), which cast a beam through the prism onto some screen. By throwing known chemical salts into the flame, different spectra were produced and through practice these could be identified. By mid-century, after much experimentation and argumentation, Gustav Kirchhoff finally showed that sodium spectra produced by salts in flames were identical with a part of the solar spectrum, thus giving the clue to chemical signatures for both earth and sun. Here my reason for calling non-isomorphic imaging "text" or code-like now becomes clear. But at the same time note that the instrument by which all this scientific knowledge is produced is simply a set of minor variations upon the antique *camera*. An antique spectroscope is simply the *camera obscura* with a slit instead of a hole (without or without lens) and a prism instead of a lens. Nor does the story end there.
- Newton firmly believed that light was made up of "corpuscles" or tiny bits of matter. Thomas Young, an exact century after Newton's *Optiks*, in 1804, (but only two years past the one slit improvement of Wollaston)

conducted an experiment which demonstrated that light had *wave properties*. To demonstrate this, he used an optical instrument called an *interferometer*, which was, again, another variant upon the *camera obscura*. The new variation produced its results through another very simple change in the setup—Young produced a simple variation by making a camera with *two slits* instead of the single slit for the spectroscope, or the single circular aperture of the camera. The result was that interfering wave patterns now were imaged on the screen. Young's analogy was that of water waves—if one throws two stones into a pond, the interacting double set of patterns of waves interfere with one another in a recognizable pattern and the same wave interference pattern in light appears from the double slit camera or interferometer.

The above examples were all developed within the period we recognize as early to late modernity. And, they are all variations upon a finite number of elements which revolve around optical or light source phenomena. A light source (sun, candle, light bulb) produces a beam by going through an aperture (hole, lensed circular opening, slit, double slit), sometimes with additional optical artifacts (lens, prism) and casts an image upon a blank screen.

This story does not end with modernity, but it does illustrate what I call instrumental phenomenological variations which produce robust scientific knowledges. [I use the contemporary term knowledges to indicate the different products resultant from the instrumental practices.] The playfulness with light is a continuous practice in many of the sciences, but in its modern sense it was limited to what I shall call "passive" light sources. I shall not, with one exception, follow the story into its twentieth–twenty-first century developments which introduce yet another set of variations into the invention of science by means of the *camera obscura*. In this perhaps "postmodern" period, it is the light itself which gets varied. In the postmodern versions of the camera, "white light" is not so frequently used at all except as a general carrier for other radiations. Instead, electrons, ions, monochromatic light (lasers), is "sprayed" or manipulated through the postmodern camera (the many varieties of spectrographs, such as mass spectrography, etc., beam splitters, and the like). I even have an astronomer friend who denies the telescope is an "instrument" since a "real" instrument is what is attached to the telescope, such as an inteferometer. The telescope simply is a gathering device for most purposes today. In short, "passive" light has been replaced by "active" beam production which is itself transformed and manipulated to be passed through the instrumental variant to produce what might be called "constructed" knowledges for contemporary science. A contemporary example of these even newer "active" technologies is the use of a laser, connected with a very recent variant upon the *camera*, which shows the "spin" of photons.

- By the end of the nineteenth century, it was discovered that finer discriminations of signatures could be obtained by using a *diffraction grating* coupled to the spectroscope (a fine lined grating which allowed the many spectral lines to have greater resolution). The device I am describing uses such a grating, but instead of natural light, laser monochromatic light is used. The laser beam passes through the grating onto a target and the stream of photons contained in the beam then produces a pattern which shows which direction the "spin" of photons has. In short, the trajectory of variants on the *camera obscura* has not ended yet.

Postphenomenological conclusions: I have now taken us through a trajectory of instrument development which has spanned several centuries. I have tried to show that what classical phenomenology needed to appreciate and understand such science practices was a sensitivity for the material—the technologies which "embody" scientific practice. Similarly, the reflexive counterpart of such sensitivity to materiality is an equal sensitivity to *bodily* materiality, active perceptual materiality, or human embodiment. I have further hinted that in imaging, science practice itself operates in a phenomenological, or better, postphenomenological sense. The instruments, here the variety of *camera obscurae*, as embodied, material instrumental variations. And in the most contemporary sense, they also become *active* or constructive counterparts to our actional human embodiment. The instruments are the means by which we engage our world, seeking to produce the knowledges needed about that which we inquire into. This kind of science remains fully within the lifeworld.

Notes

1. Carl Mitcham's *Thinking through Technology: The Path between Engineering and Philosophy* (Chicago, 1994) is the most definitive work on the history of the philosophy of technology. He claims Ernst Kapp first used the term in a book, *Philosophie der Technik*, in 1877.
2. René Descartes, *La Dioptrique* (1637, from the 1913 edition, p. 144).
3. Edmund Husserl, *The Crisis of European Sciences and Transcendental Phenomenology*, trans. D. Carr (1936, Evanston: Northwestern University Press, 1976), pp. 360–1.
4. Ibid., p. 28.
5. Maurice Merleau-Ponty, *The Phenomenology of Perception*, trans. Colin Smith (1945, New York: The Humanities Press, 1962), pp. 52–3.
6. See my *Technics and Praxis: A Philosophy of Technology* (Dordrecht: Reidel Publishers, 1979) for the first schematism of embodiment, hermeneutic and background relations.
7. Ibid., p. 143.
8. Blackwell Publishers has recently published a major anthology in the philosophy of technology, edited by Robert Scharff and Val Dusek, *Philosophy of Technology: The Technological Condition* (New York: Blackwell Publishing, 2003). There is included a major section of Heidegger texts, plus scholarship on Heidegger, an indication of his pre-eminence as a founder of this subdiscipline.

9. Martin Heidegger, *Being and Time* (1927, New York: Harper and Row, 1962), p. 97
10. Heidegger, "The Question Concerning Technology," often reprinted and included in Scharff-Dusek above (1954/2003).
11. Hans Achterhuis, Van stoommachine tot cyborg: denken over techniek in de niew wereld (Ambo, 1997), expanded an revised, translated by Robert Crease *American Philosophy of Technology: The Empirical Turn* (Bloomington: Indiana University Press, 2001).
12. Don Ihde, *Instrumental Realism: The Interface between Philosophy of Science and Philosophy of Technology* (Bloomington: Indiana University Press, 1991).

References

Achterhuis, Hans. *Van stoommachine tot cyborg: denken over techniek in de niew wereld* Ambo, 1997, translated by Robert Crease, in an expanded edition, *American Philosophy of Technology: The Empirical Turn*. Bloomington: Indiana University Press, 2001.

Borgmann, Albert. *Holding On to Reality: The Nature of Information at the turn of the Millennium*. Chicago: University of Chicago Press, 1999.

Feenberg, Andrew. *Questioning Technology*. New York: Routledge Publishing, 1999.

Heidegger, Martin. *Being and Time*, trans. J. Macquarrie and E. Robinson. New York: Harper and Row, Publishers, 1962.

Herschel, J. B. *The Analysis of Starlight: One Hundred and Fifty Years of Astronomical Spectroscopy*. Cambridge: Cambridge University Press, 1987.

——. "The Question Concerning Technology," reprinted in Robert Scharff and Val Dusek, (eds), *Philosophy of Technology: The Technological Condition*. New York: Blackwell Publishing, 2003.

Husserl, Edmund. *Cartesian Meditations*, trans. D. Cairns. The Hague: Martinus Nijhoff, 1969.

——.*The Crisis in European Sciences and Transcendental Phenomenology*, trans. D. Carr. Evanston: Northwestern University Press, 1970.

Ihde, Don. *Technics and Praxis: A Philosophy of Technology*. Dordrecht: Reidel Publishers, 1979.

——. *Technology and the Lifeworld: From Garden to Earth*. Bloomington: Indiana University Press, 1990.

——. *Instrumental Realism: The Interface between Philosophy of Science and Philosophy of Technology*. Bloomington: Indiana University Press, 1991.

——. *Bodies in Technology*. Minneapolis: University of Minnesota Press, 2002.

Merleau-Ponty, Maurice, *The Phenomenology of Perception*, trans. C. Smith. London: Humanities Press, 1962.

Mitcham, Carl. *Thinking through Technology: The Path between Engineering and Philosophy*. Chicago: University of Chicago Press, 1994.

Part III Epistemology, Identity, and Imaging

5
Concrete Images for Abstract Questions: A Philosophical View*

Patrick Grim

5.1 What does imaging give us?

What are imaging technologies *for*?

The short answer is that imaging technologies allow us to see things. But that answer is clearly *too* short; we see things all the time, without any artificial imaging and without the necessity of technological aid. What is it that imaging technologies give us, then, that ordinary vision does not?

Consider a very traditional form of imaging: photography. When you look at the image in Figure 5.1, what you see is what you would have seen had your eye been where the camera lens was when the photograph was taken. That is not quite true—yours would have been color vision, focused for depth in a way that does not occur here, and with some further differences in optics—but the characterization is true enough (see also Cohen and Meskin 2004). What the eye sees is clearly what the inventors of photography were trying to capture.

There is a sense, then, in which a traditional photograph does not show you anything that you could not have seen without it. It does, however, show it to you at a different time and place. The time and place can be *very* different; given the contingencies of history, for example, *your* eye could *not* have been where the camera lens was when this image was taken. What the camera captured, moreover, was visible to the human eye for only an instant. The photographic image, in contrast, is something we can study in great detail, far distant in space and time, and for an extended period.

There are other forms of imaging that give us more. Microscopic images allow us to see things too small for the naked eye. Telescopic images allow us to see the night sky in a detail we cannot see without them. With radio telescopes and the right forms of imaging we can do even more. We cannot see radio transmissions, not because they are too small or too faint, but because they fall outside the range of visible frequencies; radio transmissions are simply not the *kind* of thing we can see. When we code radio frequencies as colors, however, we can survey the patterns of radio frequencies as if they

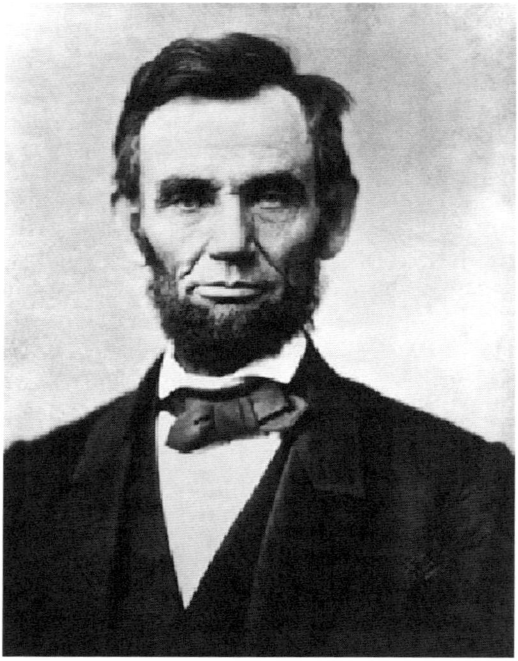

Figure 5.1 Abraham Lincoln.

were visible. The image of Andromeda in Figure 5.2, showing radio signals from hydrogen clouds and the gas disk of the galaxy, is constructed in precisely this way. Infrared images are of the same sort. Though we cannot *see* heat, infrared images can show us the heat distribution across a person's body, for example, by coding heat in terms of something we *can* see. In both cases, if we are using photographic images, we can study distributions of radio frequencies and heat in detail, at length, and at our leisure.

The work that follows represents a further step in this progression. There are imaging technologies beyond those that let us see something that we could have seen only at another time and place, beyond those that allow us to see what would otherwise be too small or too faint to see, and beyond those that let us see something that is objectively before us—like radio and infrared frequencies—but that we are unable to see without those technologies. Beyond all of these are imaging technologies that let us see something that *isn't there at all*.

My focus here is the application of imaging technologies to purely abstract questions. Let me make it clear that I am a realist about these issues; I think the abstract relations at issue are real, and that the things we say about them can be true or false. What is at issue, however, are images of things which do

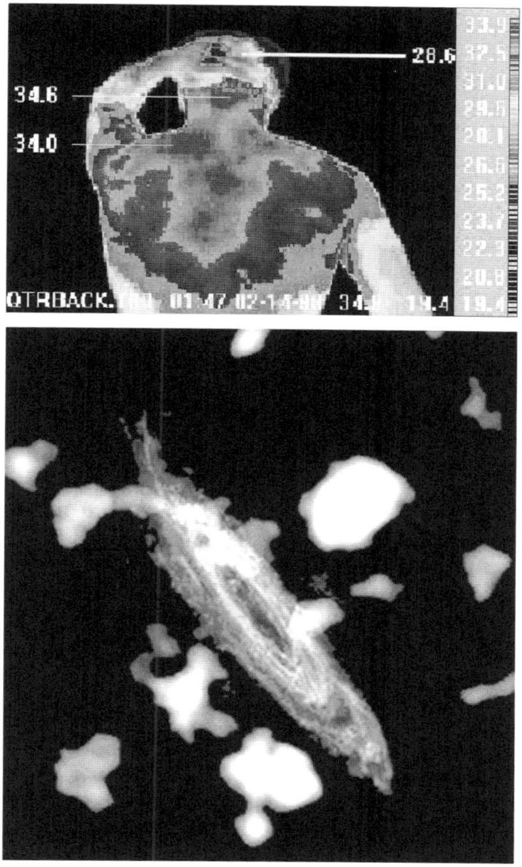

Figure 5.2 Thermal imaging of the human body (top). Radio imaging of the Andromeda galaxy (bottom). Images courtesy of Teletherm infrared at www.thermology.com; D. Thilker (JHU), Robert Braun (ASTRON), WSRT and NRAO/AUI/NSF.

not by any means qualify as physical. The reality in which I am interested here is the timeless reality of abstract concepts, sometimes said to exist "in Plato's Heaven." Using that turn of phrase, what I want to discuss are snapshots from Plato's Heaven: concrete images for abstract concepts.

5.2 What does abstract imaging give you?

The virtues of abstract imaging are the same as the virtues of some forms of non-abstract imaging. Why, for example, do we use thermal imaging to convert temperature to color? What thermal imaging does is allow us to literally *see* where things are hot and cold. The reason to convert temperature to something visual is simply this: our visual abilities are *better* than our heat-distinguishing abilities.

Let me introduce a thought experiment in terms of two maps:

(1) The first is a world map color-coded for average annual temperature. Most of Australia glows red, the Arctic is blue-black, and the temperature differential on the west coast of South America is visibly obvious. Figure 5.3 is such an image, though the thought experiment requires that you imagine it in color. (This figure and many that follow can be seen in color and with full animation at www.pgrim.org/Imaging.)
(2) The second map codes temperature not in terms of color but in terms of temperature. This map you would read with your fingers. The Sahara would *feel* warmer to the touch, while the Arctic would *feel* cooler.

The thought experiment is this: Would the second map give you better information?

The thermal map codes temperature *as* temperature; it is what Don Ihde calls an isomorphic image (Ihde 1997). The visual map is a non-isomorphic image, coding temperature as color—an arbitrary association to something entirely different. My bet, however, is that both the details of temperature distribution and their overall pattern would be more obvious to you in the visual map—the map that codes temperature as something else. The simple fact is that we are better at visual discrimination and immediate visual inference than we are in other sense modalities. We are visually advantaged because we are descendants of fruit-finding primates. If information can be converted to visual information, we are better able to exploit our inherited visual abilities.

We have all had the experience of seeing things in the shapes of clouds: giants' heads, unicorns, and beautiful women. If not in the clouds, then perhaps in swirls of color in the linoleum on a bathroom floor, or in water

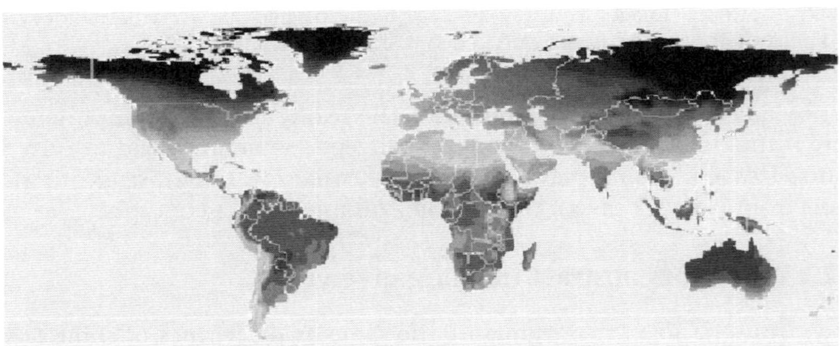

Figure 5.3 Average annual temperature.

stains in an acoustic ceiling above your hospital bed. You stare at the water stains and can see mountain ranges, centaurs, and bearded wizards leaning on their staffs. Here I offer not a thought experiment but a real one. Next time you see a centaur cantering across swirls of color in the linoleum, grab a piece of tracing paper and trace it. What you will find is that your tracing is really quite good. Indeed the result may be a much better picture of a centaur than you thought yourself capable of. (Precisely such an experiment, using swirls in bark paper, is outlined in Haseltine 2002.)

How could it be that your tracing of a centaur glimpsed in the swirls of the linoleum is better than any centaur you could have drawn unaided? Here is a piece of advice from Leonardo da Vinci's "Way to augment and stimulate the mind toward various discoveries," taken from the *Treatise on Painting*:

> [L]ook at walls splashed with a number of stains or stones of various mixed colors. If you have to invent some scene, you can see there resemblances to a number of landscapes, adorned in various ways with mountains, rivers, rocks, trees, great plains, valleys and hills. Moreover, you can see various battles, and rapid actions of figures, strange expressions on faces, costumes, and an infinite number of things . . .
>
> . . . it should not be hard for you to stop sometimes and look into the stains of walls, or the ashes of a fire, or clouds, or mud, or like things, in which, if you consider them well, you will find really marvelous ideas. The mind of the painter is stimulated to new discoveries, the composition of battles of animals and men, various compositions of landscapes and monstrous things, such as devils and similar creations.
>
> (Da Vinci c. 1500 [1956], 50–1)

The explanation of your ability to trace a centaur "seen" in swirls of linoleum, and the explanation for da Vinci's method, is that we have very well-developed visual *recognition* systems. We are able to recognize faces immediately, and are good at distinguishing familiar things even through the fog or under conditions of limited illumination. When you "see" the centaur in the swirls of linoleum it is your visual recognition system that is operative. When you trace the centaur, what you are doing is not exhibiting your centaur-drawing skills but rather exploiting your centaur-recognition skills. You are much better at recognizing centaurs when you see them than you are at drawing them from scratch. The same may even have been true of Leonardo.

Quite generally, the idea of visual imaging for theoretical purposes is to make use of our inherent visual recognition systems. This applies to coding abstract relations in terms of the visible, precisely as it applies to coding temperature as color in thermal imaging. Once an abstract relationship is rendered visible, we are able to employ visual recognition systems that are already in place. What was hard to conceptualize abstractly, difficult to grasp and manipulate formally, may be as obvious as the nose on your face when you can *see* it. Concrete images can therefore allow us to deal with abstract questions as if they were something far simpler—as if they were questions about the visible. The point is very much that which Herbert Simon and Jill Larkin make with regard to diagrams:

> The major difference in a diagrammatic representation, we believe, is difference in recognition processes. We have seen that formally producing perceptual elements does most of the work of solving the geometry problem. But we have a mechanism—the eye and the diagram—that produces exactly these "perceptual" results with little effort. We believe the right assumption is that diagrams and the human visual system provide, at essentially zero cost, all of the inferences we have called "perceptual". As shown above, this is a huge benefit. . . . It is exactly because a diagram "produces" all the elements "for free" that it is so useful.
>
> (Larkin & Simon 1987, 92)

> The advantages of diagrams . . . are computational . . . diagrams can be better representations not because they contain more information, but because the indexing of this information can support extremely useful and efficient computational processes.
>
> (Larkin & Simon 1987, 99)

What I offer in the sections that follow are a range of examples of concrete images for abstract questions. The idea of converting abstract relationships to concrete images is of course far from new. It appears clearly in Plato's "divided line," drawn in *The Republic* as an image of the relationship between the realm of perception and the realm of conception. One of our primary tools of visualization, across all fields of analysis, is the gift of another philosopher—Descartes. Cartesian coordinates are a wonderful example of precisely what is at issue: a technique for converting abstract algebraic relationships in two variables into something geometric that one can *see* in two dimensions. Why do that? Precisely because we're better in grasping and manipulating what we can see. It is much more obvious what a parabolic function really is, for example, when one can actually see the parabola.

Over more than a dozen years, the informal Group for Logic and Formal Semantics in the philosophy department at SUNY Stony Brook has been exploring a range of philosophical questions using computer modeling and computer generated images (www.pgrim.org/Imaging). That work has involved undergraduates, graduate students, visitors, and several faculty members. The work detailed in the following sections is drawn from two very different periods: (1) our early work on the semantics of paradox, and (2) our most recent work in social modeling, applying spatialized game theory to questions of prejudice reduction.

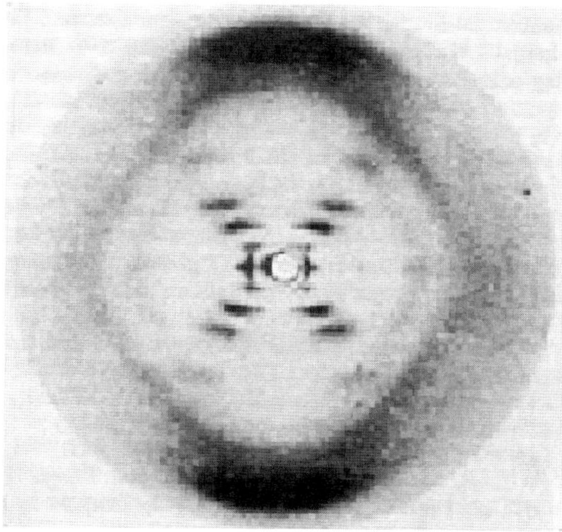

Figure 5.4 Rosalind Franklin's X-ray crystallography image of DNA.

I should also emphasize that one set of examples discussed below is more successful than the other. Not every image is equally informative. You may look at hundreds of X-Ray crystallography images of DNA before you see the one that tells you that it has to be a helix (Figure 5.4). The same will be true of concrete images for abstract concepts.

5.3 The imaging of semantic paradox

The topic of semantic paradox is already pretty abstract. In this section I want to outline some of our work that makes it even more so.

Consider the Liar paradox:

Our informal reasoning regarding the Liar may run something like this:

The boxed sentence is false.

If the sentence is true, then it's true that the boxed sentence is false. But this *is* the boxed sentence. If the boxed sentence is true, then, it is false. Suppose then that the sentence is false. That is precisely what the sentence says—that it is false. So if it is false it must be true.

But if it is true, then it is true that the boxed sentence is false. So it is false.

But it says it is false. So if false, it must be true.

We need not go on; the result is clearly an oscillation in estimated truth values. When we suppose the sentence is true, we are forced to conclude that it is false. When we suppose the sentence is false, we are forced to conclude that it is true.

If we mark "True" at the top of the vertical axis and "False" at the bottom, and if we track iterations across the horizontal axis, the semantic behavior of the Liar paradox looks something like Figure 5.5.

We can make things more abstract by converting to Fuzzy or infinite-valued logic. Classical logic is bivalent, admitting only two truth values. There are but two options for every sentence or every proposition: it is true, or it is false. Fuzzy logics, on the other hand, take seriously the idea that there may be more things in heaven and earth than are dreamt of in your Classical logic. In particular, there may be more truth-values; there may be gray areas of truth between absolute truth and absolute falsehood.

In his classic piece "Truth," J. L. Austin asks, "[i]s it true or false that Belfast is north of London? That the galaxy is the shape of a fried egg? That Beethoven was a drunkard? That Wellington won the battle of Waterloo?" He concludes that "[t]here are various degrees and dimensions of success in making statements: the statements fit the facts more or less loosely." (Austin 1950, [1979, 129–30]). George Lakoff asks, "[i]n contemporary America, how tall do you have to be to be tall? 5′8″? 5′9″? 5′10″? 5′11″? 6′? 6′2″? Obviously there is no single fixed answer. How old do you have to be to be middle aged? 35? 37? 39? 40? 42? 45? 50? Again the concept is fuzzy." (Lakoff 1973, 458). If we take these philosophical intuitions seriously, it seems natural to model "degrees of truth" on the [0,1] interval. Such an approach is formalized in multi-valued and infinite-valued logics from Lukasiewicz in the 1920 and 1930s to contemporary Fuzzy logics pioneered by Lotfi Zadeh (Rescher 1969, Zadeh 1975, 1972).

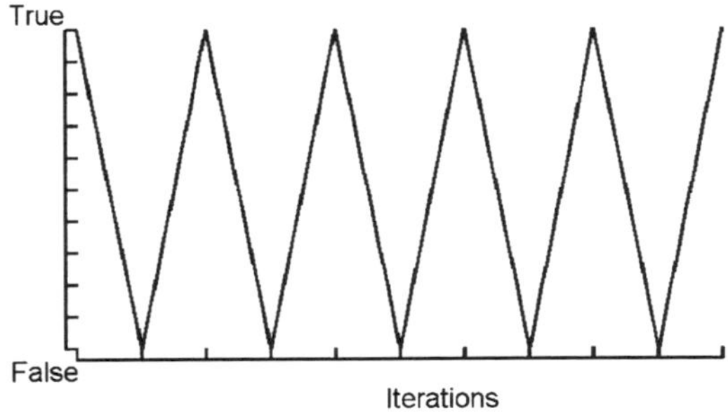

Figure 5.5 Semantic behavior of the Classical Liar (Animated versions of this and later illustrations are available at www.pgrim.org/Imaging.).

Here is a standard treatment of the logical connectives. We take /p/ to indicate the degree of truth of a proposition p. In Fuzzy logic, this can be any value between 0 and 1. A sentence might take a truth value of 1/2, for example, of 2/3, or of .54321.

If the value of p is 2/3, the value of its negation is 1/3. In general,

$$/{\sim}p/ = 1 - /p/.$$

In Classical logic, the value of a conjunction is that of its least conjunct: the conjunction of true and false gives us false, for example. This is generalized in Fuzzy logic using "minimum" operator:

$$/p \,\&\, q/ = \min(/p/,/q/)$$

In Classical logic, the value of a disjunction is that of its greatest disjunct: the disjunction of true and false gives us true. The Fuzzy logic generalization uses a "maximum" operator:

$$/p \vee q/ = \max(/p/,/q/).$$

The fundamentals of Fuzzy logic, then, are an assignment of truth-values anywhere between 0 and 1 and a fairly straightforward generalization of the logical connectives to the infinite-valued case.[1]

There is one more tool we need: Nicholas Rescher's Vvp schema. Suppose that a sentence has a particular truth-value—it really does have that truth value—but someone claims that it has some *other* truth value. How true is that second-order assertion about it? Suppose, for example, that we have a sentence like:

Jack is tall.

Suppose that sentence is pretty true—a good .9 true, perhaps. But now suppose that someone says that it is very untrue, that "Jack is tall" is only .1 true.

"Jack is tall" is .1 true.

How true is what *they* have said?

The Vvp schema calculates the truth-value of this second-order proposition in terms of how far off it is. The closer the truth-value it attributes to the sentence is to its real truth-value, the truer our second-order sentence is. The farther the attributed truth-value is from real truth value, the falser the second-order sentence is. In general, where Vvp is a sentence that attributes a value v to proposition p, and where /Vvp/ is the value of such a sentence:[2]

$$/Vvp/ = 1 - \mathrm{abs}\,(v - /p/)$$

In this case, we have said, the original sentence is .9 true. The person who claims it is only .1 true is .8 off the mark. The value of his statement, then, is 1 – abs (.1–.9), or .2.

With those basics, we have an infinite-valued or Fuzzy logic that can handle the truth of assertions about assertions. What this gives us is an infinite-valued logic that can model self-reference.

Suppose a sentence attributes *itself* a certain value—the Liar, for example, which attributes itself the value "false." Such a sentence amounts to a higher-order statement about its own value. If we start with an estimate of the sentence's truth-value, then, the Vvp schema will force us to a series of revised values. We are able to model the effects of self-reference in terms of an iterated function.[3]

What happens to the Liar when we move from a Classical logic to an infinite-valued logic? If we say the Liar is "true," we get an oscillation between "true" and "false." If we say it is "false," we get the same oscillation. That much is clear from the fact that Classical values are preserved in the infinite-valued logic we have outlined—for the special cases of "pure truth" and "pure falsity" the two logics agree. But what happens if we assign the Liar some fuzzy value in between?

If we estimate the value of the Liar as not 1, not 0, but 1/4, we once again get an oscillation—but this time an oscillation between 1/4 and 3/4 (Figure 5.6).

If we estimate the value of the Liar as 2/3, we get a similar oscillation between 2/3 and 1/3. If we estimate the Liar as precisely 1/2 true, we get a fixed point. Here for the first time we can get a stable and consistent truth-value for the Liar paradox. What is important for present purposes, of course, is that all of this allows us to *see* how values assigned to the Liar will have to be revised over time. What these techniques give us is a concrete image of abstract logical relationships.

Once we have these tools in hand, there is no reason to stop with the Liar; a range of interesting self-referential sentences lie open for examination.

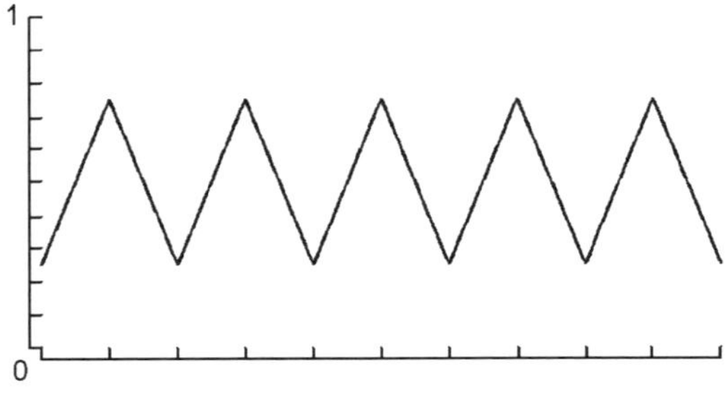

Figure 5.6 The infinite-valued Liar with an initial value of ¼.

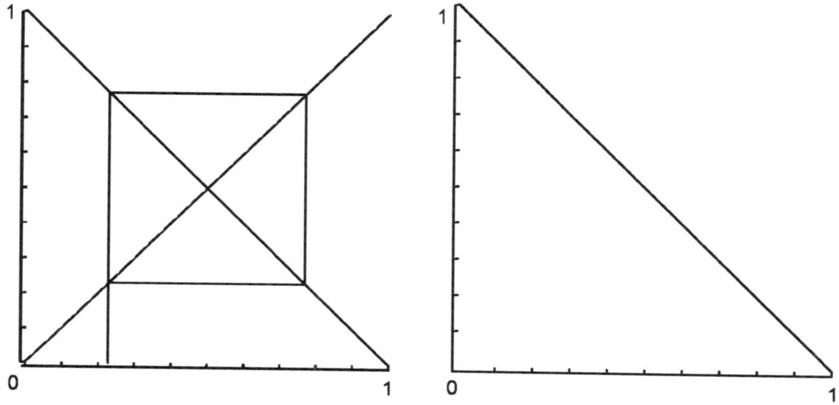

Figure 5.7 Web diagram for the Liar.

Here a further shift in imaging helps as well. The "bounce" or "time-lapse" diagrams we have used so far have a built-in limitation: they show only a finite stretch of the iterated pattern. Web diagrams are an elegant imaging solution to this problem. Thanks to Descartes and Cartesian coordinates, we can operate in a two-dimensional space. All of our semantic values lie within the [0,1] interval, so we let both our axes run between 0 and 1. Within this semantic space, we can map what self-referential sentences do as graphed functions. For any value you assign to the Liar, for example, the revised value calculated through the Vvp schema comes out as 1 minus your original value. The line $y = 1-x$ line descending from upper left to lower right therefore maps the value returns of the Liar paradox. Put in a value on the x-axis and one can read off the revised value on the y-axis (Figure 5.7a).

In dealing with self-reference, however, we are dealing with iterated values. We want to put in a value, get a revision, put that in as a further value, get a further revision, and so on. We want to convert the y-values we get from the calculation in Figure 5.7 into x-values again. We can do that by bouncing them off the $x = y$ line. For an infinitely iterated function, we put in a value on the x-axis, take its value at our function read on the y-axis, bounce that off the $x = y$ line to convert it to an x-axis value again, and repeat. Figure 5.7b shows what the Liar paradox looks like in a web diagram for an initial value of .23. Here the Liar appears as a box because it gives a repeated oscillation between values of .23 and .77.

Consider now a different self-referential sentence:

This sentence is exactly half as true as you think it is.

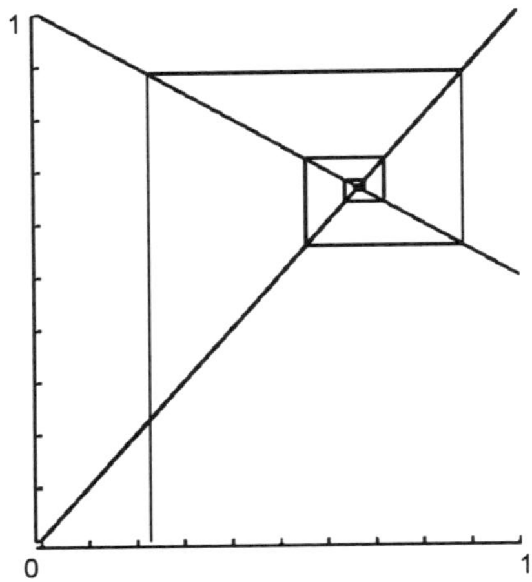

Figure 5.8 The Half-Sayer.

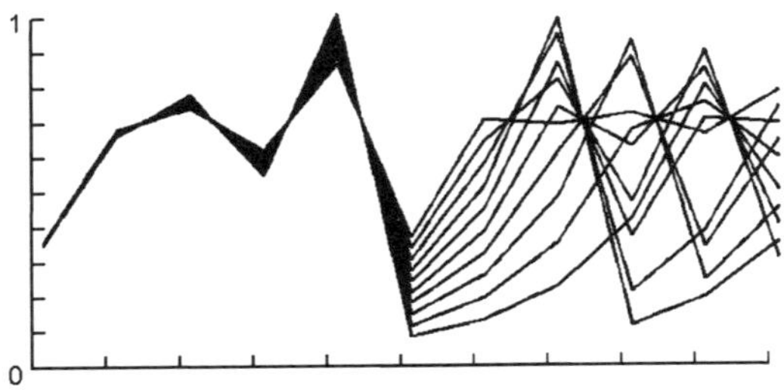

Figure 5.9 The Chaotic Liar: Sensitivity to initial conditions.

For any estimate you give, what this sentence asserts is that its value is precisely half of that. This too will force a series of revised estimate. Figure 5.8 shows what this "Half-Sayer" looks like in a web diagram, starting with an initial estimate of .23.

Here revised values converge to a truth-value of 2/3. The interesting fact is that the same thing happens with *any* initial value. For any value you start with, the Half-Sayer will force you through Vvp recalculation to an attractor point of 2/3.

Here is the Chaotic Liar:

This sentence is as true as you think it is false.

The semantical dynamics of the Chaotic Liar is provably chaotic on the unit square. There are a number of mathematical definitions of chaos.[4]

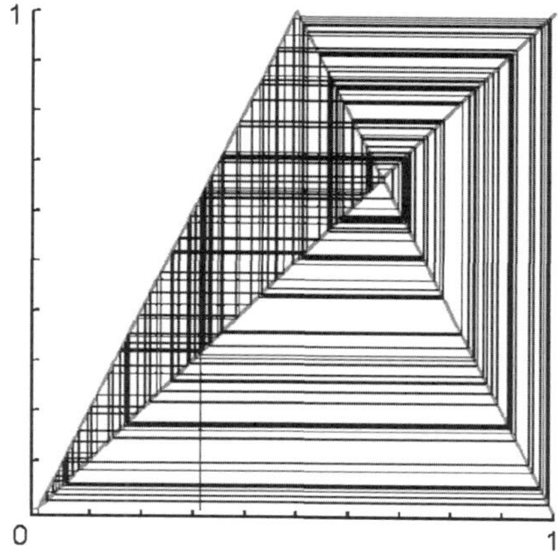

Figure 5.10 The Chaotic Liar.

What they have in common is "sensitivity to initial conditions": within the constraints of your domain, arbitrarily small differences in initial input produce arbitrarily large differences on iteration. Figure 5.9 shows initial inputs for the Chaotic Liar separated by only thousands, with their divergent courses on iteration. Figure 5.10 shows a web diagram for the Chaotic Liar.

5.4 Semantic paradox in two and three dimensions

In this section I want to take a further step in abstract imaging for semantic paradox. To this point it is clear what we are looking at: the semantical dynamics of self-referential sentences in an infinite-valued logics. Though questions of semantical dynamics are fairly abstract, concrete images allow us to track them visually.

To this point we know how to read our images; I want now to outline some further work in which we have images of the abstract that we do *not* know how to read. These images certainly appear to be showing us something; indeed they seem to be showing us something important. But it is not clear what.

Let us move from single self-referential sentences to pairs of sentences that talk about each other. The medieval philosopher Jean Buridan (Buridan c. 1340) has a Dualist form of the Liar:

Plato: *The next thing Aristotle says will be false.*
Aristotle: *Plato has spoken truly.*

In our infinite-valued logic, we might consider two sentences like these:

The Chaotic Dualist: X: X is true to half the extent that Y is true.
Y: Y is as true as X is false.

Here again we will have a dynamics that operates in terms of initial estimates and forced revisions. In this case, however, we start with two values—our estimates for each of our two sentences. We are forced to revise each, giving us a set of two revised values. We are then forced to revise those in turn.[5]

Suppose that we start with a guess of .1 for the first sentence above, and .9 for the second. We are now working with two variables—one for the value of sentence X, one for the value of sentence Y—and can thus can envisage our initial estimates as a point (.1, .9) in the Cartesian plane. On iteration we will be forced to revise values, which we can envisage plotting as further points in the plane.

If we start with values of .1 and .9 for sentences X and Y, the pattern that appears for Vvp revision points in the Chaotic Dualist is shown in Figure 5.11. If we do this for a whole range of initial points from 0,0 to 1,1, amazingly enough, we get exactly this same image.

What is clear is that the successive values for the Chaotic Dualist form an intriguing strange attractor. Across the range of points from 0,0 to 1,1, series of revision values are drawn into this same folded pattern. What is far from clear, however, is what that *means*. This is a snapshot from Plato's heaven, giving us a concrete image for some abstract semantic phenomenon. But it remains an image that we cannot read.

There are further images in the same category. An "escape-time diagram" is a way of capturing a snapshot of the motion of various points—like those in Figure 5.11—over time. For a pair of sentence such as X and Y we can, for example, draw an escape-time diagram indicating the number of iterations before a certain point "escapes" from a uniform distance of a bit over 1 unit from the origin.[6] An escape-time diagram for the Chaotic Dualist appears on the right in Figure 5.12. If we do something still more abstract—if we expand our basic values beyond the unit square—we get the image on the left in Figure 5.12.

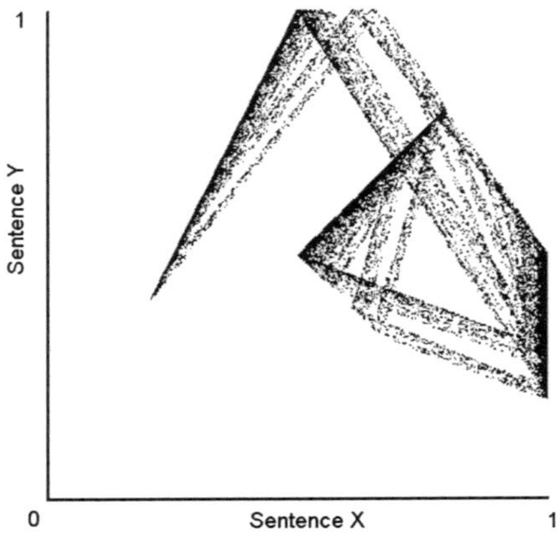

Figure 5.11 Revision points for sentences X and Y.

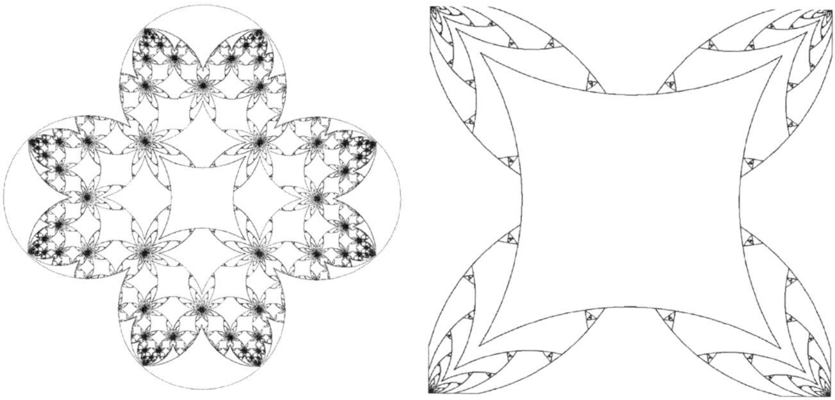

Figure 5.12 Escape-time diagrams for the Chaotic Dualist, for the unit square (right) and for values beyond the unit square (left).

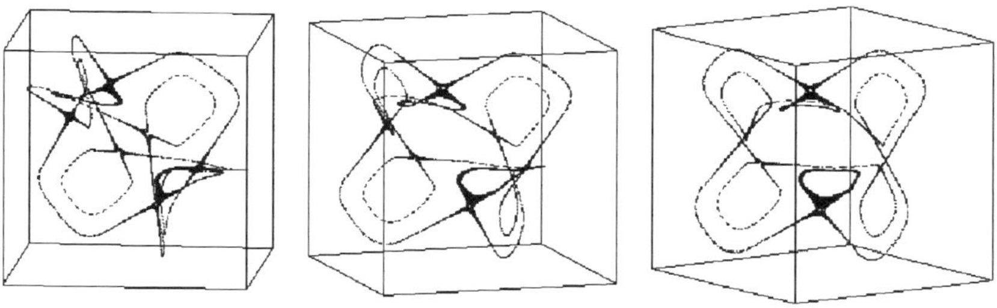

Figure 5.13 Triplist attractor.

These images are beautiful, they are clearly fractal, and they seem to be telling us something important about the semantics of paradox. Unfortunately, it is a message that we cannot yet read.

Once we have gone to Dualists, it is tempting to go to Triplists—triplets of sentences that talk about each other. Here again we can plot the movement of revised estimates of truth-value, but in three dimensions rather than two, and we can construct escape-time diagrams.

Consider, for example, three sentences each of which says it is one-fourth as true as the difference between the other two:

X: X is ¼ as true as the difference in truth-values between Y and Z.
Y: Y is ¼ as true as the difference in truth-values between X and Z.
Z: Z is ¼ as true as the difference in truth-values between X and Y.

The attractor for these sentences, rotated in three dimensions, is shown in Figure 5.13.

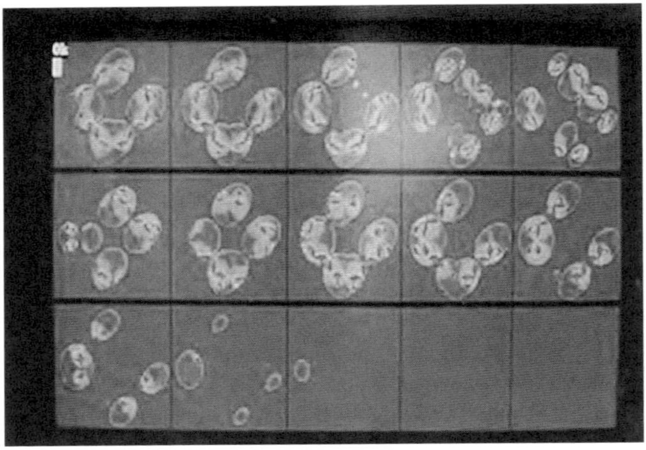

Figure 5.14 Two-dimensional slices of a three-dimensional semantic solid.

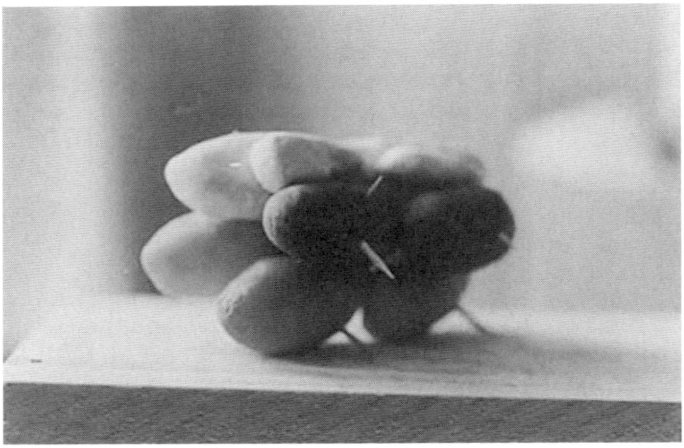

Figure 5.15 First attempts at a three-dimensional model (images in color at website).

The escape-time diagram for three mutually referential sentences will be a three-dimensional object: an escape-time solid in three dimensions. Once we began working with Triplists, we naturally wanted to know what their escape-time solid would look like. Our first attempts consisted of two-dimensional slices through the object along the x-axis and z-axis. Figure 5.14 shows a photograph from the archives, taken directly from a monitor using a 35mm camera. It is amazingly hard to envisage a three-dimensional solid

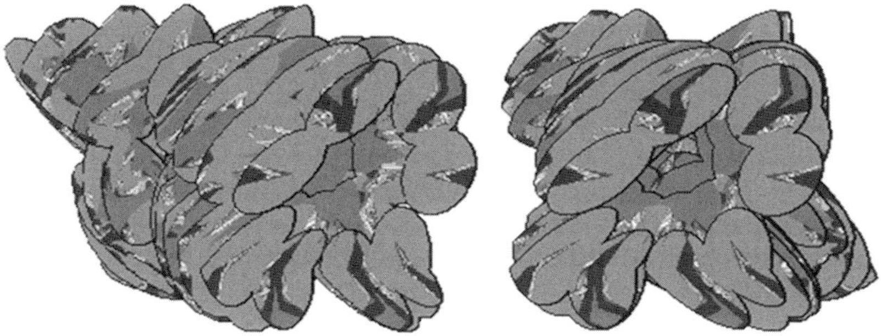

Figure 5.16 A Triplist solid.

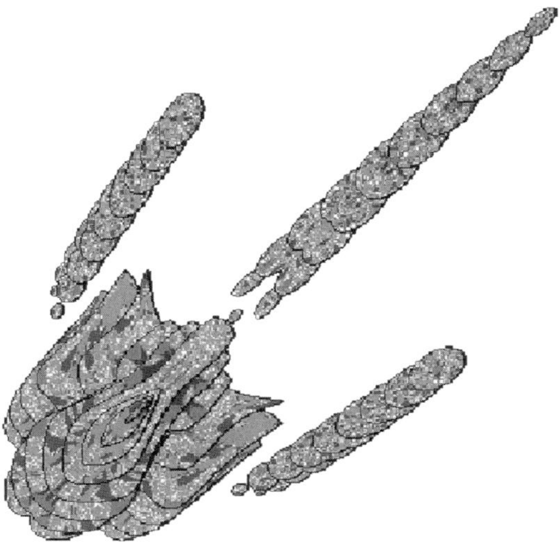

Figure 5.17 Another Triplist solid.

from merely a series of two-dimensional slices, however. Our first attempts at imaging were made with Playdough (Figure 5.15).

In the end, however, we were able to manipulate our sliced images in a way that gave us a better glimpse of the abstract object itself. The solid for the trio of mutually referential sentences above appears in Figure 5.16. Figure 5.17 shows an escape-time solid for a further Triplist.

These all qualify as concrete images for abstract questions. Sometimes the answers they give us are clear and immediate—answers to questions about whether the dynamics of a particular sentence will converge on a final value, for example. Sometimes, however, the images are such that we cannot read the answers, and indeed may not be sure that we can properly frame the questions these images would answer. It is said that a man's reach should exceed his grasp. Sometimes our imaging reach exceeds our intellectual grasp. In the abstract realm, as in other realms, we may sometimes be unsure of precisely what it is that we are seeing.

5.5 Imaging, game theory, and prejudice

Because the investigation of semantical paradox by means of concrete images gives us images we do not know how to read, it cannot claim to be a complete success story. The second example I want to offer is a success story all the way through. What we have developed is concrete imaging designed to answer a very particular range of abstract questions regarding game theory: a new measure of robustness for game-theoretic results.

For the past 25 years or so, the formal tool of choice regarding cooperation, generosity, and altruism has been the Prisoner's Dilemma. Figure 5.18 shows the standard 2×2 matrix, meaning just that we are dealing with two players, each of which has two options for play. Each player's payoff from an interaction is dictated by choices on both sides. Changing the values in a 2×2 matrix gives one different games, but by far the most studied game in game theory is the Prisoner's Dilemma, using the standard values shown here.

In a one-shot Prisoner's Dilemma, each player has two options: to cooperate or to defect. Should both players cooperate, each gets 3 points; should both defect, each gets 1 point. Should one defect and the other cooperate, the defector gets 5 points while the cooperator gets 0. The dilemma is which option you should choose when playing against an unknown opponent.

		Player B	
		Cooperate	Defect
Player A	Cooperate	**3**,3	**0**,5
	Defect	**5**,0	**1**,1

Figure 5.18 The standard matrix for the Prisoner's Dilemma.

Perhaps you should defect. That could get you the highest payoff—5 points, if your opponent chooses to cooperate. If he defects as well, however, you will get stuck with only 1 point each. So perhaps you should cooperate. There you could both get 3 points. If your opponent defects against you, however, you will end up with 0.[7]

The Prisoner's Dilemma is most interesting in an iterated form.[8] Here there isn't a single game; one faces the same opponent repeatedly. What you do on this round may therefore influence what your opponent does on later rounds, and vice versa. In the Iterated Prisoner's Dilemma there will be strategies for play over time. There are, for example, 8 simple "reactive" strategies, which gear play to whatever it is that the opponent did on the last round, shown in Figure 5.19. Using 0 for defect and 1 for cooperate, the first digit of each 3-tuple indicates the option chosen at the very beginning of the game, the second digit indicates the option played when the opponent has cooperated on the previous round, and the third the option played when the opponent has defected on the previous round.

In 1980, Robert Axelrod staged the first computer tournament for the Iterated Prisoner's Dilemma. Strategies submitted were played in a round robin competition against all other submitted strategies, against themselves, and against a random strategy. That strategy which garnered the most points over all was declared the winner (Axelrod 1980, 1984). Despite the ingenuity and complexity of some of the submissions, the winner of that first tournament was one of the simplest: Tit for Tat, thought of as a "cooperative" strategy. Tit for tat dictates that you play this round just like your opponent played against you on the previous round. It returns defection with defection, repaying cooperation with cooperation in kind.

Axelrod announced Tit for Tat as the winner of the first tournament and solicited strategies for a second competition. Tit for Tat was clearly the

<0,0,0>	All-Defect (All-D)
<0,0,1>	
<0,1,0>	Suspicious Tit for Tat
<0,1,1>	
<1,0,0>	Cooperate-then-All-D
<1,0,1>	
<1,1,0>	Tit for Tat (TFT)
<1,1,1>	All-Cooperate (All-C)

Figure 5.19 The 8 "reactive" strategies.

strategy to gun for. But when the smoke had cleared, the winner in this second tournament was again Tit for Tat (Axelrod 1980b, 1984).

Working with William Hamilton, Axelrod switched from a round robin tournament to a biological model using replicator dynamics: winners in a given generation had a reproductive advantage and thus occupied a larger proportion of the population on the next round. Strategies were pitted against all others in terms of their proportion in the population, but proportions of each strategy in the population changed over time. Using the strategies from the second tournament in this more biologically realistic model, the winner was once again Tit for Tat (Axelrod and Hamilton 1981).

In our earlier work on cooperation, we used a spatialized version of the Iterated Prisoner's Dilemma. Here individuals are embedded as cells in a two-dimensional cellular automata array. We start with a randomized array, using different colors for the strategies at issue. Unlike Axelrod's tournaments, however, individuals in the Spatialized Prisoner's Dilemma are *not* pitted against all other players in the field; here competitive action is local rather than global. Each cell plays a succession of games with only its immediate neighbors—the eight cells touching it on each side and diagonally. It then totals its points and compares its score with that of its immediate neighbors. If a cell has a neighbor that has been more successful—if a neighboring cell has a higher total score—the cell changes its strategy to match that of its most successful neighbor.[9]

A typical run of this sort, starting with a randomized array of the 8 reactive strategies, is shown in Figure 5.20. Given the set up outlined, those strategies that prove more successful should progressively spread to occupy the entire array. This is indeed what happens. All-Defect and Cooperate-then-All-Defect show early growth, but the eventual winner is Tit for Tat.

Here dynamic imaging makes it clear—immediately and visually obvious—that Tit for Tat triumphs in a spatialized context as well. Tit for Tat has a well-founded reputation as a "robust" strategy, triumphing repeatedly despite important changes in the details of the competition. It is an abstract question regarding "robustness" that will be at issue in what follows.

In very recent work, we have applied tools of spatialized game theory to the topic of social prejudice. The tools are formal and abstract, and many of the questions we are asking are abstract as well. The target phenomena that we want to understand, however, is all too familiar: the social phenomena of racial prejudice.

There are a number of theories in the social psychological literature regarding the nature and sources of prejudice. Personality theories characterize prejudice in terms of personality types, with Adorno's work in the 1950s as a prime example (Adorno, Frenkel-Brunswick, Levinson, and Sanford 1950). Social Identity theory and Self-Categorization theory depict prejudice as a natural result of categorization and group identification (Tajfel and Turner 1986; Turner and Oates 1989). The "Realistic Conflict" model portrays prejudice as a consequence of group conflict over limited resources

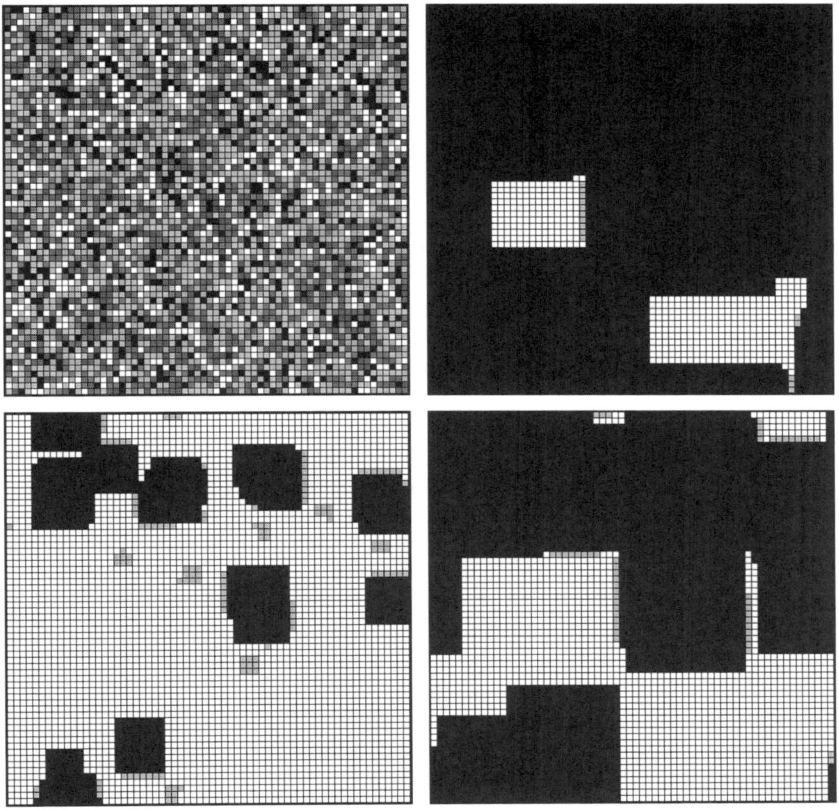

Figure 5.20 Conquest by TFT in a randomized environment of 8 reactive strategies. Complete development of the array can be seen at www.pgrim.org/Imaging.

(Campbell 1965), while Relative Deprivation theory emphasizes perceived resource differences (Gurr 1970).

Though there are many theories of prejudice, there is only one real theory of prejudice *reduction*. This is the contact hypothesis, tracing back to the work of Gordon Allport. According to the contact hypothesis, prejudice against members of one group by members of another will be reduced with increased social contact between the groups. Allport qualifies the hypothesis immediately with a set of further conditions; that the contact at issue be between participants of equal status, that the groups share common goals, that they participate in intergroup cooperation, and that prejudice reduction receives support from authorities (Allport 1954). Details and elaborations of those qualifications continue to be debated (Pettigrew 1998).

The basic contact hypothesis, however, is simple: one can reduce prejudice by increasing contact. It underlies a number of important social policies, including, for example, the desegregation of US public schools under *Brown v. Board of Education*.

What we have done recently is to construct game-theoretic models that match the basic conditions of the contact hypothesis (Grim, Selinger, Braynen, Rosenberger, Au, Louie, and Connolly 2004, 2005). Ours is work in "simulational sociology" or "artificial societies." What we wanted to see was how much of the contact effect would appear—and thus could be explained—purely in terms of spatialized game-theoretic advantage.

Though the social psychological literature contains a wealth of evidence for the contact hypothesis, it contains little regarding an explanatory mechanism. If contact reduces prejudice, precisely how does it do so? The attempts at explanation that have been made, moreover, tend to rely on high-level cognition in the agents involved. With increased contact, it is hypothesized, one's categories are restructured and one's stereotypes fall away. One makes friends with members of the other group, and the virtues one sees in new friends neutralize earlier prejudice. What we wanted to explore was whether the contact mechanisms of prejudice reduction might not be significantly simpler than that—mechanisms captured in the simple gains and losses of spatialized game theory, for example. Our game-theoretic agents are *extremely* simple, far below the cognitive level demanded by explanatory hypotheses that turn on acquaintance, friendship, or the reorienting of conceptual categories.

Our simulations regarding prejudice are based on the model for cooperation outlined above. To that model we add another parameter: our agents also come with a particular color. Each cell in our array is either green or red; unlike strategies, these background colors cannot be changed. We also add a further strategy

<0,0,0>	All-Defect (All-D)
<0,0,1>	
<0,1,0>	Suspicious Tit for Tat
<0,1,1>	
<1,0,0>	Cooperate-then-All-D
<1,0,1>	
<1,1,0>	Tit for Tat (TFT)
<1,1,1>	All-Cooperate (All-C)
<1,1,0 / 0,0,0>	PTFT: Plays TFT against its own color, All-D against others

Figure 5.21 The 8 reactive strategies plus PTFT (Prejudicial Tit for Tat).

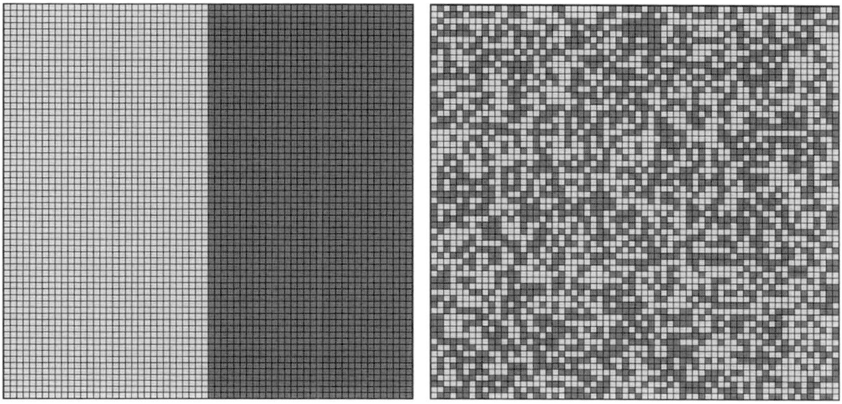

Figure 5.22 Segregated (left) and mixed patterns of background color.

(Figure 5.21). Our eight reactive strategies are color-blind; their play is geared to the opponent's last move, but does not depend in any way on the opponent's color. To these we add a ninth strategy, designed precisely to be Prejudicial with regard to the color of its opponent: PTFT, for "Prejudicial Tit for Tat." PTFT plays Tit for Tat with opponents who happen to share its color: green if it is green, red if it is red. With players of the other color, however, it plays All-Defect.

In order to model the conditions of the contact hypothesis, we compare two different arrays—a segregated array, in which our colors are split right down the middle, and a mixed array, in which they are arranged randomly (Figure 5.22). If the contact hypothesis holds, and holds even in this simple game-theoretic instantiation, a mixed environment should result in a lower level of prejudice than the segregated environment.

What happens if we start with a randomization of our nine strategies in a context segregated by color? The result is shown in Figure 5.23: a dynamics resulting in a roughly 50/50 split between TFT and our Prejudicial PTFT. A typical result is graphed in Figure 5.24. In a segregated environment, Prejudicial PTFT proves as successful as color-blind TFT.

What of a mixed environment? Here there is a very different dynamics, shown in Figure 5.25. In a mixed environment the result is standardly and robustly a complete takeover by TFT alone (Figure 5.26).

What our simulations show is a game-theoretic instantiation of precisely the result that the contact hypothesis would predict. Our agents, however, are far below the cognitive level of people; they operate purely in terms of advantage and disadvantage, dictated by the gains and losses of the Prisoner's Dilemma matrix. Strategy change in the model is simply a matter of copying one's most successful neighbor. What our results seem to indi-

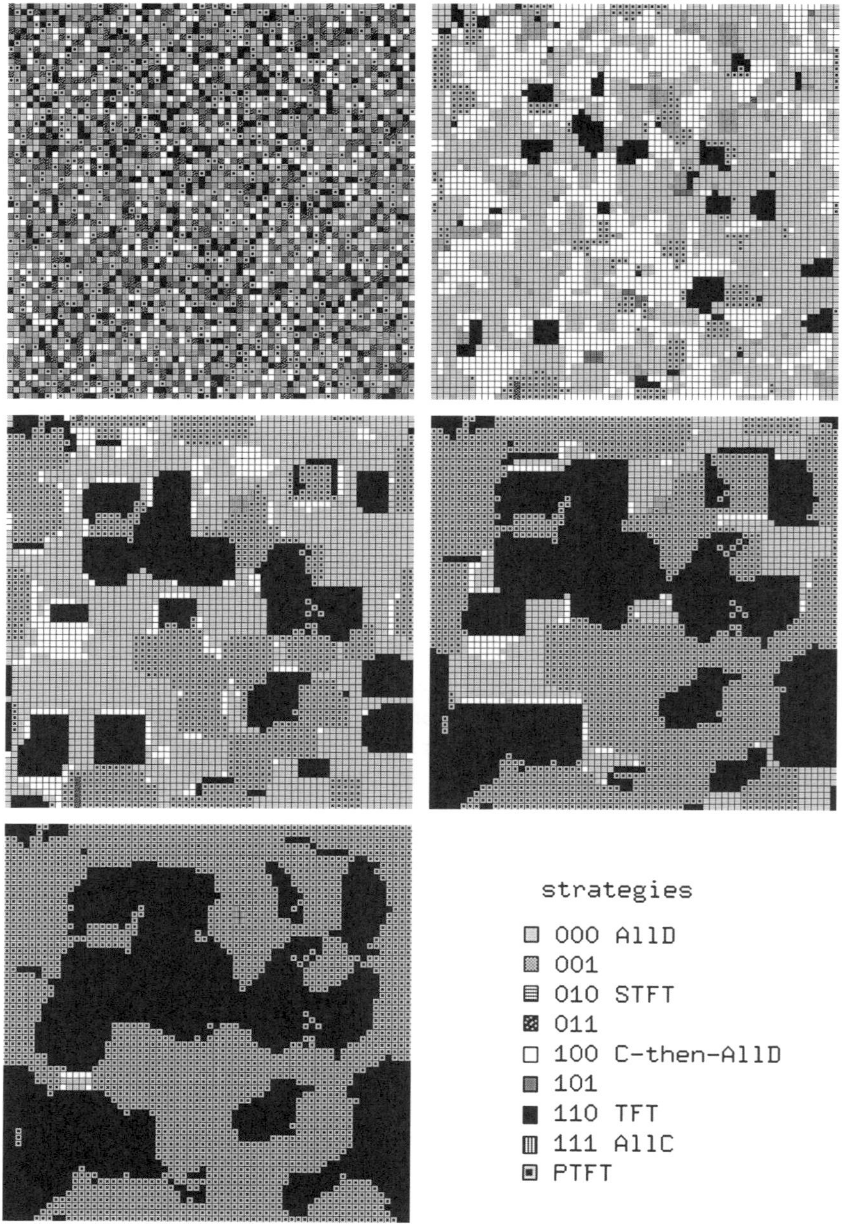

Figure 5.23 Evolution of randomized strategies to shared dominance by TFT and PTFT in an array segregated by color. Generations 0, 2, 4, 6, and 10 shown.

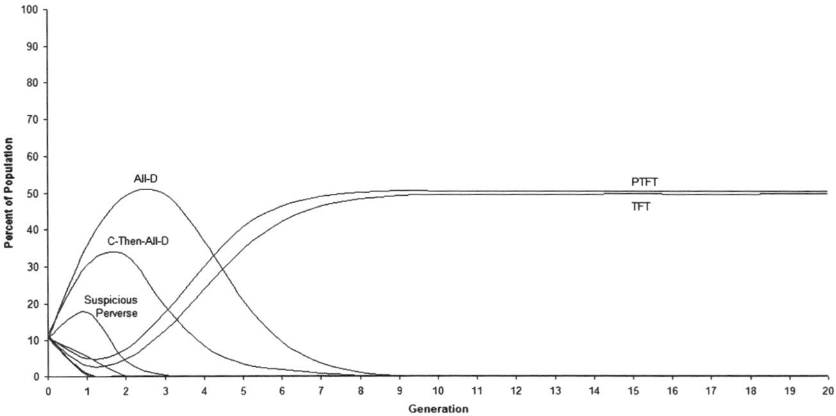

Figure 5.24 Percentages of the population for 9 strategies in an array segregated by color. 20 generations shown.

cate, then, is that a very simple mechanism may explain at least a large measure of the contact hypothesis effect.

This result can be strengthened by incorporating an element from Social Identity theory (Tajfel and Turner 1986). Because PTFT is the only strategy that gears its play to color, let us add a single point for PTFT—its "Social Identification point"—when it is playing with another of its own color. This is a tiny addition; a green PTFT cell playing a green all-C, for example, will be assigned 601 points rather than 600 over a round of 200 games.

The result in a segregated array with that slight change is shown in Figure 5.27. Here the segregated environment leads not to a 50/50 split but to total dominance by Prejudicial PTFT. Still keeping the social identification point for PTFT, the very different evolution of a mixed array is shown in Figure 5.28. With a single "Social Identification point," then, a segregated environment goes entirely to the Prejudicial strategy. A mixed environment goes almost entirely to color-blind TFT.

5.6 The game-theoretic cube: An image of robustness

The previous sections have employed imaging to show the dynamic development of cellular automata arrays, used as models for cooperation and prejudice. In what follows I want to outline a further imaging tool for answering an abstract question regarding these models. Here the central question is a question of robustness.

As indicated in Figure 5.18, the standard matrix used for the Prisoner's Dilemma is rendered in terms of four specific values: 0, 1, 3, and 5. A player gets 3 points in the case of mutual cooperation, 1 point in the case of mutual

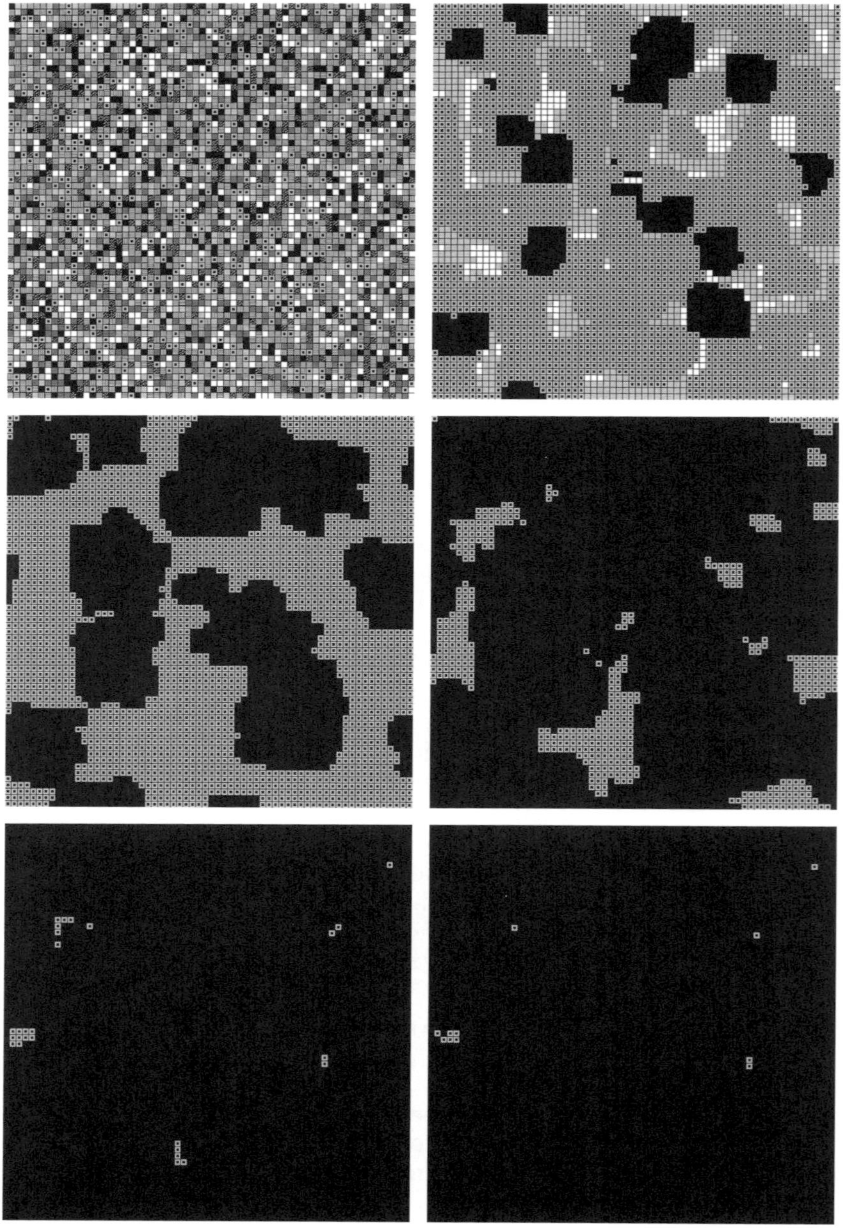

Figure 5.25 Evolution of randomized strategies to dominance by TFT in an integrated (randomized) color array. Generations 0, 4, 8, 12, 16, and 20 shown.

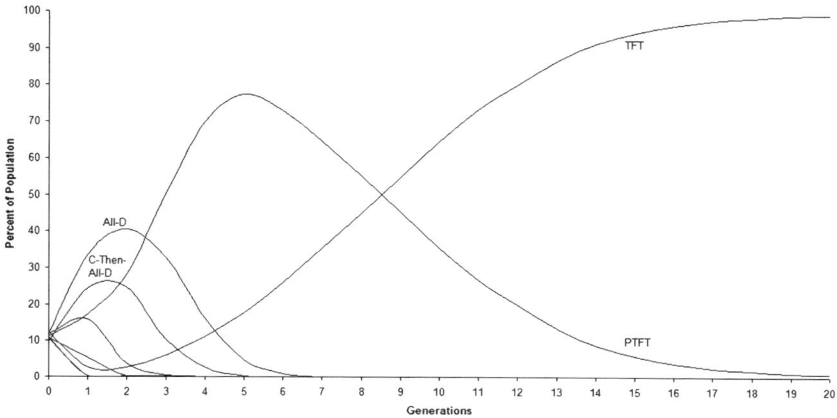

Figure 5.26 Percentages of the population for 9 strategies in an array randomized by color. 20 generations shown.

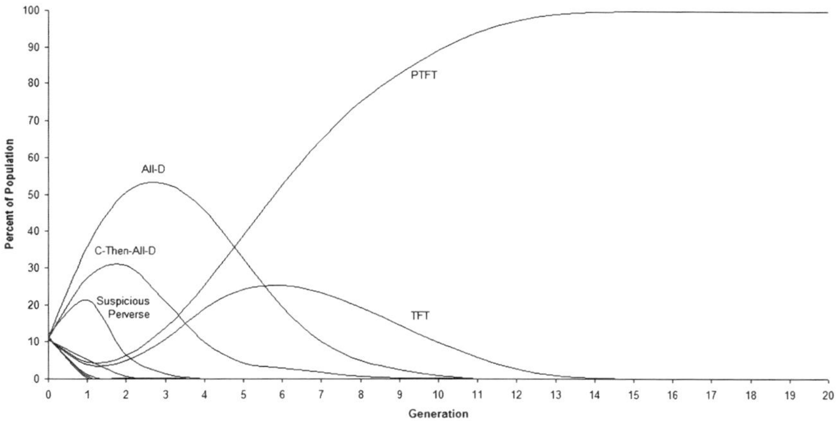

Figure 5.27 Percentages of the population for 9 strategies in an array segregated by color, with one extra "social identification" point for PTFT playing a cell of its own color. 20 generations shown.

defection, 0 points if he cooperates with a defector but 5 points if he defects against a cooperator. The technical definition of the Prisoner's Dilemma is broader, however. What qualifies something as one game rather than another is how the payoffs are organized: it is defection against cooperation that gets the highest payoff, for example, cooperation against cooperation the second, and so forth. Where DC represents defection against cooperation and CD represents cooperation against defection, what is necessary for a Prisoner's

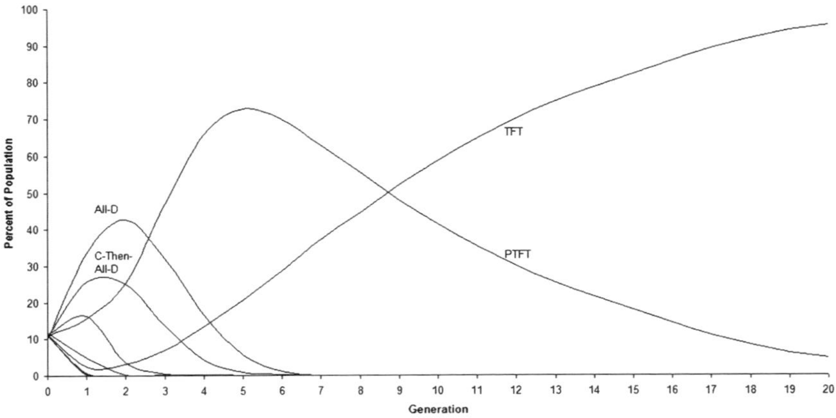

Figure 5.28 Percentages of the population for 9 strategies in an array randomized by color, with one extra "social identification" point for PTFT playing a cell of its own color. 20 generations shown.

Dilemma is that the ranking of payoffs is DC > CC > DD > CD. A second condition is also standard for the Prisoner's Dilemma—that CC > (DC + CD) / 2—but this is simply to disqualify cases in which strategies might do better by alternating defections and cooperations.

Different arrangements for our values give us different "games." DC > DD > CC > CD gives us Deadlock. DC > CC > CD > DD gives us Chicken. CC > DC > DD > CD gives us Stag Hunt, named for a discussion in Rousseau (Rousseau 1754 [1994]). Despite this breadth of possibilities, almost all of game theory to date has concentrated on a single limiting case. Virtually all of game theory is 2-person game theory. The overwhelming bulk of the literature on 2-person games, both theoretical and applied, is work on the Prisoner's Dilemma. The overwhelming bulk of *that* work concentrates on just four values: 0, 1, 3, and 5. What is of importance here is that the universe of game theory is clearly much larger; the Prisoner's Dilemma is by no means limited to those particular values, and 2-person game theory is by no means limited to the Prisoner's Dilemma.

An issue of increasing interest in game theory is how *robust* certain effects are. Here we will concentrate on robustness across changes in matrix values. In asking how robust an effect is in this sense, we are asking whether the effect is tied to very particular matrix values, or whether it has a wider impact across a larger portion of the game-theoretic universe. Some specific questions with respect to the work outlined above are these: How robust a result is conquest by TFT in a spatialized environment? How robust a result is the PTFT effect we've demonstrated? Is one of these results more robust across changes in matrix values—if so, which one? These are abstract questions for which we have developed a new form of imaging: a visual representation of the spread of an effect across the universe of game theory.

All matrices, for all 2 × 2 games, are specified in terms of 4 relative values: CC, CD, DC, and DD. At first blush, then, one might conceive of the game-theoretic universe as a 4-dimensional domain. Each point in the domain—each set of 4 values—could be specified in terms of coordinates x, y, z, and w. The universe of game theory could be represented as a hypercube.

Unfortunately, however, it is very difficult to think in terms of 4-dimensional geometry. Homo sapiens are used to a mere 3 dimensions; the step to conceptual work in 4 dimensions is hard. I have tried to argue earlier that imaging has the power it does because it allows us to exploit our immediate perceptual inferences, derived from our evolutionary heritage. But those perceptual inferences are tied to our familiar 3 dimensions. Once we step into 4 dimensions, for the most part, this power of imaging is lost.

Luckily, however, the universe of game theory is written in terms of four *relative* dimensions; what matters is not absolute but relative values. If there are 4 relative values for any game—CC, CD, DC, and DD—we can plot 3 of them relative to a particular value for one (it doesn't matter which) as our "normalization." We normalize our 4-dimensional space to a more manageable 3 dimensions by mapping the relative status of the other values to our chosen one. Here, for example, we will set CC at a constant value of 5. We plot the relative CD value on the x-axis, the DC value on the y-axis, and the DD value on the z-axis (Figure 5.29).[10]

When we first tried to envisage the universe of game theory, we didn't have the computer-generated images in Figures 5.29 and 5.31 through 5.33. We were limited to the tools at our fingertips. Two of the working

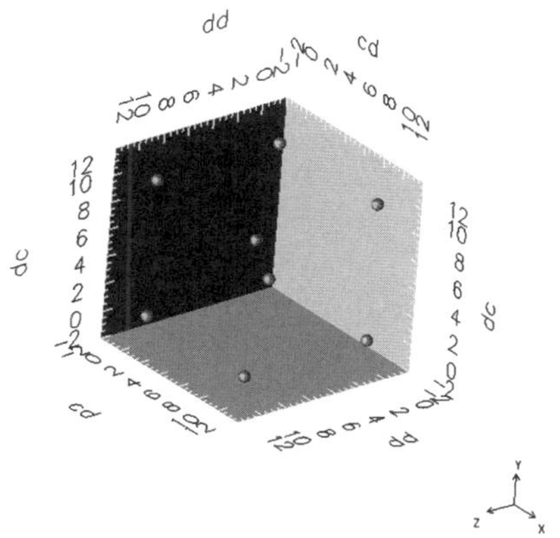

Figure 5.29 The universe of 2 × 2 game theory. Spheres indicate vertices.

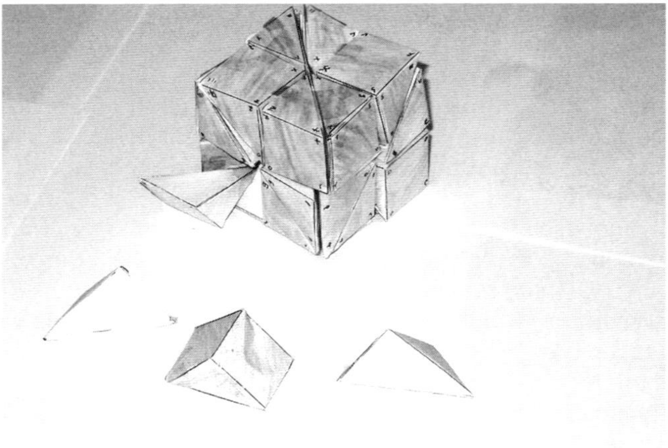

Figure 5.30 First paper mock-up of the cube.

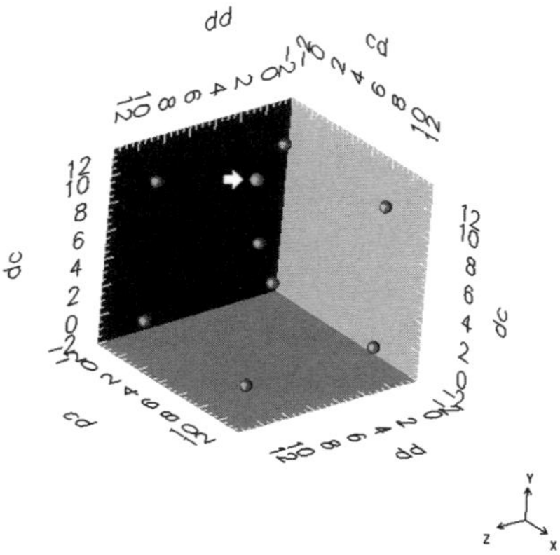

Figure 5.31 The single most studied point in game theory.

group—Randy Au and Nancy Louie—fashioned a first model of the universe of game theory in paper, including removable slices that represented particular games (Figure 5.30).

Later we were able to construct the computer image of the cube. Figure 5.31 shows the most studied point in game theory in that cube: the point that represents the Prisoner's Dilemma matrix with values 0, 1, 3, and 5. Figure 5.32 shows the larger area that represents the Prisoner's Dilemma as a whole. Animated rotations of these and later figures can be found at www.pgrim.org/Imaging.

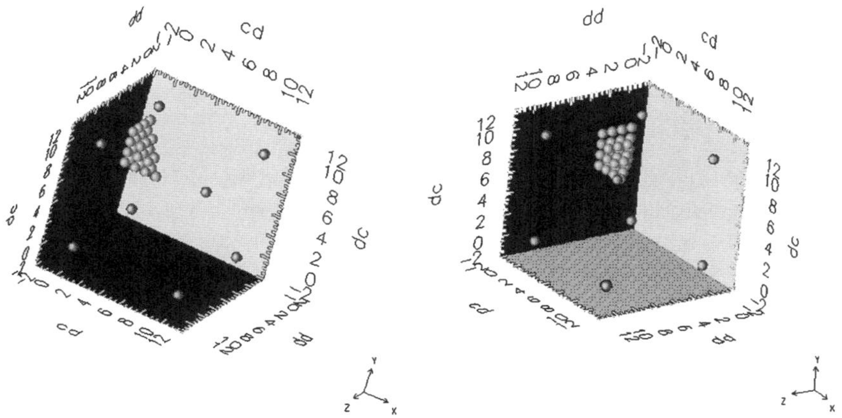

Figure 5.32 The area of the Prisoner's Dilemma.

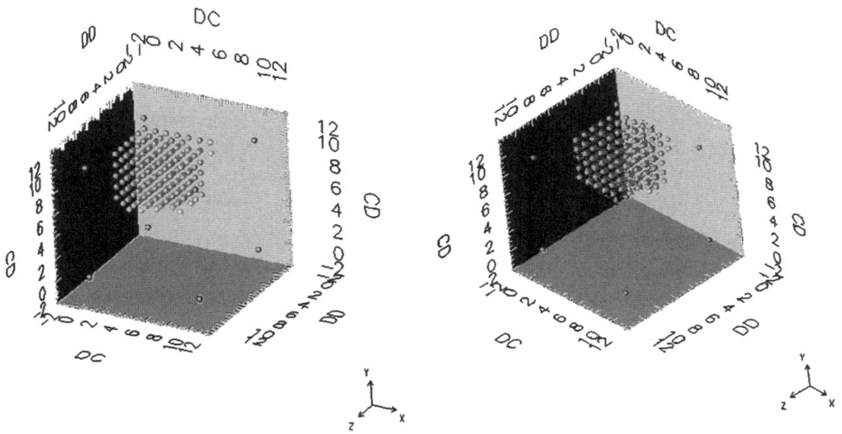

Figure 5.33 Robustness of the TFT effect.

Let us return, however, to our most studied point in game theory (Figure 5.31). We know two things that happen at that point. At those matrix values, TFT conquers in a spatial array of 8 strategies. It is also true that at those matrix values the contact hypothesis PTFT effect holds—that PTFT is as successful as TFT are codominant in a segregated environment, but that TFT triumphs in a mixed environment.

With imaging for the game-theoretic universe as a whole, we can use survey techniques to obtain a measure of robustness across changes in matrix values. With CC fixed at 5, we survey all combinations of values for CD, DC,

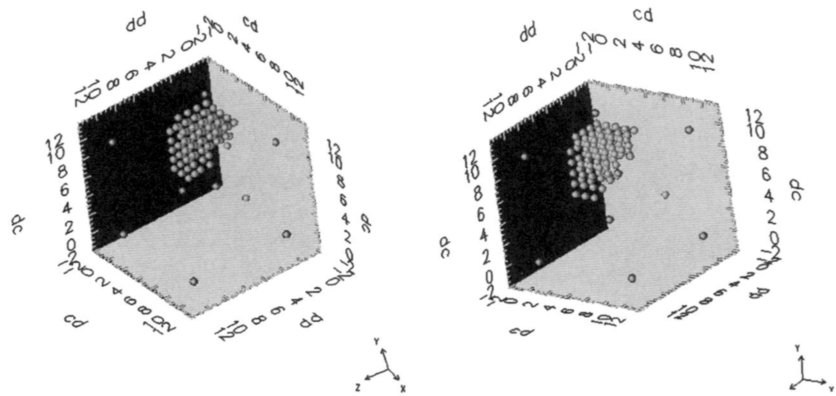

Figure 5.34 Robustness of the PTFT effect.

and DD. We can, for example, do a sampling across the universe of game theory in order to see where the TFT effect appears. The result appears in Figure 5.33. Here we show those areas in which, after 100 generations, TFT shows a conquest of 90 percent of the array or better after 100 generations. What we see is a very robust effect across changes in matrix values: TFT conquest holds not only in the territory of the strict Prisoner's Dilemma but far beyond it.

A similar survey for the PTFT effect, incorporating data from both mixed and segregated arrays, is shown in Figure 5.34. Here again we have a robust effect in the same area, and once again the effect occupies a greater range than merely the Prisoner's Dilemma. For the parameters we have used (conquest of 40 percent or greater by each of TFT and PTFT in a segregated array, conquest of 90 percent or greater by TFT in a mixed array), it is clear that the effect is comparable but not quite as robust as that of TFT.

Unlike some of the Fuzzy logic work, these images are well behaved, staying neatly within the bounds of clear interpretation. In that regard, this is a case of successful imaging, with all the virtues of real applicability. We know exactly what it is telling us. I must confess, however, that these lack some of the tantalizing character of the earlier images, such as those in Figures 5.16 and 5.17. There we didn't always know what we were looking at, but there seems to be something wonderful in that too. Perhaps our imaging grasp *should* exceed our intellectual grasp, else what is Plato's Heaven for?

5.6 Conclusion

What I have tried to emphasize here is the role that concrete images can play in approaching abstract questions. Traditional imaging allows us to see (roughly) what we would have seen at a particular time or place. Microscopic and telescopic imaging allows us to see things that are too small or too faint to see without them. Thermal, radio, and X-ray imaging allow us to see things that are genuinely there before us, but that we cannot *see* without their aid. Concrete images for abstract questions, in a way both similar and distinct, allow us to see things that are not there before us at all. They allow us to literally *see* the abstract dynamics of self-reference in fuzzy logics, for example, or the relative robustness of phenomena in spatialized game theory.

The claim is that we seek out and develop all of these kinds of imaging because we are particularly adept at pattern recognition in a particular sensory mode—the visual. There is, of course, an evolutionary history behind those visual capabilities. Our ability to detect tactile heat is not nearly so refined, and we are no good at all at detecting radio waves. If we can convert these to something visual we can exploit the ease with which we recognize and manipulate visual information. That is also, I have tried to suggest, the motivation for seeking concrete images in approaching the abstract.

Notes

*This chapter is transformed from a presentation for a conference on Rethinking Theories and Practices of Imaging: Technology, Representation, and the Disciplines, Rochester Institute of Technology, April 2004. Some of the informality of the presentational form undoubtedly remains, though hopefully some of its accessibility does as well. Thanks to Jack Sanders, my co-presenters, and the conference audience for helpful discussion. Research by Patrick Grim, Gary Mar, Paul St. Denis, Evan Selinger, William Braynen, Robert Rosenberger, Randy Au, Nancy Louie, and John Connolly.

1. For purposes of simplicity, I have here omitted the vexed question of how to treat the conditional. A more complete treatment appears in Grim, Mar, and St. Denis 1998.
2. Rescher 1969.
3. For details see Grim, Mar, and St. Denis 1998.
4. See for example Devaney 1989.
5. For details see Grim, Mar, and St. Denis 1998.
6. Here again full details appear in Grim, Mar, and St. Denis 1998.
7. The popular story that gives the Prisoner's Dilemma its name—involving two criminals deciding whether or not to turn State's evidence against each other under certain conditions—is credited to Rand consultant Albert W. Tucker. See Poundstone 1992.
8. In the one-shot game just outlined, it is clear that defection is strongly dominant: Whatever the opponent does, one will be better off by defecting. It doesn't follow at the level of strategy-selection, and it isn't true, that playing All-Defect is dominant in an iterated Prisoner's Dilemma.
9. Should two neighbors have identical scores higher than that of a cell, the strategy of one is chosen at random. For details see Grim 1995; Grim, Mar and St. Denis 1998.
10. For details see Grim, Au, Louie, Braynen, Rosenberger, and Selinger 2005.

References

Adorno, T. W., E. Frenkel-Brunswick, D. J. Levinson & R. N. Sanford. *The Authoritarian Personality*. New York: Harper, 1950.

Allport, G. W. *The Nature of Prejudice*. Cambridge, MA: Addison-Wesley, 1954.

Austin, J. L. "Truth," *Proceedings of the Aristotelian Society* (1950), Supplement 24, reprinted in *Philosophical Papers of J. L. Austin*. Ed. by J. O. Urmson and G. J. Warnock. Oxford: Clarendon Press, 1979, pp. 117–33.

Axelrod, Robert. "Effective Choice in the Prisoner's Dilemma," *Journal of Conflict Resolution* 24 (1980a): 3–25.

Axelrod, Robert. "More Effective Choice in the Prisoner's Dilemma," *Journal of Conflict Resolution* 24 (1980b): 379–403.

Axelrod, Robert. *The Evolution of Cooperation*. New York: Basic Books, 1984.

Axelrod, Robert & William D. Hamilton. "The Evolution of Cooperation," *Science* 211 (1981): 1390–6.

Buridan, John, c. 1340. *Sophisms on Meaning and Truth*. Tr. by Theodore K. Scott. New York: Appleton-Century-Crofts, 1966.

Campbell, D. T. "Ethnocentric and other Altruistic Motives," in *Nebraska Symposium on Motivation*. Ed. by D. Levine. Lincoln, NE: University of Nebraska Press, 283–311, 1965.

Cohen, Jonathan & Aaron Meskin. "On the Epistemic Value of Photographs," *Journal of Aesthetics and Art Criticism*, vol. 62. no. 2 (2004): 197–210.

Da Vinci, Leonardo, c. 1500. *Treatise on Painting [Codex Urbinas Latinus 1270]*. Tr. by A. Philip McMahon. Princeton: Princeton Univ. Press, 1956.

Devaney, Robert L. *An Introduction to Chaotic Dynamical Systems*, 2nd edition. Menlo Park: Addison-Wesley, 1989.

Grim, Patrick. "Greater Generosity in the Spatialized Prisoner's Dilemma," *Journal of Theoretical Biology* 173 (1995): 353–9.

Grim, Patrick. "Spatialization and Greater Generosity in the Stochastic Prisoner's Dilemma," *BioSystems* 37 (1996): 3–17.

Grim, Patrick, Gary Mar, & Paul St. Denis. *The Philosophical Computer*. Cambridge: MIT Press. 1998.

Grim, Patrick, Evan Selinger, William Braynen, Robert Rosenberger, Randy Au, Nancy Louie, and John Connolly. "Reducing Prejudice: A Spatialized Game-Theoretic Model for the Contact Hypothesis," in *Artificial Life IX: Proceedings of the Ninth International Conference on the Simulation and Synthesis of Artificial Life*. Ed. by Jordan Pollack, Mark Bedau, Phil Husbands, Takashi Ikegami, & Richard A. Watson. Cambridge, MA: MIT Press, 244–9, 2004.

Grim, Patrick, Evan Selinger, William Braynen, Robert Rosenberger, Randy Au, Nancy Louie, & John Connolly, 2005. "Modeling Prejudice Reduction: Spatialized Game Theory and the Contact Hypothesis," forthc*oming in Public Affairs Quarterly* 19 (2005): 95–126.

Grim, Patrick, Randy Au, Nancy Louie, Robert Rosenberger, William Braynen & Evan Selinger. "A Graphic Measure for Game-Theoretic Robustness," Research Report #05-01, Group for Logic & Formal Semantics, Dept. of Philosophy, SUNY at Stony Brook, 2005.

Gurr, T. R. *Why Men Rebel*. Princeton, NJ: Princeton University Press, 1970.

Haseltine, Eric. "Apparitions: Did you Just See a Ghost?" Neuroquest column in *Discover*, (February 2002).

Ihde, Don. "The Structure of Technology Knowledge," *International Journal of Technology and Design Education* 7 (1997): 73–9.
Lakoff, George. "Hedges: A Study of Meaning Criteria and the Logic of Fuzzy Concepts," *Journal of Philosophical Logic* 2 (1973): 458–508.
Larkin, Jill H. & Herbert A. Simon. "Why a Diagram is (Sometimes) Worth Ten Thousand Words," *Cognitive Science* 11 (1987): 65–99.
Pettigrew, T. F. "Intergroup Contact Theory," *Annual Review of Psychology* 49 (1998): 65–85.
Poundstone, William. *Prisoner's Dilemma*. New York: Anchor Books, 1992.
Rescher, Nicholas, *Many-valued Logic*. New York: McGraw-Hill. 1969.
Rousseau, Jean-Jacques, 1754. *Discourse on the Origin of Inequality*. Tr. by Franklin Philip. New York: Oxford University Press, 1994.
Tajfel, H., and J. Turner. "The Social Identity Theory of Intergroup Behavior," in *Psychology of Intergroup Relations*. Ed. by S. Worchel & W. G. Austin. Chicago: Nelson, 7–24, 1986.
Turner, J. C. and P. J. Oates. "Self-Categorization Theory and Social Influence," in *The Psychology of Group Influence*. Ed. by P. B. Paulus. Hillsdale, NJ: Erlbaum, 233–75, 1989.
Zadeh, L. A. "Fuzzy Logic and Approximate Reasoning," *Synthese* 30 (1975): 407–28.
Zadeh, L. A., "A Fuzzy Set-theoretic Interpretation of Linguistic Hedges," *Journal of Cybernetics* 2 (1972): 4–34.

6
The Epistemic Status of Brain Images

Megan Delehanty

6.1 Introduction

Functional brain images are very interesting types of evidence and have recently drawn the interest of some philosophers.[1] They look—to a greater or lesser degree—somewhat like a picture of a brain, but just how it is that they can be claimed to represent an active brain is far from obvious. While they are presented as if they are the product of a perceptually available process, they are very different from any sort of photographic image[2] since they are neither produced by any kind of optical imaging technique, nor are the quantities they detect accessible to human senses. In addition to this, and more importantly, they are produced only through the use of intensive mathematical and statistical processing. As a result of these features, they are difficult to accommodate into any broadly empiricist view according to which observation and (direct) perceptual evidence are epistemically privileged ways of providing justification for our beliefs about the world.

In this chapter I will look at the mathematical and statistical procedures that are required to produce these images and suggest that there are very general epistemic difficulties that attach to functional brain images because of the nature of these computations. So far there has been little attention paid to how the heavily mathematized nature of these images affects the kind of evidence we take them to be, but this is where the real epistemological interest lies. It is precisely the statistical processing that is involved in producing these images that creates the most serious difficulties in applying any sort of empiricist account of observation to these technologies since it seems to blur the distinction between producing data and interpreting it. This, in turn, suggests that the epistemic privilege that often gets attached to perceptual evidence may not apply to these images. Furthermore, if we examine how functional brain images represent their object (the brain in action), we must conclude that PET and fMRI images are properly thought of as a hybrid between a mathematical model or computer simulation and a standard photograph-like image. Moreover, they are far closer to being a mathematical model than they are to being a photograph. A full investigation of the epistemic status of these images, therefore, requires a determination of precisely how they are like photographs, how they are like models or simulations, and how the interaction between these various modes of representation might produce novel epistemic considerations which are not present in any mode individually.

6.2 Problematic features of functional images

Empiricism was traditionally linked with the idea that unaided human perception is the standard for what is observable.[3] Furthermore, perception and sense experience—as direct and unmediated as possible—have an epistemically privileged status with respect to the justification of beliefs about the natural world. The best ground for a belief is that one perceives it to be the case. Thus Carnap defined observable as "directly perceived by the senses" (Carnap 1966, 225) and van Fraassen defined it as "what the unaided eye discerns" (van Fraassen 1980, 59). Functional images obviously fail to meet this criterion, but critics of empiricism have argued that this criterion is misguided. It has been claimed that this criterion is too narrow and that observation should not exclude the use of various sorts of instruments (microscopes, telescopes) to assist human perception (Maxwell 1962; Hacking 1983) as well as that all observation is theory laden and is no more epistemically secure than any theoretical assumption (Hanson 1958; Kuhn 1970; Feyerabend 1975). Thus that these imaging systems fail the strict criterion may not tell us much. The crucial issue is not the fact that functional images are not produced via direct, unmediated perception but rather what gives perception its epistemic credentials and whether the production of functional image shares the criteria that are required to gain these credentials.

The available options, then, seem to be (1) to deny perceptual knowledge its traditional privilege and allow that nonperceptual processes can, in principle, produce evidence with equal justificatory strength, or (2) to extend the boundaries of perception to include other processes that produce evidence that is epistemically equally good. Which of these options we choose is, I will claim, related to how the final data (image in this case) represents the object/phenomena it represents. This is a feature of the causal and other processes involved in data production. A difference in the causal process of image production makes an epistemic difference.

6.3 Mathematization of imaging technologies

While I will address only functional brain imaging in this chapter, the argument I will develop for the epistemic status of these images can be generalized to many varieties of scientific images. The history of visual imaging technologies in biology and medicine in general is one of increasing mathematization.[4] Part of this mathematization reflects an increased desire and ability to provide accurate, precise measurements of objects or phenomena, but an equally important component of this development is the contribution of mathematical and statistical techniques in allowing the visualization of fundamentally *non-visible* phenomena. This is not simply quantification of the image after it has been produced but mathematization that is required to even *generate* the image. So, for example, the production of images by standard light microscopy can be given a mathematical description, but this refers only to the operation of the microscope (its optics). The mathematical description explains how the image is produced, but it does not contribute to the production of the image. The now standard use of image analysis software in conjunction with photography or other digital data collection techniques allows more information to be extracted from various modern forms of microscopy but still starts with optical data about an essentially visual object. Brain imaging technologies, however, allow us to accomplish something fundamentally different: they make visible patterns of neuronal activity that would not otherwise be visible at any level of resolution and, in so doing, enable the mapping of brain function. In contrast to the case of microscopy, imaging the brain *requires* mathematization to even get started. Moreover, at key points later in the history of brain imaging technologies mathematical developments were crucial in permitting continued development of brain mapping strategies with the result that mathematical techniques and their attendant assumptions and limitations became embedded deeper and deeper within the images that are presented as data about (or as models of) the human brain.

There are four key problems that arise in attempting to produce functional maps of the brain and for which a mathematical solution is required. The first problem arises from the inaccessibility of brain activity, even after X-rays opened up the interior of other parts of the body to view. To investigate the living human brain, researchers needed to find some physiological variable that was correlated with local brain activity but, at the same time, could be used to study the whole brain. The slowly established connection between blood flow and brain activity provided such a variable.[5] However, this was simply a prerequisite to the more explicitly mathematical project of identifying a specific aspect or physical characteristic of this variable that could be measured with a sufficient degree of precision and designing an appropriate signal detection system. Positron emission tomography (PET), magnetic resonance imaging (MRI), and functional MRI (fMRI) all are connected to blood flow but use a different physical characteristic of blood flow to detect patterns of brain activity. I will discuss the signal detection aspects only in the case of PET as the basic issues are similar in all these technologies. The second problem is finding a way to generate an image of the object, in this case the functionally active brain, from the non-optical data acquired by the detectors. The development of image reconstruction algorithms in the 1950s and 1960s provided the solution to this problem.[6] The third problem that arose was how to adequately model noise as the spatial and temporal resolution of brain imaging technologies increased and standard statistical models proved to require assumptions which were incompatible with contemporary understandings of how the brain works. The final problem was how to map an abstract entity—an average human brain. This required a method for comparing images from different individuals. Numerous sources of variation between individual brains make this a difficult task requiring complex mathematical techniques—warping algorithms—to map images from different individuals onto a standard reference space.

In all of the above problems, both aspects of mathematization are essential, but I will concentrate primarily on its ability to make the non-visible visible. Improving accuracy and precision are, in all of these cases ongoing projects that continue far beyond the original provision of a solution to a general problem. While this sort of tweaking of the original solution often has important consequences for the epistemic status of the images produced,[7] for the purpose of this chapter it will be more important to focus on what the problems that require mathematical solutions are and the general form of the solutions rather than intricacies of particular solutions.

6.4 The epistemology of images and models

6.4.1 Images

What kind of evidence is an image? Galison (1997) identifies the visual similarity of the image with the represented object as the source of their utility as evidence. This general claim has been echoed by many authors in aesthetics and philosophy or sociology of science (cf. Gombrich 1960; Mitchell 1986, 1994; Arnheim 1969; Latour 1988; Tibbetts 1988; Peacocke 1998). It is important, however, to separate two aspects of this question. The first is made up of issues surrounding the cognitive accessibility of evidence in visual form and the impact of aesthetic values on the selection or acceptance of visual evidence. The second involves questions about what the image represents. These are quite distinct subjects of investigation and while there are many interesting questions that arise in the first, I will be concerned here only with the latter.

Functional images are hybrid epistemic objects. They combine features of two distinct types of representation: pictures (photographs, naturalistic illustrations) and graphical displays of statistical information. These types of representation—and their means of production—correspond to the distinction that Galison has drawn between "picturing machines" and "electronic counters coupled with electronic logic circuits" (1997, 19–20). The central difference between the two is that "picturing machines" produce images of particular events or objects that bear a close visual resemblance to the entity represented while "electronic counters" collect data on particular features of large numbers of events or objects and produce evidence which is statistical in nature. The virtue of one of these means of data production is the weakness of the other. The key advantage of the former is that it preserves specific features of the particular object represented. This may be valuable in many contexts, but can also be problematic when random individual variation deflects attention from—or even makes it impossible to detect general patterns that hold across individuals. "Electronic counters" solve this problem through statistical manipulation of data, but in so doing can obscure individual differences that *are* important, particularly when these manipulations are based on false or exaggerated assumptions about stability or homogeneity of certain features of the object under study.

Functional images are hybrid epistemic objects in that they purport to be images of an individual object (though this may be either an "individual" person's brain or an "individual" average human brain), but are clearly produced in the manner of the "electronic counter" as will be discussed in section 6.5. Given their means of production, it might seem tempting to classify them simply as an unusual type of graphical display of statistical data. But to do this would be to ignore the fact that an enormous amount of work goes into mapping this data onto the naturalistic space of an individual brain. This includes not only the recovery of spatial information from the data, but the often extensive transformations that are required to combine data from different individuals. Furthermore, single events simply cannot be mapped. Thus that these images are images is not an accidental feature of their production.

If an image taken by itself is evidence, in virtue of what can it be considered good (or bad) evidence? Since images are nonpropositional, they cannot be true or false. As Gombrich states in *Art and Illusion*, "a picture ... can no more be true or false than a statement can be blue or green."[8] Gombrich goes on to say that "in scientific illustrations, it is the caption [the figure legend] which determines the truth of the picture."[9] Yet even if images cannot be true or false in this strict sense, there are other senses in which we might consider them to be true or false. There is a range of meanings of "true," and the point is not so much whether the labels true or false do or do not apply, but how and on what basis they are arrived at. What sort of relationship must an image stand in with reference to its "object" for it to be faithful or informative? Here it will be important to specify that an image may be used in the same general way as convention-governed forms of representation such as maps; they may be used to designate or denote only certain features of an object. A map or globe, for instance, may denote the fact that the world is round, but not that North America is pink. While it is probably true of all images, it seems especially relevant to scientific images that they are produced for a certain purpose and with a certain audience (familiar with certain conventions of representation) in mind.

The essential feature of an image, then, is perhaps not its truth or approximate truth, but its objectivity and its resemblance/similarity to its subjects in the relevant respects. What, then, are these relevant respects? For the most part, this will be an empirical question. For Galison's example of bubble chamber experiments, for instance, the relevant aspects of visual similarity are the number and orientation of lines. For a black and white photograph taken of a field in a fluorescence microscope, it will be the relative position and intensity of white areas with respect to black or gray. The key determinant of what visual features can bear relevant similarity to the represented object is the process by which the image is produced.[10] To take an obvious case, a black and white photograph cannot bear resemblance in terms of color (differences in hue at least; tonal gradations can be represented). By focusing on the process of image generation, it will be possible to identify which of the similarity relations that PET and fMRI images stand in with respect to the brain share features with photography (or other techniques of mechanical reproduction) and which are shared with mathematical models.[11]

6.4.2 Models

Empiricists have traditionally worried about the need to have sensory evidence—whether delivered by the unaided senses or via magnifying or amplifying devices—in order for a scientific claim to be well-grounded. The central issue is the idea that there is a crucial gap, which must be maintained, between producing data and interpreting it. While empiricists may be able to respond to the challenge of instruments that register properties such as radioactivity and electrical currents which are not detectable (at any level) to our human senses by extending the notion of empirical evidence to apply to what registers on observational equipment rather than human sensory systems, the problems raised by functional brain imaging are not so neatly dispensed with. The mathematical and statistical processing which go into production of functional brain images blur the line between producing empirical data and interpreting the data to draw conclusions about the system under investigation. In doing this, it takes these images away from the realm of empirical evidence into the realm of evidence gained from a mathematical model or simulation.

The literature on models and computer simulations is far too large to deal with comprehensively here; I will address only the question of the sort of relationship that simulations must bear to their subjects in order to provide satisfactory evidence.[12] There are, in fact, many similarities between images and simulations. Like images, simulations (or their products) are not true or false, but can be judged more or less realistic, more or less useful, more or less successful for the uses to which they are put. They fall somewhere between theory and experiment, thus raising the empiricist's red flag.

As a form of representation, they typically represent either some aspect of the world, or some aspect of our theories about the world, or both at once. To examine the epistemic status of a model/simulation, then, requires that we know what is involved in its functioning so that we can tell how much and what kind of resemblance to expect. A model/simulation is able to serve as a tool for understanding, describing, and exploring a physical system insofar as it is an analog of that system. Just as an image is useful in virtue of having some relevant visual similarity to the object it represents, a model/simulation must *behave* in a relevantly similar way as the physical system in order for it to serve as evidence and support certain inferences. The model/simulation is set up such that an input to the model—input from the detector of a PET scanner, direct human input—prompts the derivation of mathematical results that may then be subject to further mathematical manipulation or may be taken directly as the output of the model. What we need to know about a model in order to judge the status of its output as evidence, then, is what and how good the relevant similarity is; how well the model can be said to represent the behavior of the system. In order to know this, we need to look at the mathematical similarity or isomorphism between the model and the system.

Assessment of the mathematical similarity relations that hold between a model and the system it is claimed to be a model of are, of course, specific to a particular model. In general, however, we need to examine both the components of the equations as well as the complex chain of inferential steps that generate the results or output of the model. The components of an equation can only rarely be perfectly paired with obvious objects in the world. In the case of functional brain imaging, for example, statistical strategies for dealing with noise refer to characteristics of the data rather than variables that correspond (if indirectly) to brain function. The choice of parameters with which to represent an object as well as the boundary conditions will affect the correspondence between the model and the physical object as will the choice of statistical model. The use of mathematical styles or statistical techniques that use aggregates or larger-scale relations instead of individuals is extremely common, for example, and may warrant more or less epistemic concern depending on the particular system being modeled.[13] The series of inferential steps that make up the mathematical transformations performed by the model may also approximate more or less well the behavior of the physical system. With these concerns in mind, let us now turn to the processes involved in the production of functional brain images.

6.5 Mathematical aspects of brain images

The first point to note is that functional brain images are not produced by optical imaging techniques. Empiricists generally take the evidential status of photographs to be related to the reliability and simplicity of optical imaging systems—light is emitted, reflected, or refracted and then strikes a detector where incoming light can be measured—as well as to the similarity between this type of system and our own optical system.[14] On this understanding, the types of similarity relations that are given evidential status in a photograph are simply those that can be represented by the optical properties of the imaging instrument in question. While similar structural similarity relationships are (or may be) present in a PET or fMRI image, that the resemblance is not generated by an optical technique means that this visual similarity cannot serve to warrant inferences in the same way that a homologous similarity in a photograph could. There is a crucial causal difference in how the similarity is produced and it is the causal chain that generally underlies any epistemic privileging of perceptual evidence. The naturalistic form of functional brain images functions to provide cognitive manageability and is, in this sense, epistemically important, but this is not what I am concerned with here. These images may be the most effective means of identifying characteristic features out of complex data sets, but the epistemic considerations that apply to these images as data do not emerge from their naturalistic form. Instead, when we examine the means by which these images are produced, we see that the fact that they even *are* images is accidental. While a photograph may be measured and subjected to quantitative analysis subsequent to its production, functional brain images require that extensive mathematical transformation occur to produce the data that can *then* be represented in the form of an image.[15] For the remainder of this chapter I will consider the image to be identical to the final numerical output of the mathematical transformations since the translation between the two is purely conventional.

To get from an active human brain as the object of investigation to the data in its final form a series of mathematical and statistical processing steps must be performed. Importantly, many of these are not simple number-crunching techniques. If this were the case, we might make the case that these images really do bear a greater epistemic similarity to photographs than I allowed above. We could argue that, although the detection system is non-optical, the straightforward conversion of the detected signal into an image (together with a good understanding of the nature of the detected signal and detection system) is epistemically analogous to the case of, say, a scintillation counter. However, the nature of the mathematical transformations which are actually required prevent us from drawing this conclusion. To get started, the non-visual phenomena must be detected and the data stored in such a way as to allow recovery of spatial information. This much may be allowed to raise no epistemic worries beyond those that might be present for photographic systems. However, transforming this data into an image requires that mathematical techniques of enormous complexity be used. Techniques are needed to: (1) reconstruct the 2-D (or 3-D) image from the data collected by the PET or fMRI detector, (2) adequately model the noise which is a significant problem (one can normally expect 1 percent change in signal due to increased brain activity and a 2 percent change due to noise) for these techniques, and (3) make comparisons between individual brains or between imaging modalities. Each of these contains several sources of epistemic concern of the sort discussed above for mathematical models.

6.5.1 Image reconstruction algorithms

The task of image reconstruction arises because of the nature of the detection system used in PET and fMRI. There are slight differences between the two, but since the issues relevant to this chapter are the same in both cases, I will consider only the case of PET. PET scanners use a system of scintillation counters arranged in a 2-D or 3-D array to detect radiation from a source internal to (normally injected into) the subject. Scintillation counters work by coupling a crystalline material known as a scintillator with a photomultiplier tube. When a gamma ray hits the scintillating crystal, some of the energy deposited is converted into a flash of visible light. The amount of light produced is low so to produce a measurable signal a photomultiplier tube is optically coupled to one face of the scintillator in order to detect the light and amplify the signal.[16] In addition to the elimination of events which do not have a coincident event at 180° (this eliminates random events since real events will emit two 511keV photons at 180° to each other), some additional data correction is required to account for slight differences in counting efficiency between detectors and for the time each detector is "dead"—processing of a detected event takes a finite amount of time and while this occurs the detector cannot "see" another event. The exact nature that these corrections take depend on the type of scintillating crystal used as well as the way in which detectors are arranged.[17]

The result of using this form of detection system (and of detecting positrons to begin with) is that raw PET data, unlike an autoradiograph or an X-ray, is not an image of the object that was scanned: it is a set of numbers. The data is stored as a 2-D matrix known as a sinogram. The vertical axis represents the angle of the line of response and the horizontal axis represents the displacement from the center of the field of view (Figure 6.1). Each element in the sinogram represents the number of counts detected by a particular detector pair (members of a pair are located at 180° from each other). Thus any point in the (r, ϕ) coordinate system of the sinogram represents the count density at a particular point in the (x, y) plane being scanned. The sonogram is 2-D but the brain is 3-D, so somehow the spatial information needs to be recovered from the data. To do this—and to produce an image of the brain from the sinogram—requires a reconstruction algorithm.

The basic problem is how to reconstruct a two- or three- dimensional image of the interior and exterior aspects of an object from data that consists of a large number of projections through the object. The mathematical

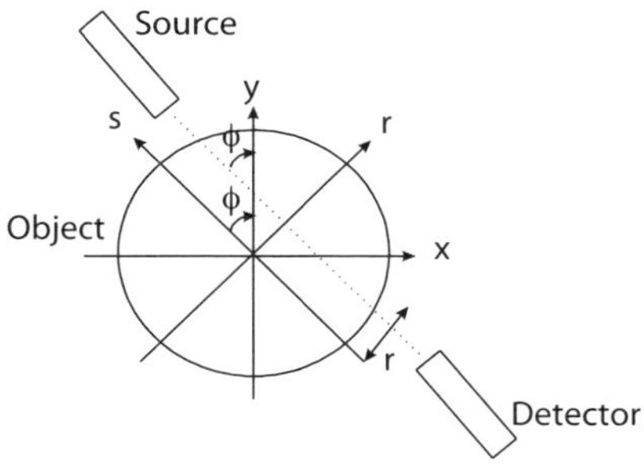

Figure 6.1 Coordinate system for describing ray paths and projections (Adapted from Brooks and DiChiro (1976))

techniques that were developed to accomplish this task are known as reconstruction algorithms. The primary problem is the superimposition of multiple planes in each data point (point on the photographic film or, later, the numerical figure representing a projection in a computer) resulting in blurring of the structure of interest by structures on front and behind it. Resolution of this problem thus required a strategy to eliminate the effect of radiodensity variations in planes other than the one of interest (or the one for which data is explicitly being sought). While I want to minimize any mathematical formalism as much as possible, it is necessary to give a brief summary here in order for it to be clear what reconstruction algorithms need to do. As described above, the process of data collection must involve both translational and rotational scanning. Each plane of the object which is scanned, then, can be represented by an (x, y) coordinate system (Figure 6.1). The contribution of each point to the total detected signal is represented by the density function $f(x, y)$. "Density" is not used in a technical sense here, but is used as a general term that covers both emission (PET) and transmission (CT) scanning. For PET, $f(x, y)$ is proportional to the radioisotope density. Ray paths are described by an (r, s) coordinate system which is rotated by the same angle as the ray (see Figure 6.1). Each ray is specified by coordinates (r, ϕ), where ϕ is the angle of the ray with respect to the y-axis and r is its distance from the origin. The coordinate s represents path length along the ray. The integral of $f(x, y)$ along a ray (r, ϕ) is called the ray-projection p:

$$p(r,\phi) = \int_{r,\phi} f(x,y)ds$$

For PET, p is directly proportional to the total detector signal. A complete set of ray-projections at a given angle is called a projection. Ideally $f(x, y)$ is a continuous function and the number of projections is infinite. In practice, of course, $f(x, y)$ is calculated at a finite number of points from a finite number of projections. This is the starting point for the image reconstruction algorithms. These algorithms can be roughly divided into three classes based on the type of strategy they use to solve equation (1). They are: (1) summation or back-projection, (2) iterative reconstruction, and (3) analytic reconstruction. There are interesting and important differences between the classes of algorithm, particularly with respect to the spatial resolution that they can provide, the artifacts or noise they introduce into the data, the assumptions they either require or are incompatible with, and the computational power they need to be useable in practice. I will not discuss them here.[18] For the purposes of this chapter, it is that a mathematical solution was required for this problem and that the solution is complex and involves trade-offs between many approximations and assumptions that matter.

6.5.2 Modeling noise

The ability to distinguish signal from noise, while important in PET and earlier imaging technologies, becomes a crucial issue in fMRI and requires improved statistical techniques for modeling noise. fMRI is distinguished from other brain imaging techniques by its combination of superior temporal resolution[19] with the excellent spatial resolution (1–2 millimeters) of MRI. The high degree of spatial and temporal resolution provided by fMRI introduces complications because of potentially high correlations among the observations. There is a lot of structure inherent in the data, but it is a kind of structure that standard statistical models are not well adapted for. Most classical statistical methods make assumptions about independence and identical distribution that are not warranted in fMRI. In addition, the very low level of the BOLD signal presents numerous statistical challenges since the change in BOLD signal due to brain activity is only about 1 percent and is, in fact, smaller than the noise (which is generally on the order of 2 percent).[20] To even begin to separate the signal from the noise, then, requires that a large number of images must be averaged in order to detect the regions of activation. Since the data sets are very large to begin with, the sheer amount of data that must be dealt with poses a challenge. This has resulted in two-stage statistical processing of fMRI data; an initial preprocessing (using a software package such as Statistical Parameter Mapping (SPM), or other commercial programs) which includes image reconstruction and removal of some sources of noise such as motion of the subject,[21] followed by a second stage of analysis which allows for modeling of other sources of noise and determination of active voxels (Lazar et al. 2001).[22]

The basic problem is that it is impossible to get away from the spatial dependence of statistical models (assume if data points are close, then they are more likely to be connected), but with brain we expect distant regions also to be connected. While scientists currently believe that elementary operations (e.g., component operations of selective attention or of visual word processing) are localized to discrete areas, cognitive tasks are carried out by networks of widely distributed neural systems. Many current techniques are based on the idea of classification, that is, deciding which voxels[23] are active and which are not. This involves, for example, calculating t-tests on individual voxels for comparing a baseline condition with an experimental one and classifying as active all those voxels with a t-statistic above a certain threshold. But a t-test assumes each voxel is independent—it doesn't capture the way brains work in that it does not take account of the spatial correlations between voxels that are close to each other. It also relies on adjustments for the multiple comparisons that are made. Because the signal is smaller in relation to the noise, the use of standard multiple comparison techniques can unnecessarily eliminate activity. As a result, voxels that perhaps should be classified as active may appear to be inactive.[24] One would like to be able to incorporate the assumption that voxels next to one another might be related. It is possible to do this by searching for clusters of active voxels in order to determine activity, but it is not known what size of clusters are relevant for brain function (and they may vary). Incorporating the wrong size of expected cluster can result in falsely identifying an area as active (if cluster size is set too small noise will be counted as activity) or as inactive (if the cluster size is too large, smaller regions of potentially real activity will be counted as inactive). As a result of these problems, the mathematical modeling of noise must be taken as just that—a model. Since the application of a particular noise model can make a significant difference to the final image, the epistemic concerns that arise for various models of noise must also be taken to affect the final brain image.

6.5.3 Warping algorithms

In order to make a map of *the*, rather than a single individual's brain, it is necessary to make intersubject comparisons. Application of some sort of averaging procedure in order to produce a composite image requires that we first be able to represent the data from each individual in a standard brain space. That this task would not be straightforward was foreshadowed long ago when simple anatomic studies revealed that not only are individual humans' brains of different sizes, but that the relative sizes of specific areas of the brain also vary. Additional complications have been introduced as cognitive studies have revealed that individuals may use different strategies, and thus different functional areas of the brain, to perform an identical task.[25] Furthermore, some familiar anatomical landmarks may be missing in the brains of even normal individuals, making their use as reference points on which to base a standard map impossible.[26] The proliferation of imaging techniques and their separate virtues and weaknesses served to increase the need to be able to make such comparisons both between individuals and between images of the same individual produced using different techniques.

The first systematic attempt at providing a general atlas (or set of maps) of the brain to which all other images could be compared was made by

Jean Talairach and Michel Tournoux. Their *Co-Planar Stereotaxic Atlas of the Human Brain*, published in 1988, remains the standard reference system today, despite acknowledged inadequacies.[27,28] The Talairach atlas is based on postmortem measurements of serial slices (with variable slice separation) through the brain of a 60-year-old female subject and allows only a limited set of linear transformations: rigid body rotations, translation, and linear scaling along pre-specified axes. Even after these transformations there often remain extreme intersubject differences.[29] Comparing images in different locations, in different orientations, and often with different geometries requires more complex transformations. These more complex geometric transformations of brain imaging data are known as warps. The geometrical transformations classed as warps include global scaling, affine transformations, linear, nonlinear, and local deformations.[30] Each of these transformations produces sets of images that can be compared both statistically and visually. In addition, the transformations themselves contain a significant amount of useful information about regional similarities and differences in the data. Thus visualizing the result of a warp, the mapping of one image onto another or onto a reference atlas, can be displayed either in the form of the image itself or as the deformation map.

A critical part of warping is the production of an image relocated to the pixels of the fixed target image, i.e., the resampling of the data onto the appropriate pixel grid. As was briefly mentioned with regard to reconstruction algorithms, the choice of mathematical procedure for carrying out such transformations has consequences for the degree to which the original information is conserved during resampling. In general, the goal is to conserve as much of the information present in the data acquired by the detectors as possible and, ideally, to enhance it.[31] The same goal applies to the more complex transformations carried out with warping algorithms. As with reconstruction algorithms, warping algorithms can be broken down into different types based on the mathematical strategies they use. The two basic categories are intensity based and geometrically based approaches. In general, intensity based approaches are less dependent on spatial features of the data and include warps employing elastic, mechanical, multiresolution, Bayesian, and fluid methods. Geometrically based approaches are those that use spatial information in the form of features, templates, and patterns. Information about anatomic landmarks is used to drive as well as to constrain the warp. It is not necessary to attempt a more detailed categorization here,[32] nor would any further discussion of specific warping algorithms help to clarify my point. Instead, I will simply conclude by reiterating that warping of images across imaging modalities and across subjects has made possible the production of comprehensive reference systems that fully describe brain structure and functions within an individual as well as within whole populations. This is required to get useful information in a clinical context as well as for generalizable information about location and ordering of cognitive functions.

6.6 Conclusion

Functional maps of the brain, whether they represent a particular individual or some sort of population average, are far from simple images. Many layers of mathematical processing are embedded within the final image. Each layer introduces, to a greater or lesser degree, assumptions about the nature of the data and imposes constraints on how such images can be interpreted. Yet without this extensive process of mathematization, there could be no access to any but extremely partial representations of human brain function. Despite this, the role of these techniques and their impact on the epistemic status of brain images has been largely ignored. Indeed, it continues to be ignored in current practice since the continued development of improved statistical and mathematical techniques is increasingly the exclusive domain of statisticians, mathematicians, and computer scientists while many brain researchers (in fields such as cognitive psychology, psychiatry, and neuroscience) treat imaging and statistical software as black boxes between the experiment and the final output of the data in the form of an image[33] and philosophers concentrate on the relationship between cognitive science and brain imaging.

The depth to which mathematical processing is embedded in these images, however, has profound importance for the evidential status of these images. They do not fit into any broadly empiricist account of evidence since they are neither perceptual nor do they clearly maintain the gap between producing data and interpreting it. The standards for assessing the reliability of evidence provided by these images must be tied not only to the physical processes involved in signal detection but to the mathematical and statistical processes and inferences involved. These images are, then, hybrid epistemic objects that share features of both photographic images and mathematical models. Furthermore, it is the mathematical aspects of image production that are especially problematic and on which we must focus in order to assess what sorts and what quality of similarity relations hold between the causal input and the final output data.

Notes

1. See, for instance, Bogen (2000, 2002), van Orden and Paap (1997), Stuffelbeam and Bechtel (1997).
2. I will use "photograph" or "photographic image" to refer to any sort of image—an actual photograph, an X-ray, etc.—that is produced by optical means. This class of images are heterogeneous in many interesting ways, but can be considered identical for the questions addressed in this chapter.
3. There is a significant amount of diversity among positions that are referred to as empiricist, but however they may differ they at least share this principle. Other aspects of different empiricist positions (such as the British Empiricists' view that all ideas come from experience) are not central to this chapter.

4. I intend by this term to capture the embedding of mathematical or statistical algorithms within imaging technologies. Quantification and measurement are also somewhat of a different issue than mathematization as used in this sense (although there is a significant amount of literature which uses the term "mathematization" to refer to what I would consider to be the development of quantitative methods in the social science). I am not concerned with how or why the need or desire to provide precise quantitative measurements arose, but with the role(s) played, given the existence of quantitative measurements of a phenomenon in some form already, by the continued development of mathematical tools for detecting and transforming phenomena of interest.
5. For a summary of this work, see Raichle (1998 a, b).
6. While research on refinements of these algorithms continues today, I will focus on the basic mathematical solution rather than the complexities of various algorithms.
7. Bogen (2000) discusses some of the epistemic problems associated with the extensive mathematical processing of functional brain images but does not identify particular forms of or stages in this processing with which the specific problems are associated.
8. Gombrich (1980, 67–8).
9. Gombrich (1980, 68).
10. Not all the similarity relations that are possible given a particular process of image production may apply in a particular case. This is most evident in painting or illustration but is also true in the case of mechanical reproduction techniques.
11. Looking at the mechanism of image production is also important in allowing epistemic distinctions to be made between identical images that may have been produced by different means. In her 2002 book on explanations in developmental biology, Evelyn Fox Keller suggests that some of the recent (partial) acceptance of theoretical biology by experimental ("wet") biologists may be due in part to the fact that the output of advanced modeling and simulation programs can be highly naturalistic images of the sort that experimental biologists are used to working with. This may be so, and it may be interesting data for the questions about the first set of questions (cognitive, sociological, etc.) about how images function as evidence, but it entirely ignores the crucial differences I am concerned with here.
12. Additionally, I will be concerned with mathematical models rather than with scale models, descriptive models, or model organisms. Certain issues specific to these other types of models do, in fact, arise with functional brain images, but it is impossible to deal with them in a chapter of this length.
13. In the case of functional brain imaging, this is a major source of concern.
14. See, for instance, Van Fraassen (1980) and Hacking (1983).
15. Essentially, the image is just a translation of a matrix of numerical data that is the endpoint of successive mathematical and statistical processing steps. Each voxel is assigned a color according to the numerical value obtained for that point in the matrix together with a conventional scale.
16. Knoll (1989).
17. Further research into PET detection systems has produced superior scintillators as well as better arrangements of detectors. Faster computers have also allowed more detectors to be incorporated into the system.
18. An excellent review of these differences is provided by Brooks and DiChiro (1976). Gordon, Herman, and Johnson (1975) provides a good, non-technical summary.

19. PET permits measurements in the scale of minutes, fMRI allows for measurements to be made in the scale of seconds to hundreds of milliseconds (Toga and Mazziotta 1996, 391).
20. Eddy et al. (1999).
21. The other main components of the preprocessing are baseline correction, deghosting, mean correction, outlier adjustment, and detrending (Lazar et al. 2001).
22. Because of epistemic difficulties which affect these programs (interestingly, these worry the statisticians who develop the software but not the end users of the software), a great deal of current research is directed toward developing statistical models that are more flexible and able to accommodate the spatial and temporal correlations that are to be expected in data that represent the functioning brain as well as being able to incorporate into a single step more sources of noise (see, for example, Eddy et al. 1999, Bandettini et al. 1993, Forman et al. 1995, Kherif et al. 2002).
23. A voxel is the basic volume element with which 3-D data is represented. It is the 3-D equivalent of a pixel.
24. Author interview with William Eddy, April 2002.
25. Posner and Raichle (1997, 233).
26. Ibid.
27. Interview with W. F. Eddy (April 2002), Toga and Thompson (1999), Roland and Zilles (1994).
28. The Montreal Neurological Institute created a composite MRI dataset from 305 young normal subjects in order to compensate for some of the limitations of the Talairach atlas (Evans et al. 1994). It still uses the Talairach coordinate system. The widely used Statistical Parametric Mapping (SPM) package incorporates this average MRI dataset.
29. Toga and Thompson (1999, 10).
30. Toga and Thompson (1999, 1).
31. Eddy and Young (2000).
32. For one such attempt, see Toga and Thompson (1999, 8).
33. William Eddy, interview with author (April 2002) and Julie Price, interview with author (March 2002).

References

Arnheim, R. *Visual Thinking*. Berkeley: University of California Press, 1969.

Bandettini, P. A., Jesmanowicz, A., Wong, E. C., and Hyde, J. S. "Processing Strategies for Time-Course Data Sets in Functional MRI of the Human Brain", *Magnetic Resonance in Medicine*, vol. 30, no. 2 (1993): 161–173.

Bogen, J. "Functional Imaging Evidence: Some Epistemic Hot Spots," in *Theory and Methods in the Neurosciences*. Eds P. K. Machamer and R. Grush. Pittsburgh: Pittsburgh University Press, 2000.

——. "Epistemological Custard Pies from Functional Brain Imaging," *Philosophy of Science* 69 (Suppl.) (2002): S59–S71.

Brooks, R. A. and G. Di Chiro "Principles of Computer Assisted Tomography (CAT) in Radiographic and Radioisotopic Imaging," *Physics in Medicine and Biology* 21 (1976): 689–732.

Carnap, R. (1965). *Philosophical Foundations of Physics*. New York: Basic Books.

Eddy, W. F., M. Fitzgerald, C. Genovese, N. Lazar, A. Mockus, and J. Welling. "The Challenge of Functional Magnetic Resonance Imaging," *The Journal of Computational and Graphical Statistics* vol. 8, no. 3 (1999): 545–58.

Eddy, W. F. and T. K.Young. "Optimizing the Resampling of Registered Images," in *Handbook of Medical Imaging: Processing and Analysis*. Ed. I. N. Bankman. San Diego: Academic Press, 603–12, 2000.

Forman, S. D., Cohen, J. D., Fitzgerald, M., Eddy, W. F., Mintun, M. A., Noll, D. C. "Improved Assessment of Significant Activation in Funtional Magnetic Resonance Imaging (fMRI): Use of a Cluster-Size Threshold." *Magnetic Resonance in Medicine* vol. 33, no. 5 (1995): 636–47.

Feyerabend, P. (1975). *Against Method*. New York: Schocken.

Galison, P. *Image and Logic*. Chicago: University of Chicago Press, 1997.

Gombrich, E. *Art and Illusion*. Kingsport, TN: Pantheon Books, 1960.

Gordon, R., G. H. Herman, and S. A. Jornson. "Image Reconstruction from Projections," *Scientific American* vol. 233, no. 4 (1975): 56–68.

Hacking, I. (1983). *Representing and Intervening: Introductory Topics in the Philosophy of Natural Science*. Cambridge: Cambridge University Press.

Hanson, N. R. (1958). *Patterns of Discovery*. Cambridge: Cambridge University Press.

Kherif, F., Poline, J.-B., Flandin, G., Benali, H., Simon, O., Dehaene, S, and Worsley, K. J. "Multivariate Model Specification for fMRI Data." *NeuroImage* 16 (2002): 1068–83.

Keller, E. F. *Making Sense of Life: Explaining Biological Development With Models, Metaphors, and Machines*. Cambridge, MA: Harvard University Press, 2002.

Kevles, B. H. *Naked to the Bone: Medical Imaging in the Twentieth Century*. New Brunswick, New Jersey: Rutgers University Press, 1997.

Kuhn, T. (1970). *The Structure of Scientific Revolutions* (Second, Enlarged ed.). Chicago: University of Chicago Press.

Latour, B. "Drawing Things Together," in *Representation in Scientific Practice*. Eds M. Lynch and S. Woolgar. Cambridge, MA: MIT Press, 19–68, 1988.

Lazar, N. A., W. F. Eddy, C. R. Genovese, and J. Welling. "Statistical issues in fMRI for brain imaging," *International Statistics Review* 69 (2001): 105–27.

Maxwell, G. "The Ontological Status of Theoretical Entities," in *Minnesota Studies in the Philosophy of Science*. Eds H. Feigl and Maxwell, G. Minneapolis, MN: University of Minnessota Press, vol. 3, 3–15, 1962.

Mitchell, W. J. T. *Iconology: Image, Text, Ideology*. Chicago: University of Chicago Press, 1986.

——. *Picture Theory*. Chicago: University of Chicago Press, 1994.

Ogawa, S., T. M. Lee, A. R. Kay, and D. W. Tank. "Brain Magnetic Resonance Imaging with Contrast Dependent on Blood Oxygenation," *Proceedings of the National Academy of Sciences USA* 89 (1990): 9868–72.

Peacocke, C. "Nonconceptual Content Defended." *Philosophy and Phenomenological Research* vol. 63, no. 2 (1998): 381–88.

Raichle, M. E. "Imaging the Mind," *Seminars in Nuclear Medicine* vol. XXVIII, no. 4 (1998a): 278–89.

——. "Behind the Scenes of Functional Brain Imaging: A Historical and Physiological Perspective," *Proceedings of the National Academy of Sciences USA* 95 (1998b): 765–72.

Stufflebeam, R. S. and W. Bechtel. "PET: Exploring the Myth and Method," *Philosophy of Science* (Supplement) vol. 64, no. 4 (1997): S95–S106.

Talairach, J. and P. Tournoux. *Co-Planar Stereotaxic Atlas of the Human Brain*. Trans. M. Rayport. New York: Thieme Medical Publishers, Inc., 1988.

Ter-Pogossian, M. M. "The Origins of Positron Emission Tomography," *Seminars in Nuclear Medicine* vol. XXII, no. 3 (1992): 140–9.
Tibbetts, P. "Representation and the Realist-Constructivist Controversy," in *Representation in Scientific Practice*. Eds M. Lynch and S. Woolgar. Cambridge, MA: MIT Press, 69–84, 1988.
Toga, A. W. and J. C. Mazziotta. *Brain Mapping: The Methods*. San Diego: Academic Press, 1996.
Van Fraassen, B. *The Scientific Image*. Oxford: Oxford University Press, 1980.
Van Orden, G. and K. R. Paap. "Functional Images Fail to Discover Pieces in Mind in the Parts of a Brain," *Philosophy of Science* (Supplement) vol. 64, no. 4 (1997): S58–S94.
Wagner, H. N. "A Brief History of Positron Emission Tomography (PET)," *Seminars in Nuclear Medicine* vol. XXVIII, no. 3 (1998): 213–20.
Worsley, K. J. "Local Maxima and the Expected Euler Characteristic of Excursion Sets of χ^2, F, and t Fields," *Advances in Applied Probability* 26 (1994): 13–42.
Worsley, K. J., A. C. Evans, S. Marrett, and P. Neelin. "A Three-Dimensional Statistical Analysis for rCBF Activation Studies in the Human Brain," *Journal of Cerebral Blood Flow and Metabolism* 12 (1992): 900–18.

7
Technologies to Bond With*

Andy Clark

Heavy metal

Los Alamos National Laboratory, New Mexico, occupies the high ground both physically and technologically. I am here, this hot and sunny day in May 1999, to deliver a talk on the interactions between mind and technology. Getting in is not easy. There have been security scares, and my permanent resident alien card is deemed insufficient proof of identity. After a flurry of panic, my secretary somehow manages to fax them a copy of my UK passport. At last—and just in time for the talk—I am issued with the inevitable plastic photo ID. Soon I find myself deep in the radiation-proof concrete bunkers that currently serve as home to my hosts, the Complex Systems Modeling Team.

Walking around this eerily silent, windowless, underground laboratory, I am struck by the stark contrast between old technology and new. The massive concrete bunkers and reinforced floors of these old buildings were designed both to resist nuclear attack and to support heavy, in-your-face technology: giant mainframes, immense monoliths of dials, lights, and levers. Yet today's action, in the Complex Systems Laboratory at least, usually requires little more than a few potent laptops and some fiber-optic links to massive databases. The heaviest piece of real, working machinery that I encounter is a somewhat sick old printer whose wheezing vibrations occasionally disturb the tomb-like silence.

The talk safely delivered, my hosts suggest a meal. Los Alamos' best restaurant turns out to be Japanese, an irony I decide not to pursue. But my political coyness proves unfounded. In fact, over the course of the meal it is decided that we will next visit one of Los Alamos' best, if lesser known, attractions—the Black Hole.

The Black Hole is the shop-cum-soapbox of peace protester Edward Groshus.[1] To visit it is to step into a retrotechnological Aladdin's Cave. Housed in a rambling, hangar-like complex on the edge of town, the Black Hole is a stunning repository of ex-National laboratory equipment and scientific junk. The stuff was purchased (by the pound!) direct from the laboratory during its postwar sell-off period. The buyer was the same Ed Groshus, one-time national laboratory employee-turned peacenik, anti-war campaigner, and retrotechnology entrepreneur. As I walk toward the installation, a billboard on the roadside catches my attention. It reads:

OMEGA CHURCH OF PEACE
BOMB UNWORSHIP CEREMONY
CRITICAL MASS EVERY SUNDAY

This is Groshus's doing. A man with an agenda, to be sure, but one nicely tempered by an enduring sense of fun. This shows, too, in his relationship with the goods in his store. Groshus despises the technologies of warfare, but he clearly sees the beauty as well as the absurdity of all that in-your-face technology. The Black Hole manages, incredibly, to be both shrine and protest. And it is a retro-techno addict's dream come true. It feels like a vast and ill-organized hardware store. The hardware here is not screws, nails, and duct tape so much as bank upon bank of imponderable valve electronics, heavyweight first generation calculating machinery, fragments of complex control panels bristling with hundreds of tiny lights and switches, filters, fans, cathode-ray tubes, testers, probes, bomb casings, wires, screens, and dials. My personal favorite was a variety of gray, heavy, metal boxes (rather like office filing cabinets) with enormous single red buttons, labeled EMERGENCY, slap-bang in the middle, items seemingly straight out of Tom and Jerry, but in fact straight out of Uncle Sam, *ca.* 1960.

What we have here is an elephant's graveyard of Un-transparent, In-Your-Face Technology. Most of this stuff was not built to fade into the background of anyone's life or work. It made few efforts to configure itself to better suit the user. It was, in many ways, the strict antithesis of Weiser's vision of ubiquitous computing. Heavy, enormous, almost maximally resistant to easy human use, such technologies ran little risk of blurring the boundaries between machine and human, between biological user and technological tool. Naturally, I bought as much of it as I could possibly carry! I came away with a large vacuum tube with shining copper coil and a heavyweight electromagnet at the base. This was the so-called triggertron, once used to discharge a large bank of capacitors in order to implode trial atomic devices. I also succumbed to the siren call of two black boxes full of inscrutable, but wisely glowing, valve electronics. To complete the order I added a few substantial fragments of complex "Bat-Cave" control panels, featuring hundreds upon hundreds of tiny red and green lights and switches. To my eternal regret I could not carry the heavy box of metal with the big red emergency button in the middle. Walking all this through airport security at Albuquerque was surprisingly easy. "What's that, sir?" "It's an antique triggering device for an atomic weapon." "That's fine—come on through."

My own suspicious eroticization of retro-technology aside, the real point of this little reminiscence was just to begin to cement the contrast between two types of technology: "transparent technologies," and what might contrariwise be dubbed "opaque technologies." A transparent technology is a technology that is so well fitted to, and integrated with, our own lives, biological capacities, and projects as to become (as Mark Weiser and Donald Norman have both stressed) almost invisible to use.[2] An opaque technology, by contrast, is one that keeps tripping the user up, requires skills and capacities that do not come naturally to the biological organism, and thus remains the focus of attention even during routine problem-solving activity. Notice that "opaque," in this technical sense, does not mean "hard to understand" as much as "highly visible in use." I may not understand how my hippocampus works, but it is a great example of a transparent technology nonetheless. I may know exactly how my home PC works, but it is opaque (in this special sense) nonetheless, as it keeps crashing and getting in the way of what I want to do. In the case of such opaque technologies, we distinguish sharply and continuously between the user and the tool. The user's ongoing problem is to successfully deploy and control the tool. By contrast, once a technology is transparent, the conscious agent literally sees through the tool and directly confronts the real problem at hand. The accomplished writer, armed with pen and paper, usually pays no heed to the pen and paper tools while attempting to create an essay or a poem. They have become transparent equipment, tools whose use and functioning have become so deeply dovetailed to the biological system that there is a very real sense in which—while they are up and running—the problem-solving system just *is* the composite of the biological system and these nonbiological tools. The artist's sketch pad and the blind person's cane can come to function as transparent equipment, as may certain well-used and well-integrated items of higher technology, a teenager's cell phone perhaps. Sports equipment and musical instruments often fall into the same broad category.

Often, such integration and ease of use require training and practice. We are not born in command of the skills required. Nonetheless some technologies may demand only skills that already suit our biological profiles, while others may demand skills that require extended training programs designed to bend the biological organism into shape. The processes by which a technology can become transparent thus include both natural fit (it requires only modest training to learn to use a hammer, for example) and the systematic effects of training. The line between opaque and transparent technologies is thus not always clear-cut; the user contributes as much as the tool. But there is a real and important sense in which some technologies are immediately better candidates for ultimate transparency than others. Most of the old heavyweight technology in the Black Hole remained eternally opaque, even to trained operators. While a very few devices are so well suited to the biological user that we either know at once how to use them, or quickly find out by an intuitive process of trial and error.

Transparent tools

Donald Norman—cognitive scientist and contemporary guru of the age of "information appliances"—describes the Rubicon between opaque and transparent technologies in terms of a historical progression from "technology-centered" to "human-centered" products.[3] Human-centered products wear their functionality on their sleeve and exploit the natural strengths of human brains and bodies. These are the kinds of products where, Norman insists, the user almost never needs to open the manual. Yet vanishingly few of our high technology, information-based products are like that today. Sadly, a thousand cases of highly opaque, run-daily-to-the-manual products spring all-too-readily to mind. Examples range from VCRs to photocopiers to personal organizers and laptops.

The trouble with technology-centered products is that, as Norman's label suggests, they answer only to the need to do things (often, many *different* things) that previous products didn't do, or that they didn't do to the same degree. What they *don't* answer to is the need to enable those things to be done fluently, reliably, and with a minimum of learning and effort on the part of the user. And the reason, as Norman notes, is simple enough. At first, creating a product that can do the job is hard enough, let alone aiming for products nicely fitted to brains like ours. As time goes by, however, the vendors must seek to extend their market beyond the gung ho early adopters and technophiles. They will need to sell to the average user who simply wants a cheap, reliable, and easy-to-use tool. The technological product then comes under cultural-evolutionary pressure to increase its fitness by better conforming to the physical and cognitive strengths and weaknesses of biological bodies and brains. In quasi-evolutionary terms, the product is now poised to enter into a kind of symbiotic relationship with its biological users. It requires widespread adoption by users if its technological lineage is to continue, and one good way to achieve this is to provide clear benefits at low cognitive and economic costs.

There are, of course, many rather less appealing routes to technological (and biological) survival: products, for example, which survive simply because they do the job, however opaquely, or products that depend on the incompatibility of alternatives with some popular platform or protocol. There is also a large gray area in which technologies actively create the very needs (e.g. "more memory") that they then rush to fulfill. Our immediate task, however, is to get a more concrete sense of some of the complex ways in which technologies simultaneously shape and adapt to the cognitive profiles of biological users. With that in mind, let's look briefly at a familiar item, one that long ago passed from the realm of opaque technology into that of transparent symbiotic partner—the humble wristwatch.[4]

We humans didn't always keep precise, objectively measured time. Before the dawn of the city, the factory, and the organized religious order, human beings used natural cycles to prompt daily activities. The sun rises and farming begins, interrupted only by a brief break when the sun is high in the sky. Darkness signals food and sleep. Today, a great many humans are not like this. We work all hours. We plan to meet friends for coffee at 11:45 a.m. We make a date for supper at 10:00 p.m. and a film at midnight, and so on. The transition from a natural-time society to our present arrangements for work and play was mediated by a long thread of technological evolution: a thread that leads from heavy, fixed, unreliable sundials and water clocks, through the development of early oscillating-element-based timekeeping, right up to cheap, accurate, personal quartz crystal wristwatches. But the technological story, though fascinating, pales beside the human-centered story. In a mere 500 years, the opaque, unreliable, fixed-location tower clocks of the Middle Ages gave way to the reliable, cheap, personal timekeepers that we now take so much for granted. Along the way our relationship to time itself was irrevocably changed and transformed.

Once the average city worker was awakened by the call of the night watch, a living person whose task was to patrol the streets shouting the time. A little later the tolling of a bell, either owned by the town or perhaps by a specific employer, woke the townspeople. These measures instilled a degree of what David Landes nicely calls "time obedience." But with the availability of personal timepieces, in the form of chamber clocks or (ultimately) wristwatches, came the possibility of something new and different—"time discipline." The presence of easily accessible, fairly accurate, and consistently available time-telling resources enabled the individual to factor time constantly and accurately into the very heart of her endeavors and aspirations. This made possible ways of thought, and cultural practices and institutions, which were otherwise precluded by our basic biological nature. Landes makes the point well:

> The public clock could be used to open markets and close them, to signal the start of work and its end, to move people around, but it was a limited guide to self-imposed programs. Its dial was not always in view, its bells not always within hearing. Even when heard, hourly bells are at best intermittent reminders. They signal moments. A chamber clock or watch is something very different: an ever-visible, ever-audible companion and monitor . . . a measure of time used, time spent, time wasted, time lost. As such it was prod and key to personal achievement and productivity.[5]

Notice that what counts here is not always consciously knowing the time. None of us, I suppose, looks constantly at his or her watch. Rather, the crucial factor is the constant and easy availability of the time, *should we desire to know it*. Therefore, a prime characteristic of transparent technologies is their *poise* for easy use and deployment as and when required. Daily, unreflective usage bears this out. As you walk down the street, you are accosted by the familiar cry of the temporarily watchless. "Excuse me, sir, do you happen to know the time?" Asked this question on a busy street, most of us will unhesitatingly reply, *even before consulting our wristwatches*, that yes, we surely do.[6] Grasping the request hidden in the formulaic question, many of us will also, and without further request, share our knowledge with the time-challenged supplicant. As we do so, we may find ourselves producing one of the characteristic body motions of the modern world. In the suited male or female, this takes the form of a controlled, punch-like extension of the arm, a clockwise half-rotation of the emerging wrist, and a slight lowering of the gaze. This knowledge-retrieval tropism serves, of course, a single practical function—it permits you to focus your gaze briefly upon the face, dial, or display of your watch, that humble example of cyborg technology.[7]

Now compare a superficially similar case. Your houseguest has encountered a word he does not know. To be concrete, let the word by "clepsydra." At some appropriate conversational juncture, the question is raised: "Good host, do you know what the word 'clepsydra' means?" Perhaps you are like me. I only learned this word a few days before writing this paragraph, until then, it wasn't part of my working vocabulary at all. But perhaps, like me, you keep a medium-size version of the *Oxford English Dictionary* somewhere in your house. So you know you have the wherewithal to resolve the matter. But what do you say? You surely won't say, "Yes, I know what that word means" and only then proceed to consult the dictionary. Yet this is precisely what usually happens when we are asked the time!

An easy dismissal of this discrepancy is, of course, to simply lay everything at the accommodating feet of convention. When we answer that we know the time, all we mean is that we have the information readily at hand. And to be sure, several cultural variants of the request exist. My wife, a native Spanish speaker, might ask me "*Tienes hora?*" literally, "Have you got the time?" with the emphasis on possession rather than knowledge. All this notwithstanding, I think the ease with which we accept talk of the watch-bearer as one who actually knows—rather than one who can easily find out—the time is suggestive. For the line between that which is *easily and readily accessible* and that which should be counted as *part of the knowledge base* of an active intelligent system is slim and unstable indeed. It is so slim and unstable, in fact, that it sometimes makes both social and scientific sense to think of your individual knowledge as quite simply whatever body of information and understanding is at your fingertips; whatever body of information and understanding is right there, cheaply and easily available, as and when needed.[8] According to one diagnosis, then, you are telling the literal truth when you answer "yes" to the innocent-sounding question, "Do you know the time?" For you *do* know the time. It is just that the "you" that knows the time is no longer the bare biological organism but the hybrid biotechnological system that now includes the wristwatch as a proper part.

To make this just a little more palatable, consider the parallel case of biological memory. Suppose I ask you whether you know the year of the first walk on the moon. You might answer "Yes, 1969." In answering "yes" you do not mean to imply that this date was present to your conscious awareness all along. You do not walk around all day mentally rehearsing "1969," "1969," "1969." Rather, your "yes" signifies that the information was indeed there, poised for easy access and retrieval from your biological memory. The informational poise of the wristwatch (and, as we'll later see, of the visual scene in front of your own eyes) may sometimes be relevantly similar. Perhaps, then, you may be properly said to know the time even before you actually look at your watch—just as you can be said to know the date of the moon landing even before actually retrieving it from your biological memory.

If this way of looking at things still strikes you as outlandish, you are in good company. Most people find such a diagnosis strange, unnecessary, and (thus) unconvincing. But this reaction is unprincipled. It rests not upon any deep fact about the nature of knowledge or the preset bounds of persons but on a simple prejudice: the contemporary version, as it happens, of the old and discredited idea of the mind as a special kind of spirit-stuff. The idea of "mind as spirit-stuff" is no longer scientifically respectable. Instead, mind is seen as the working of a purely physical device. In identifying that physical device solely with the biological brain, we again make a leap of faith, depicting the biological brain itself as the sole and essentially *insulated* engine of mind and reason. This conception is the old idea of special spirit-stuff in modern dress. A thoroughgoing physicalism should allow mind to determine—by its characteristic actions, capacities, and effects—its own place and location in the natural order. We should not, at any rate, simply assume that it is correct to identify and locate the individual *thinking system* by reference to the merely *metabolic* frontiers of skin and skull.

We can, in any event, take away two somewhat less contentious lessons from our discussion of modern timekeeping. The first is that transparent (nonopaque, human-centered) technology is by no means a new invention. It is with us already in a wide variety of old technologies, including pen, paper, books, watches, written words, numerical notations, and the multitude of almost invisible props and aids that scaffold and empower our daily thought and action.[9] The second is that the passage to transparency often involves a delicate and temporally extended process of co-evolution. Certainly, the technology must change in order to become increasingly easy to use, access, and purchase; but this is only half the story because at the same time, elements of culture, education, and society must change also. In the case at hand, people had to learn to value time discipline as opposed to mere time obedience, and this transition itself, Landes tells us, took over a hundred years to fully accomplish.

Smart worlds

What happened with timekeeping is now happening with the flow of information itself. Mark Weiser's vision of ubiquitous computing is finding concrete expression in attempts to design and market what Norman calls "information appliances."[10] We have met this phrase once or twice already, and it is time to try to pin it down. Information appliances are characterized by three central features:

1. An information appliance is geared to support a specific activity, and to do so via the storage, reception, processing, and transmission of information.
2. Information appliances form an intercommunicating web. They can "talk" to each other.
3. Information appliances are transparent technologies, designed to be easy to use, and to fade into the background. They are *poised to be taken for granted*.[11]

Weiser's vision of the home and workplace is filled with small, intercommunicating, unobtrusive intelligent devices was a vision of a world of such appliances, but Norman offers several rather more restricted, less futuristic examples. He imagines the use of inexpensive, tiny cameras to beam information (the shape of the coffee table, the color of the sweatshirt) directly to family and friends while shopping. This is a natural extension of the current use of cell phone technology. (In fact, as I write the second draft of this text, I note that several cell phone companies offer "picture messaging" with attendant ability for personal digital input.) Norman goes on to imagine houses with permanent wall-mounted weather displays, constantly showing the local forecast and conditions, to imagine (echoing Weiser) applications embedded in walls and furniture, and supermarkets where you simply wheel the leaden shopping cart through a sensor, which scans each item and debits your bank account accordingly. He imagines devices embedded in our clothes, eyeglasses capable of comparing a currently presented face to a database and retrieving name and details (again, I lately discover that such glasses now exist and are being marketed as aids for mild Alzheimer's sufferers) And he imagines—the inevitable final step—similar devices implanted in our own bodies, monitoring the world, communicating with other such devices, and enabling us to manage, recognize, store, and compare information quite effortlessly as we go about our daily business.

Such is Norman's vision: a vision of a world in which "information is more available to all of us, no matter where we are, whenever we need it."[12] Such technologies, to support the kind of profound integration into human life here envisaged, need to be just about maximally nonopaque. They should contribute nothing to the complexity of the tasks they support: "the complexity of the appliance is that of the task, not the tool."[13] That does not mean, of course, that the technology itself needs to be simple. Quite the contrary. It often takes highly complex (but robust and special-purpose) technology to create a device, which can simply be taken for granted by the user in pursuing her goals and projects. What matters is that as far as our conscious awareness is concerned, the tool itself fades into the background, becoming transparent in skilled use. In this respect the technology becomes, to coin a phrase "pseudo-neural." In childhood we learn how to use our various neural circuits to guide actions (learning to read, to walk, to talk, to write) and, later on, we simply take those capacities for granted as we confront the problems of adult life (preparing the business presentations, going out for groceries, etc.).

Personal information appliances, functioning robustly, transparently, and constantly, will slowly usher in new social, cultural, educational, and institutional structures. Perhaps we will one day live in a world in which, thanks to some easy-to-access implant or wearable device, your answer to the clepsydra question—like the one about the time—is simply, "Yes, (tiny delay) it means 'water clock,'" rather than "No, hold on while I go and look it up." For our sense of self, of what we know and of who and what we are, is surprisingly plastic and reflects not some rigid preset biological boundary so much as our ongoing experience of thinking, reasoning, and acting within whatever potent web of technology and cognitive scaffolding we happen currently to inhabit.

That web is already beginning to include a varied and mutually empowering matrix of human-centered technologies. Other elements in the near-future matrix include the development of lightweight, constantly running, personal computing appliances and of new techniques for rendering the informational substantial, thus blurring the boundaries between the virtual and the physical ("tangible computing," more on which is discussed later).

To see why we need this even-richer web of support, reflect that the original vision of Ubiquitous Computing, with its image of a smart world populated by semi-intelligent desks, doors, freezers, and coffeemakers, aims to put all the computational work out of sight. It seems unlikely, however that we will do away with all need for personal data storage and knowledge access. As a result, we cannot really off-load all the computational work onto some fixed environment. Some of it should be, as Norman realized, linked more directly to a specific user. Wearable Computing, by attaching (quite literally) certain resources directly to the biological agent, offers a nonpenetrative means of catering to just this need. Instead of seeing Wearable and Ubiquitous Computing as competing approaches, then, it is much more fruitful to consider their large potential for harmonious interaction.[14]

A Wearable Computer is an information-processing tool that is, in a deep but noninvasive sense, integral to the user. It is portable, constantly running, and may be used while the agent is in motion or otherwise engaged. As such, it should support hands-free use and be capable of presenting data unobtrusively to the user whenever it sees fit. Such devices are "designed to be useable at any time with the minimum amount of cost or distraction from the wearer's primary task (which is) not using the computer (but) dealing with the environment."[15] Wearable Computing is thus, in a very broad sense, another instance of what Norman called a human-centered technology; it belongs, or aims to belong, to that species of technology that fades into the background in use, providing support while allowing us to focus not on the technology but on the task at hand. It achieves this not by embedding the computing into the world but by affixing it to the agent.

An early example is Bradley Rhodes' "wearable remembrance agent."[16] This is described as "a continuously running proactive memory aid." The device comprises a commercially manufactured heads-up display, which presents an 80 × 25-character screen in the upper visual field via a slightly clumsy "hat-top" mounting.[17] More discreet technologies, including Eye Glass displays and laser-based retinal displays, could also be used.[18] The Microvision firm has piloted a device that uses safe laser technologies to scan images directly onto the user's retina.[19] Ultimately, one can imagine direct electronic input into the main visual processing gateway to the brain, somewhat along the lines of the cochlear implants described in Chapter 1 of *Natural Born Cyborgs*. In Rhode's device the heads-up display is combined with a special one-handed keyboard for input, known as a Twiddler keyboard (made by Handykey in New York) and a special software package, the Remembrance Agent (RA) itself. The software is designed to run constantly, and to respond to inputs by intelligently searching through the agent's local or distal file spaces for items whose contents match the current probe. Think Google, but imagine the resource roving over your own personal file spaces, looking for the notes you yourself entered last time you encountered such-and-such a person or situation. In principle, search-initiating probes could also originate from signals continuously broadcast by local devices (here is one potential source of synergy with Ubiquitous Computing approaches). A typical pattern of use, as imagined by the designer, might go like this:

> Say the wearer of the RA system is a student headed to a history class. When she enters the classroom, note files that had previously been entered in that same classroom at the same time of day will start to appear . . . when she starts to take notes on Egyptian hieroglyphics, the text of her notes will trigger suggestions pointing to other readings and note files . . . when she later gets out of class and runs into a fellow student, the identity of the student is either entered explicitly or conveyed through an active badge system or automatic face recognition. The RA starts to bring up suggestions pointing to notes entered while around this person, including an idea for a project proposal that both students

> were working on. Finally, the internal clock of the wearable gets close to the time of a calendar entry reminding the wearer of a meeting.[20]

The idea is thus to combine the advantages of personal, agent-specific information, storage, and retrieval with input from a variety of fixed, environmentally distributed resources providing the wearable device with a stream of useful context-fixing information, helping it to guess where the agent is and what she is probably doing. The user is at once a mobile locus of highly personalized resources and a useful interface for local, embedded computational devices. She is also a kind of automatic electronic trail-leaver, whose movements and choices can be tracked—for good or for ill—by the devices she passes near (for much more on this, see Chapters 6 and 7). Wearable Computing and Ubiquitous Computing are natural allies whose full synergetic potential has yet to be explored.

There is, however, another problem lurking in the general move toward ever-more-integrated, invisible, automatic, pseudo-neural technologies. The danger is one of loss of control.[21] Opaque technologies were, of course, hard to use and control; that's what made them opaque. But truly invisible, seamless, constantly running technologies resist control in a subtler, perhaps even more dangerous, manner. How then can we alter and control that of which we are barely aware? Suppose, for example, I am unhappy with the performance of my biological memory regarding names and dates. There is nothing very direct that I can do about this. I might engage in memory-training exercises; I might augment my biological memory with new resources (Palm Pilot, remembrance agent); I might try some neurotropic substances like Ginkgo biloba. But—considered as a piece of "cognitive kit"—my biological memory is pretty hard to get at and reconfigure. It is too far along the spectrum that leads to fully invisible computing.

Proponents of what has become known as tangible computing take this kind of worry very seriously indeed. In making our technologies truly, permanently invisible to the user, we may similarly limit our own capacities for creative intervention. The philosopher, Heidegger, writing in 1927, distinguished between a tool's being "ready-to-hand" and its being "present-at-hand." The hammer, while in use, is ready-to-hand. It is not an object of conscious reflection. We can, in effect, "see right through it," concentrating only on the task (nailing the picture to the wall). But, if things start to go wrong, we are still able to focus on the hammer, encountering it now as present-at-hand, that is, as an object in its own right. We may inspect it, try using it in a new way, swap it for one with a smaller head, and so on. The effective use of tools thus often involves a kind of flipping between invisibility-in-use and availability for thought and inspection. Paul Dourish, a leading proponent of tangible computing, thus reminds us that "the effective use of tools inherently involves a continual process of engagement, separation and re-engagement."[22] Dourish, a one-time colleague of

Mark Weiser, invented the term "tangible computing." Tangible computing maintains key elements of the invisible computation model but seeks to do so without allowing the tools and technologies to become permanently invisible, available solely as ready-at-hand. In common with the work discussed earlier, however, the interactions between the user and the tool are meant to be as natural and easy as possible, and to make the most of our basic skills and knowledge.

An appealing example of tangible computing is the Marble Answering Machine designed by Durrell Bishop at London's Royal College of Art.[23] Standard digital answering machines can be hard to use and control; they may have multiple functions hidden in menus and useable only via complex sequences of button-pushings and key-holdings. By way of contrast, Bishop's design, Dourish tells us, works like this:

> [The] answering machine has a stock of marbles. Whenever a caller leaves a message . . . it associates that message with a marble from the stock, and the marble rolls down a track to the bottom, where it sits along with the marbles representing previous messages. When the owner of the machine comes home, a glance at the track shows . . . how many messages are waiting: the number of marbles arrayed at the bottom of the track. To play a message, the owner picks up one of the marbles and drops it in a depression at the top of the answering machine; because each marble is associated with a particular message, it knows which message to play. Once the message has been played, the owner can decide what to do: either return the marble to the common stock for reuse (so deleting the message) or returning it to the track (saving it to play again later).[24]

Now imagine a good spy-movie scenario. You receive a long and vital message on the home machine, but you have just one minute until your nosy roommate gets home. You don't want her to see or hear the message, but you cannot delete it yet. What do you do? I am willing to bet that every single reader of this text immediately thought, "Take the marble and put it in your pocket, then later, in private, drop it into the machine." The situation was unusual, yet you at once knew how to proceed—but what would you have done on your usual digital device?

The point is that the Marble Answering Machine, by giving a familiar kind of physical presence to what is really a digital abstraction (the message) allows us to use our well-developed intuitions about physical objects to interact with the virtual/informational realm. As Dourish explains, the problem of interacting with the virtual is thus transformed into the more familiar one of interacting with concrete, movable objects. Instead of pushing technology to become totally invisible, the idea is to make it extra-visible: to take digital abstractions and data-flows and make them as solid and manipulable as rocks and stones. In so doing, it is hoped, we provide for the kind of easy flippability (between ready-to-hand and present-at-hand) characteristic of many of our favorite tools.

Our biologically given neural "tools" typically lack this characteristic flippability and it may be thought that this is sufficient reason to reject the idea of a potential symmetry between neural structures and any nonbiological equipment that is *ever* encountered as an object in its own right. This cannot be quite right, however, since we can easily imagine encountering some of our own neural mechanisms as objects too. Biofeedback techniques already allow an indirect form of this, as when we learn (by way of audible signals) to induce neural alpha rhythms at will, to lower our blood pressure, and so on. More direct forms of encounter may become commonplace as neuron-imaging techniques allow us to watch our own brains as they process information. In so doing, we surely do not render these aspects of our neural functioning less part of ourselves. The increasing visibility of our biological information-processing routines is, however, an interesting counterpoint to the increasing transparency of our best nonbiological props and aids.[25]

The Tangible Media Group at MIT Media Lab is also in pursuit of this vision of embodied digitality. Their goal is to create a new generation of interfaces that increasingly blur the distinction between the virtual/informational and the tangible/physical. A typical project is the aptly named Sensetable, a tabletop display that uses electromagnetic sensing to determine the position of a variety of physical objects (placed on the tabletop), which the user can then move around so as to amend and alter the information displayed.[26] One example is a chemistry-teaching package in which a variety of atoms and molecules are displayed on the desktop screen. On top of the screen (i.e., on the desktop) there is also an array of small, puck-like objects. Each puck, if moved on top of a specific atom or molecule, becomes "bound" to that item (much as a marble became bound to a specific message). The puck is now a physical embodiment of that information. Moving the puck around on the display, or bringing it into contact with another puck, now causes the system to simulate the effects of bringing that atom or molecules into contact with others. Chemical reactions can be investigated, and new molecules built and examined. In addition, modifiers can be attached to the puck via a surface slot, in order to change the charge on that atom or molecule, and so on.

Sensetable is a descendent of a system called metaDesk, which used cameras and computer vision techniques (instead of electromagnetic sensing) to allow a variety of physical icons ("phicons") to interact with a tabletop display.[27] Common to all these projects, then, is the use of what the group calls Tangible User Interfaces (TUIs) in which familiar physical objects, instruments, surfaces, and spaces are used to mediate our exchanges with digital information systems. Such interfaces aim, also, to erode the gap between input systems and output systems.[28] When I write on a piece of paper, the input space and the output space are one and the same; the stored item appears exactly where it was *input*. Standard Graphical User Interfaces (GUIs) pull the spaces apart: you type on a keyboard and the information is stored somewhere else, and displayed on a screen. TUIs use displays, which are themselves "aware" of the user's activities, and which act as input to the system (the puck-sensitive tabletop screen, for example).

One promising idea is to exploit the kinds of interface we find familiar in the noncomputational world to better mediate our contact with digital and informational resources. In this vein, Neil Gershenfeld and his colleagues produced a bow-using interface to mediate the contact between a world-class cello player (Yo-Yo Ma) and an electronic cello. The bow provides a superbly sensitive, delicately nuanced, feedback-friendly means of continuously controlling the musical ebb and flow. It is an interface that has been tuned and adapted over centuries of use, and to which the human cellist has devoted a lifetime of study. Why throw all that away in favor of a few buttons and a mouse? If synthesized music can sometimes sound cold and lifeless, might that have more to do with the use of such stale interfaces, rather than the potential of the digital medium itself? Recently, the bow-based interface was used to great effect by Yo-Yo Ma in a Tokyo performance. The digital media allowed the artist to create new sound combinations beyond the reach of any normal cello, while the familiar interface allowed him to explore these new possibilities with all his characteristic flair and insight. Ma himself, according to Gershenfeld, is engagingly enthusiastic and unsentimental, treating both his original cello (a Stradivarius) and the new array as just two technologies: simply the means to his musical ends.

Another area in which the notion of the interface is being reinvented is in work on Augmented Reality. In this work, the interface is nothing more than your own view of the world as you look around, but the view is augmented using some kind of heads-up or eyeglass style display system. The display might use video systems to mix computer graphics and input from cameras aimed at the scene before you, or take direct optical input and overlay it with computer graphics. The idea, in each case, is to overlay our experience of the physical world with layers of personalized digital information. This kind of work uses many of the same techniques and technologies as work on Virtual Reality and Wearable Computing. But instead of trying, as with standard Virtual Reality approaches, to re-create a simulation of the real physical world entirely *inside* some computer-generated realm, the goal of Augmented Reality is to *add* digital information to the everyday scene. Think of it as a kind of digital annotation and enhancement regime, with the specific annotations and enhancements being tailored to the needs and desires of different users passing through the (real-world) scene.

For example, combining Global Positioning information with locally poised digital resources makes it possible to associate specific items of information with geographical locations. Such information could be picked up using special eyeglasses, or via handheld or other wearable devices. Thus imagine you are lost on a university campus.[29] To find the library, you simply enter the name "library" in a handheld local guide-box and don a pair of special eyeglasses. As you look around, you see a giant green arrow take shape in the sky, pointing at the roof of the library! Looking down at the path, you see smaller arrows indicating the best route. Hanging in the air around your body you notice a variety of small icons offering you other local services. To use them, you just reach out and "touch" them, sending positions and motion information through sensors in your clothing.

The term "Augmented Reality" was first used by a group of Boeing engineers and scientist in the early 1990s.[30] Their idea was to use such systems to help workers install complex wiring harnesses in aircraft. The workers would see the desired positioning superimposed upon the actual physical structure of the plane. In a similar vein, engineers seeking to repair broken equipment might soon see the innards of the machine alongside specific repair instructions highlighting the elements to be removed and replaced. Surgeons seeking to repair human brains or bodies could benefit in the same way, seeing ultrasound scans or brain imaging information projected onto the appropriate areas. Researchers at the University of Central Florida have overlaid a model of a knee-joint on a woman's leg. Using infrared LED's (Liquid Electronic Displays) to inform the system about current leg position, the Augmented Reality interface allows onlookers to see just how the bones move while the woman walks and bends. The use of overlaid digital resources to enhance our ordinary daily experience of the world and to provide new means of physical–virtual interaction is likely to play a major role in the next decade. Very soon we may expect to see various kinds of personalized electronically overlaid information, from advertising to information about incoming cell phone calls or even about ourselves apparently suspended in the air as we roam about. Such information might appear attached to the space around an individual, or a shop, or a designated electronic advertising area. Once again, the key innovation is to allow the physical and the informational realms to seamlessly merge and mingle, in ways that unobtrusively support daily activity and that make maximum use of our normal means of embodied, socially embedded activity. It is worth repeating that such work in Augmented Reality, though it uses some of the same technologies as Virtual Reality, is really quite different as root. The aim here is not to create a richly detailed version of the daily world inside the machine, but to use the machine to add new layers of meaning and functionality to the daily world itself. Some theorists thus speak not of Virtual Reality but Real Virtuality—a kind of deliberate blurring of the boundaries between physical and informational space.

This kind of blurring has educational importance too. If we are indeed becoming complex biotechnological hybrids, a major challenge for the future will be to train young minds to think well about a world in which the physical and the informational/digital are densely and continuously interwoven. To that end, researchers are developing forms of so-called mixed reality play.[31] In mixed reality play, the virtual/informational is made tangible, the physical made virtual, and the two realms interwoven in single play-based experiences. One of the keys to such experience is the development of what the team calls "traversable interfaces between real and virtual worlds." A traversable interface creates the impression of a seamless join between the real and the virtual, and encourages users to frequently and naturally cross over between the two realms.

Versions of such techniques can allow children to engage in "mixed reality play," in which a coherent play space is created with characters who are able to cross over between the physical and virtual realms. In one of the last papers written before his untimely death, my colleague and friend Mike Scaife, working as part of a multiuniversity interdisciplinary research collaboration, helped design a mixed reality adventure game called "Hunting the Snark."[32] The game targets children in the 6 to 10 age group, and the idea is to locate an elusive entity (the Snark) that can live in both the physical and the digital world, and whose activities seamlessly crisscross the two realms. This game also relies on traversable interfaces that give the illusion of joining the two worlds. For example, physical items bearing electronic tags can be used to trigger events in the digital world when placed into some key locations including a well, a wardrobe, a cave, and a mouth. When placing a real object in the well, the children could see the object in their hand disappear in the real world, emerging at once in the digital one, or to be eaten, or rejected, by the digital manifestation of the Snark. The use of simple forms of wearable computing also allowed some of the children's real movements to be coordinated with the movements of the digital Snark. This kind of playful technology provides novel and exciting experiences, which should help young brains learn better how to make the most of a world in which the physical and the digital are ever-more-closely intertwined—worlds in which everyday objects (medicine cabinets, coffeemakers, refrigerators) have informational state, and informational phenomena have much more tangible physical presence.

Nurtured by such experiences, and living and moving in a world populated with ubiquitous computing devices, augmented reality displays, and various kinds of tangible computing, next-generation human minds will not invest very heavily in the virtual/physical divide. Instead, these minds will focus on activity and engagement, seeing both the virtual and the physical as interpenetrating arenas for motion, perception, and action.[33] Mixed reality play intends to block the stale opposition between the real and the virtual, or the bodily and the informational, revealing each for what it is: just one more aspect of a larger world in which hybrid selves live, move, work, and play.

Moving on

Time to take stock. We have now met several visions of the near future. The visions of Invisible Computing, of Tangible Computing, of Wearable

Computing, and of Augmented Reality. Of these, Invisible Computing and Tangible Computing at first seem like diametrically opposed research programs, but this is not really the case. The differences are real, but easily overplayed. Is the wristwatch an example of invisible or tangible technology? Norman and Weiser's original vision does not require total invisibility so much as invisibility-in-use, and on this, both models converge. The Marble Answering Machine, to take another case, is every bit as good an example of an Information Appliance as it is an instance of Tangible Computing. And it is the very same features that make it "tangible" (the way it exploits our ease and familiarity with everyday objects) that allow it to become invisible in daily use.

The differences between the two visions thus show up only, if at all, at the very extremes, where some Information Appliances will indeed be designed to remain firmly out of sight and out of mind. But sometimes, as Norman stresses and Dourish would surely admit, this is the best way. The engine management system of a modern automobile is a case in point, and the intelligent coffeepots and carports imagined by Norman and Weiser may be others. The question we really should *not* ask is, may be, "Which way is best?" That is rather like asking whether our best tools should be more like hands, hammers, or the hippocampus. The question is misguided, because each of these tools is specialized for different purposes and (hence) needs to be accessed, used and/or reconfigured in very different ways. For certain purposes we want tools that we can step back from and think about. For other purposes, we want tools that function continuously and quasi-independently, requiring little or no conscious attention and that resist easy reprogramming. The various kinds of transparent, human-centered technologies that we have so far imagined are, however, typically restricted in one specific way. All the *fitting*, the adaptation of the technology to the needs and capacities of the biological user, is done by the slow cultural process of design and redesign; the final dovetailing of biology and technology is achieved courtesy of individual human learning. This neglects the important—perhaps ultimately transformative—potential of information appliances that, in use, actively work to learn about and better fit the user. I dub such appliances "dynamic appliances." Not all dynamic appliances are transparent and unobtrusive. Existing speech-to-text software is quite hard to set up and use, but it is dynamic in that it learns about specific users and adapts to *their* voices and vocabularies. The combination of dynamic appliances and transparent technologies is surely a match made in cyborg heaven. Imagine information appliances that actively learn about the user. First, you take the word-processing function out of the PC and lodge it in a special, dedicated writing and composing platform—an information appliance using a nice Direct Manipulation Interface. Next, you allow the machine to monitor its own use. After a while, unused functions can be temporarily disabled, and frequently used functions are given a more efficient implementation. Response times speed up. The appliance becomes *skilled* at doing just what the user requires, no more and no less. This is what we already find in our own neural structures, and pseudoneural technologies need to aim for the same kind of effect. There is a cost of course: the *immediate* functionality is reduced. But this is a trade-off the biological brain makes all the time. You become skilled at driving a car with a certain configuration of controls. After a while, the cognitive effort needed to drive that kind of car drops, and the capacity to rapidly and effectively avoid collisions increases. The cost of this is felt if driving demands change—say you rent a car with a very alien configuration. But the process then repeats. The biological brain is constantly striving to streamline, chunk, compile, and automate, and it does so by attending to repeated patterns of activity and use. Dynamic information appliances would, when appropriate, do just the same. The combination of brains that learn about technologies, with ubiquitous, increasingly transparent technologies that "learn" about individual brains sets the scene for a cognitive symbiosis whose full potential and implications none of us can yet fully appreciate.

The technological present, then, is a shifting kaleidoscope of visions of the future. The smart world full of invisible technologies; the world of constantly running, easily deployed wearable computers; the world of neuroelectronic implants; the world of tangible computing and real virtuality; the world of dynamic, self-reconfiguring wearables and information appliances. Which is it to be? The visions jostle for space, but they are not truly competing. Quite the reverse. These are, as we will see, complementary threads in an emerging biotechnological fabric. Our cyborg future, like our cyborg present and our cyborg past, will depend on a variety of tools, techniques, practices, and innovations. What they will increasingly have in common is that deep human-centeredness that Norman so powerfully celebrates. These will be technologies to live with, to work with, and to *think through*. Such technologies are apt for the most profound and enduring kinds of interweaving into our lives, identities, and projects, and into our constantly constructed sense of place, presence, and self.

Notes

*This chapter is an edited version of Chapter 2, *Natural-Born Cyborgs: Minds, Technologies, and the Future of Human Intelligence*, Oxford University Press, 2003. The author and editors are grateful to Oxford University Press for allowing the reprint to occur here.

1. You can find a nice description on the web at www.wps.com/about-WPS/personal/black-hole/.
2. See Mark Weiser, "The Computer for the 21st Century," *Scientific American*, September 1991, 94–110, and Donald Norman, *The Invisible Computer* (Cambridge, MA: MIT Press, 1999). Similar ideas are found in the philosophical works of Maurice Merleau-Ponty, such as *The Phenomenology of Perception*, trans. Paul Kegan (1945; reprint, London: Routledge, 1962) and Martin Heidegger, *Being and*

Time (1927; reprint, New York: Harper and Row, 1961). An early appearance of the idea of "transparent tools" in the context of human–computer interaction is found in Jacob Mey's 1988 paper "CAIN and the Transparent Tool: Cognitive Science and the Human-Computer Interface," presented at the 3rd Symposium on Human Interface in Osaka, Japan, in *Journal of the Society of Instrument and Control Engineers* 27: 1 (1987): 247–52.

3. See Donald Norman, *The Invisible Computer* (Cambridge, MA: MIT Press, 1999).
4. There is some discussion of this case in Norman, but what follows is based upon David Landes's wonderful tome *Revolution in Time: Clocks and the Making of the Modern World* (London: Viking Press, 2000).
5. Ibid., 92–3.
6. I came upon this example several years ago but cannot seem to recover the original source. To my best recollection, the example was then deployed in the service of a philosophical argument concerning the nature of implicit beliefs. If any reader can supply the full reference, I would be most grateful.
7. A tropism is a kind of automatic, hardwired response. I use the term, however, to mean a learned but now largely automatic response.
8. For a sustained philosophical defense of this position, see A. Clark and D. Chalmers, "The Extended Mind," *Analysis* 58 (1998) 7–19.
9. This idea of "scaffolding" originates with the work of Soviet psychologist Lev Vygotsky, who stressed the way a child's experience with external props (especially an adult's helps and hints) could alter and inform the way the child solves a problem. Since then, the term "scaffolding" has come to mean any kind of external aid and support, whether provided by a notepad, a computer, or another human being. See L. Vygotsky, *Thought and Language* (Cambridge, MA: MIT Press, 1986).
10. Norman credits the term to Jeff Raskin and dates the coinage as 1978, in an internal Apple memo. See Norman, *The Invisible Computer*, 275, para.1.
11. This list is a distillation from Norman's work. It is also influenced by conversations with Ed Hutchins at UCSD and Mike Scaife and Yvonne Rogers at the University of Sussex.
12. Norman, *The Invisible Computer*, 59.
13. Ibid., 67.
14. For this, see Bradley J. Rhodes, Nelson Minar, and Josh Weaver, "Wearable Computing Meets Ubiquitous Computing: Reaping the Best of Both Worlds" (*Proceedings of the International Symposium on Wearable Computers,* October 1999), http://www.media.mit.edu/rhodes/Papers/wearhive.html.
15. Bradley J. Rhodes, "The Wearable Remembrance Agent: A System for Augmented Memory," *Personal Technologies Journal Special Issue on Wearable Computing, Personal Technologies* 1 (1997): 218.
16. Ibid.
17. Known as a "Private Eye," the unit is made by Reflection Technology and marketed (at last sighting) by Phoenix Group, NY. See www.reflection.com for details.
18. See T. Starner, S. Mann, B. Rhodes, J. Levine, J. Healey, D. Kirsch, R. W. Picard, and A. P. Pentland, "Augmented Reality through Wearable Computing," *Presence, Special Issue on Augmented Reality* (1997).
19. P. Scott, "Eye Spy," *Scientific American*, September 2001, News Scan sec.
20. Rhodes, "The Wearable Remembrance Agent," 219.
21. For more on this issue, see Chapter 7 of *Natural Born Cyborgs.*
22. Paul Dourish, *Where the Action Is: The Foundations of Embodied Interaction* (Cambridge: MIT Press, 2001), 141.

23. Ibid., 42–3.
24. Ibid., 42.
25. Thanks to Ron Chrisley for pointing this out.
26. J. Patten, H. Ishii, J. Hines, and G. Pangaro, "Sensetable: A Wireless Object Tracking Platform for Tangible User Interfaces" (Proceedings of the ACM CHI 2001), http://citeseer.nj.nec.com/464521.html.
27. B. Ullmer and H. Ishii, "The metaDESK: Models and Prototypes for Tangible User Interfaces" (Proceedings of the ACM UIST '97 Symposium on User Interface Software and Technology, 1997), 223–32. http://citeseer.nj.nec.com/ullmer97metadesk.html.
28. See especially David Small and Hiroshi Ishii, "Design of Spatially Aware Graspable Displays" (Short paper, during proceedings of the ACM CHI '97, Atlanta, March 22–27, 1997).
29. Thanks to Frank Biocca for helpful discussion on this topic.
30. For this information and the examples in the same paragraph, I am indebted to S. Feiner, "Augmented Reality: A New Way of Seeing," *Scientific American* 286:4 (April 2002): 48–55.
31. The idea of using mixed reality play was introduced to me by Steve Benford and Tom Rodden in a plenary session at the Fourth International Conference of Cognitive Technology, Coventry, UK, 2001. There, they described a "traveling story tent," using some of the technologies described in the text.
32. Yvonne Rogers, Mike Scaife, Eric Harris, Ted Phelps, Sara Price, Hilary Smith, Henk Muller, Cliff Randall, Andrew Moss, Ian Taylor, Danae Stanton, Claire O'Malley, Greta Corke, and Silvia Gabriella, "Things Aren't What They Seem to Be: Innovation Through Technology Inspiration"(unpublished manuscript).
33. One highly funded European endeavor, the Equator Project, is entirely dedicated to the exploration of such interpenetration. The Equator Project presently spans eight academic institutions. A typical vision statement reads: "Instead of treating these [the digital and the physical] as two different worlds, our view is of these two media as being complementary halves of the same world. They are continually interwoven by the activity of people." For more on this work, visit the Equator website: www.equator.ac.uk.

8
On the Subject of Neural and Sensory Prostheses*

Lisa Cartwright and Brian Goldfarb

> Is it really so sad and dangerous to be fed up with seeing with your eyes, breathing with your lungs, swallowing with your mouth, talking with your tongue, thinking with your brain . . .? Why not walk on your head, sing with your sinuses, see through your skin . . . . Where psychoanalysis says, stop, find your self again, we should say instead, "Let's go further still . . . . Find your body without organs. Find out how to make it. It's a question of life and death."
>
> (Deleuze and Guattari 1987, 151)

In *A Thousand Plateaus* Gilles Deleuze and Félix Guattari thought about the psyche and challenged psychiatry and psychoanalysis, fields based in neurology. But they consider only marginally the role of the spinal cord and nervous system in the production of a body without organs. During the 1990s, years the American Psychiatric Association dubbed the Decade of the Brain, neurology took seriously the production of its own material versions of a body without organs, facilitated by evidence accruing since the 1950s that the brain and nervous system were not rigid and hardwired but plastic and malleable. By the 1980s it was clear that adult nerve fiber was capable of changing form and making new connections. This sea change in neurology made it possible to imagine the willful redistribution of intrasubjective relationships among organs, the nervous system, and sensory experience through neural prosthetics designed for people with injuries or progressive diseases of the spinal cord.

The nervous system emerged with prosthesis as a scholarly trope in the 1990s. These terms often appeared with *network*, a concept that emerged across computer science and biomedicine. Variously modified by the terms neural or computer, the network, like prosthesis, signified enhanced connectivity between organisms and machines in the late-century frenzy over digital and organic systems anticipated in the cybernetic theories of Norbert Wiener and figured in his postwar cybernetic prosthesis (Serlin 2004, 47–51).[1] This chapter will chronicle some neural and sensory prosthetic technologies introduced in the 1990s. We consider material and semiotic relationships among instruments, bodies, and subjects across these concepts.

In 1998, one of us chronicled the development of ocular prosthetics in *Ocular Convergence* (Goldfarb 2001), a digital artwork that constructed an imaginary global consortium of government, university, and private ocular prosthetic research and development. The research documented was actual, the federation a fabrication meant to comment on the state of global science. In this chapter, we mine that work to identify changes in concepts of sensory prostheses and the nervous system at the century's end.

We also view prosthesis actively through the experience of spinal cord injury sustained by one of us (Cartwright) in the early 2000s and experienced by both of us as a condition requiring collaborative research, accommodation, and medical management. Prosthesis, narrowly speaking, hasn't been part of that injury's management; orthotics and surgical hardware have. We propose, however, that prosthetics exist in a continuum with orthotic and surgical hardware and software designed to facilitate nerve function, sensory function, and motor ability. Experiencing life through the dynamic neural perspective that a spinal cord injury and its accommodation in the early twenty-first century invites, we look back on the critical writings about prosthesis of the 1980s and 1990s. That discourse inspired us to write an earlier review of the prosthesis trope and the meaning of disablement (Cartwright and Goldfarb 2002). How, we ask now, do we reconceive the body and the subject when their organization is altered by changes in concepts of the nervous system introduced in the 1980s and 1990s? How do neural prostheses reorganize the body imagined across the discursive realms of biomedical research, medicine, rehabilitation, and patient discourse? What constitutes the object lost when nerves not organs are the primary entities of interest in treatment and rehabilitation? And how do the very ideas of object, organ, and part change when the senses and the nervous system are recognized to be the entities replaced or supplemented in a prosthetic system?

The image of the adult nervous system earlier in the twentieth century was that of a hardwired system with limited potential for adaptation or regeneration after injury or degenerative illness. After World War II discourses of neurology and machine computing became mutually constitutive. Descriptions of the nervous system as a model for machine intelligence were paralleled by characterizations of the nervous system as a conductive mediating network. By the early 1980s there was clear evidence that nerves were capable of change and regeneration. By 1990 the image of the nervous system had become that of a dynamic, changeable network with potential for organic regeneration and reconfiguration.[2] The nervous system had come to life, conceptually and practically, as much more than the body's hardwired interface. At the same time, the brain and spinal cord emerged as previously untapped sites of potential medical management through prosthetic intervention. Instrumentation introduced late in the century included external and implanted elements (neural stimulators) and surgical manipulation (deep brain stimulation). We propose that neural and sensory prostheses introduced at the end of the century offered a different set of meanings and relationships of body and technology than those theorized on the basis of mechanical prostheses in late-century critical writings on the concept. We emphasize overlooked changes in intrasubjective relationships of experience and meaning produced across the sensory and nervous systems. The purpose of this chapter is to introduce some thoughts about changes in intrasubjective experience with the new neural organization of the body and with the introduction of neural prostheses.[3]

Touch and intersubjectivity: Myoelectrics and the prosthetic sense of feel

Myoelectric refers to devices using electrical systems to tap the body's nervous system. In development since World War II, the myoelectric limb was still not commercially viable by the end of the century (Ott 2002, 20). David Serlin notes that in the postwar US, Norbert Wiener was commissioned to develop electronic prosthetic limbs, but the model he developed was picked up as an industrial production device first. The pricey models introduced to the prosthetics market from the 1960s forward were never widely used (Serlin 2004, 47–51).

These early models offered their wearers greater control over limb movement but not limb sensation. Jack, "the mean-looking guy with the myoelectric arm" in William Gibson's 1981 short story "Burning Chrome," reflects some of the desires driving experiments to introduce sensitivity to prosthetic and paralyzed limbs (Beeker 1967, Pfeiffer 1969). Jack describes his encounter with Rikki Wildside, a woman erotically drawn to his prosthesis. For readers of the 1980s and 1990s, Rikki and Jack embodied the erotic potential of technologies that might restore not only body parts, but also sensations like touch reciprocally shared across bodies, through virtual skin. Importantly, Jack's prosthesis replaces not just his limb but also its enveloping organ, the skin, as a medium of sensual pleasure. Histories of prosthetics emphasize the regard of the hand as functional, cosmetic, and symbolic.[4] For centuries, designers of mechanical hands have attended not just to form but to appearance, crafting glove-like covers to simulate the appearance of skin or emphasizing the aesthetics of functionality in, for example, sleek design features. The hand clearly serves a visually symbolic function that includes the hands and fingers as instruments of speech in sign language. Jack's prosthesis emphasizes the role of the hand as an instrument of sensual feeling. A sense strongly linked to sentiment

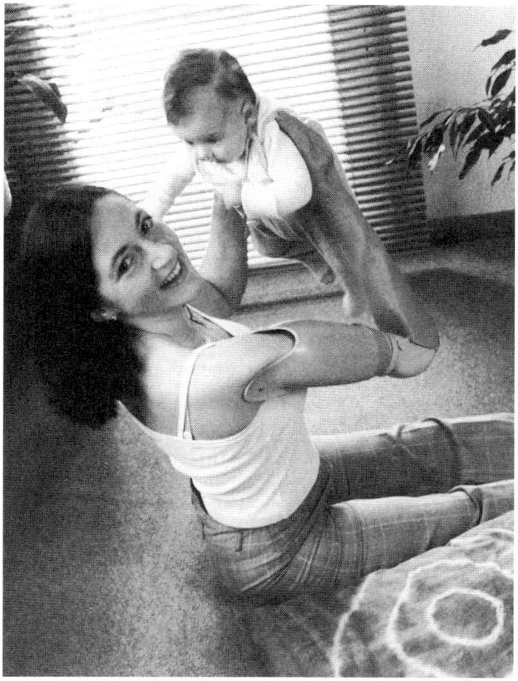

Figure 8.1 Utah Myoelectric arm from the website of Motion Control, Inc.

and the emotions and tied to language, touch is a supplement of language that conveys emotion through what we might loosely call a semiotics of feeling. In all of that term's valences, *feeling* is the object facilitated via Jack's cyberarm. Through this skin prosthesis, the meaning of feeling slips from the sense of touch to affect and emotion to communicated sentiment.

In 1980, following six years of development, researchers at the University of Utah released a nonsensory myoelectric limb offering refined control over pressure and movement (Sears, Andrew, and Jacobsen 1989). The robotic Utah arm, produced by Motion Control, Inc., places electrodes in contact with nerve endings in the amputee's residual limb. The prosthetist picks a "control site" in the amputee's residual limb by testing for areas that provide good nerve conductivity. The wearer is taught to control the device's motorized movement through nerve impulses triggered in remnant musculature. Existing nerve pathways in the body are essentially retrained to accommodate the prosthesis. The wearer of a myoelectric limb must engage in a process of reflexively retraining and incorporating, and being retrained by and, as it were, aggressively *hailed* by a device that stimulates nerve pathways.

Incorporation of a mechanical prosthesis also involves retraining of remnant musculature to control the attached device. One could argue that the wearer of the mechanical limb is also interpellated by and incorporates the prosthesis (Sobchack 1995). However, the myoelectric device requires the wearer's body physically to redeploy nerve pathways. This process suggests a system in which the body is more radically and deeply reorganized. Myoelectric devices tap the body's nervous system, and the body reciprocally extends into the device, producing a system that is interconstitutive beyond the interface between body and prosthetic part, or between prosthetist and amputee.[5]

The prosthetic facilitation of sensual feeling invoked in "Burning Chrome" appears in a brief narrative describing the Sense-of-Feel system, a prosthesis developed by John Sabolich. The surface of the prosthesis is implanted with sensors that transmit information—about temperature, for example, from a prosthetic hand, or about pressure, from the base of a prosthetic foot. Sensory information passes into the flesh through thermopiles and electrodes that press against the base of the wearer's residual limb, triggering sensations elsewhere in the wearer. NovaCare's promotional materials posted to the Web in the late 1990s offered testimony of a client fitted with an experimental Sense-of-Feel arm:

> For the first time in a decade, I could reach out and touch my wife's hand and feel the warmth. That was a very emotional moment.[6]

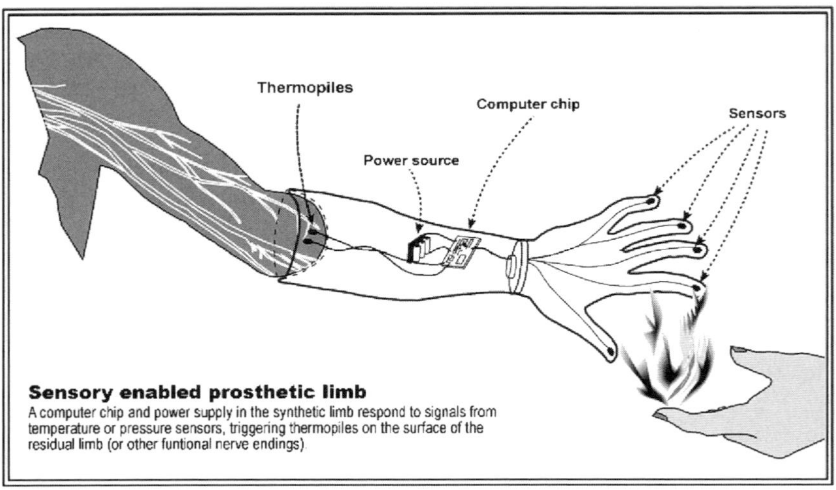

Figure 8.2 Sensory Enabled Prosthetic Limb.
Source: Brian Goldfarb, 2004.

The client's clause "for the first time in a decade" makes this scenario profoundly poignant and nostalgic. The Sense-of-Feel arm's hand, like Jack's, approximates lost manual sensation in the body of the wearer, and in so doing draws on memories of "warm feeling" (love) communicated through hand-to-hand contact. This system requires of its wearer psychological training not only in how to incorporate and control a limb with electrical components that redeploy the nervous system, but also in the orientation of where and how "warmth" enters into and is experienced in and through the body. It is tempting to say that the client quoted above has gained virtual heat sensation *in his new hand*. But it is more accurate to say he must be trained to recognize as "hand" sensations stimuli felt in another, relatively arbitrary, location in his physical body.

Abstraction is an important aspect of this process. Temperature sensation is delivered in isolation of other types or qualities of sensations conducted through touch, such as texture, tickle, itch, numbness, position sense, and pressure (the latter delivered to the Sense-of-Fell prosthesis not by thermopiles but through electrodes). Sensations are parsed in this system; touch in itself is reduced to the abstraction of one of its aspects, temperature, which is in return reduced to warmth as a synecdoche of the "feeling," a euphemism for love.[7] Warmth produces a different meaning than would extreme heat, which invokes the experience and meanings of pain. Warmth is an important signifier of life and of arousal (e.g., of desire and pleasure). Warm flesh signifies the simple fact that a body is alive. Warmth is also an involuntary response to and signifier of arousal as blood flows to the surface of the body in response to nerve stimulation by contact, and in response to the psychical chemistry of desire.

Who would deny the value of technologies that facilitate such essential psychical pleasures as being able manually to touch and feel these involuntary signifiers of "life" and "love" in another's body? But these meanings are unstable and relative, even if they are strongly supported by memory and experience. In "Burning Chrome" Jack conveys to the reader the pleasure he takes in the fact that this sensory prosthesis of desire is reciprocal, between himself and Rikki, but she is a "flame" and not someone with whom he shares a past experience of organic touch. Rikki, he lets us know, is exciting because she "feels" differently: she finds the cold metal of his prosthesis hot, preferring it to his warm flesh. We might ask, what does the Sense-of-Feel hand feel like to the wife of the Sabolich client? Had she learned to incorporate the mechanical prosthesis into their relationship as fetish? Is this new prosthesis a model that offers the surface texture of flesh to her hand, and will Sabolich design a model that can also communicate warmth back to her hand on contact? Might the simulation of warmth involve mechanisms for simulating blood flow to the surface of the "skin" on stimulation, so the hand can radiate the meaning of warmth? We might also ask, how important is it that this example relies on the beloved as the object that communicates "warmth" to the sensory hand? Can the client discern between different degrees of physical and semiotic intensity—say, between the warmth of love and the heat of anger, or between the heat of a body and that of a burning flame?

Rather than pursuing these hypothetical questions about fleshing out what it means for a partner to be incorporated in a prosthetic of feeling, we will stay with the Sense-of-Feel wearer's body and the problem of regulating intrasubjective communication of emotional feeling via sensation. We want to emphasize the extensive representational and psychical work—especially the work of spatial displacement, reorientation, and imagination of a new organization of "feeling"—that is required of the Sabolich client to make his arm a satisfactory instrument for communicating within himself "safe" degrees of feeling. There are limits to how far this prosthesis may go in providing access to feeling. The Sabolich Sense-of-Feel arm allows the wearer to use his virtual hand to feel the warmth of a loved one's body, but it does not afford him the experience of cutaneous pain and nerve trauma of being burned should his prosthesis touch a flame. There are limits on the degree of agency the assistive technology allows its user. This is in contrast with standards we set on technologies we produce for the able body, which push these limits all the time. A prime example is automotive technology (Jain 2004). The toddler learns to recoil from the heat of a fire and not to touch the invisibly hot stove burner. But the wearer of the prosthetic hand that restores the sensory capacity of skin is not re-presented with that potential to get burned. Understandably, companies producing sensory prostheses bypass sensory risks to avoid liability, and to avoid the potential for nerve damage in the subject who has already experienced injury and loss. That subject will not get "burned" twice.

Prosthetic stimulation

So far we have dwelled on the sensory prosthesis in cases where the limb is lost. Paul Schilder is widely cited for his observations on body image in war amputees. However, he also studied and published significant work on the psychology and symptomatology of progressive paralysis (Schilder 1930). Here we wish to draw out the implications of progressive paralysis for body image and as impetus for neural prosthetic support. Paralysis can be sudden and traumatic (as in stroke or an accident that damages the cord), but loss of sensation and control can also be gradual and partial as a result of degenerative changes following injury or illness. Since the spinal cord itself generally does not register localized pain with impairment, the location of impairment relative to site of pain can be remote, its connection to the site of injury obscure—hence the need for inbuilt protections from injury. In my own experience (Cartwright), an injury to the cervical spine produced minimal immediate pain and no visible injury but triggered the slow development of nerve damage and the formation of an undetected lesion in the cord that resulted in escalating chronic and dispersed pain and functional limb and organ impairment over the course of more than a year. Vivian Sobchack notes that numbness is "not not-feeling, but the feeling of not-feeling" (Sobchack 1995, 207). The gradual loss of nerve supply to muscles in the body following injury or degenerative illness can be described in a more varied sort of detail than the language of absence and presence allows. In my own experience, numbness was always arbitrary and abstract. It was a momentary stop *en route* to other phases of sensation, and was almost always partial, as in instances where the sense of the limb's weight was exaggerated in the absence of an ability to control it or feel physical stimuli through it. Numbness was profoundly semiotic, serving an important material signifying function as an indicator of degenerative processes in action. Periods of limb numbness after my injury progressively became longer, and the regions, areas, and types of numbness multiplied and expanded over months, signaling progressive nerve and muscle degeneration. "Not feeling" in my arms and legs sparked the feelings of confusion and distress, triggering anxious hypervigilance of the limbs. The numb limb, like a phantom, demands attention precisely because it is not "there." Inevitably, throughout sleep my numb hand, arm, or leg would rouse me to consciousness, sometimes a dozen times a night. Numbness disturbed my body image profoundly, requiring me to feel for, or open my eyes and look for, the "missing" limb, to locate the sleeping appendage where I imagined it should be relative to my body schema. To be roused from an unconscious state with this negative signal of the body image gone awry is an uncanny feeling that invokes death. More than once I responded with consternation at the ironic situation in which my senseless limbs would have the "nerve" to deny me much-needed sleep. The sleeping limb usually inspired me to manually lift and shake it with the intact hand, an attempt to awaken it. Over months, as I gradually lost nerve supply in various degrees, I recurrently experienced the disorienting "feeling of not-feeling" in appendages that would randomly come back to life with the unpredictable fireworks of fine internal twitching (neurons misfiring, muscle fasciculations) and radiating, cramping muscle pain, only to feel the limb periodically slip back into a *dead weight*, a term that accurately captures the sense of pressure that can be so exaggerated in numbness, and that belies the characterization of numbness as absence.

Elizabeth Grosz describes changes to the body and its organs as a result of disease as they are reflected in changes in body image. Her point is followed by reference to depersonalization, in which the subject loses interest in bodily zones or the whole of the body (Grosz 1994, 76 and 84–5). Disease typically is progressive, producing an unstable and changeable body that is always out of sync with prior schemas. This temporal mismatching of material condition and schema is partly responsible for this depersonalization. Over the course of months I was interested to find that although I did "lose interest" I was hardly inattentive to my body. Rather, loss of control generated a psychic response of fearful hypervigilance in response to the mismatch of material and psychic schemas. I did become inclined psychically to detach from my uncooperative fingers, hands, and feet. I watched and felt muscles flicker with detached fascination, unable to will them to stop as they signaled imminent failure. My hands became clumsy objects, dissociated from my controlled body as if they were the hands of another less capable subject, or as if they were parts externally attached to my body. When I dropped things or stumbled, I mockingly scolded and pitied the uncooperative appendages. To "lose" one's limbs, in this slow and gradual sense, is to experience as process the death of a part that, confusingly, remains deceptively present and visually consistent. My response was emotionally to divest myself of those parts when they performed badly. But this was a performance of alienated disinterest that accompanied the de-alienating process of accommodating the loss.

As with the loss of vision in an eye, spatial orientation and the sense of body image that comes with balance can be lost with sensory and nerve losses to the legs and feet. Randomly, I would temporarily lose the ability to control motor function in one foot to the degree that I could not always anticipate when that appendage would make contact with the floor. A new propensity to stumble and fall out of the blue eroded my sense of autonomy and control, a function of the mismatch between material body and body image. Depersonalization was something I struggled against in my attempts to consciously control not only that which was voluntary but also what was autonomic and sympathetic.

Limb or organ sensation and function may be lost in the death of nerve supply, but these are not the only objects involved. Nerves, residing out of sight and potentially undiagnosed in their pathology, are also objects wanting accommodation or prosthetic support. Yet in the management of damage to the spinal cord and nerves, this fragile object and its extensive network are carefully avoided. They are that which is hardest accurately to image and that which surgeons carefully avoid touching with their implements, except in extreme and rare cases of cord pathology (drainage of a syrinx, a fluid-filled lesion in the spinal cord, for example). In most neurosurgical procedures to address cord or nerve damage, the focus is on the hard matter, the bone and cartilage of the spine that block nerve pathways or compress the cord. Bones and cartilage are cut, removed, ground to a paste and remolded anew, and replaced with substitute hardware (plates, screws, cadaver bone) to make room for the cord and nerve fibers, and to unblock the flow of cerebrospinal fluid.

By contrast, devices to mobilize paralyzed limbs use direct stimulation of the nerves. Functional Electronic Stimulation (FES) was developed in the1980s in conjunction with research into functional electronic rehabilitation as a means of making the muscle "perform useful external work" (Phillips 1991, viii). FES was approved by the US Food and Drug Administration for various uses in 1997. One of these uses was to facilitate mobility in paralyzed limbs. Electrodes attached to non-paralyzed muscle in the body are wired to stimulators embedded in paralyzed muscle. As with myoelectric devices, the wearer's control over (in this case paralyzed, not prosthetic) limbs involves adapting to a schematically remapped nervous system pathway. In the FES mobility systems demonstrated in the late 1980s and early 1990s, electrodes are implanted or attached throughout the paralyzed legs. A computer is then activated to generate sequences of electronic impulses that jolt the paralyzed muscles into action (Phillips 1991, 12–17).

Joseph Shapiro describes the early FES user as becoming physically exhausted and performing a "jerky, tortuous version of walking." "Unlike nature," Shapiro noted, "the 'machine' overstimulates muscles, leaving users exhausted after short distances" (Shapiro 1993, 225). Indeed, one of the early researchers in this mode reported that the two main problem of this system were fatigue and inability to protect the subject from falling (Phillips 1991, 14).[8] Shapiro complained that the FES user neither learned to walk with an extraneous device, nor learned to use his or her paralyzed legs as if they were prostheses. Rather, he or she learned to anticipate the forced way of walking programmed into the computer. The FES device of the early 1990s had a mind of its own, so to speak, and the wearer had to learn to anticipate and accommodate the system.

What is most concerning about this apparatus, for Shapiro, is the degree of agency the FES system affords the technology over the user. Against the argument that the user's dignity is enhanced with the appearance and the performance of legged mobility (a position argued strongly throughout Phillips 1991), Shapiro argues that this appearance is at the expense of the dignity of agency that comes with granting the subject control over movement. "Only in a society that equates a wheelchair with tragedy," Shapiro notes, would such technological "puppetry" be viewed with the optimism that has driven continued work on FES technology (Shapiro 1993, 225).

Although we are in sympathy with his critique, we suggest that Shapiro's understanding of power as an entity that individuals, broadly conceived as persons and technologies, might possess and wield against one another is in itself problematic and misses the point of neural devices. Rather than asking where power is held between the dyad user and technology, or among user, technology, and technician, we might trace power in its multiple and dispersed functions, and investigate its effects in producing new forms of subjectivity (Foucault 1980). To do this, we would need to understand the prosthesis as neither an iconic and substitutive nor a supplemental and facilitative object. Neural prostheses suggest a different configuration of agency in the act of control, in which the device is not a medium but materially reconfigures the intersubjective unit body and technology as an intrasubjective entity. Agency is radically and synesthetically distributed and interconstitutive, within the unitary body. The status of consciousness relative to bodily control is key in this neural reconfiguration of the body.

The FES limb mobility system is kin to a variety of neural stimulators approved for the market by the US Food and Drug Administration from 1997 forward and designed to offer users control over a range of functions. Comparing some of these devices will better help us to explain our point about control and agency. In 1997 an electronic nerve stimulator dubbed the Neurocybernetic Prosthesis (NCP) was approved for use in people with seizures resistant to medication. In the earliest models, electrodes implanted in the neck in the region of the vagus nerve are wired to a small generator usually implanted in the chest cavity near the collarbone. The system delivers electronic impulses, and is controlled either through an external device or through the application of a magnet to the area above the implant when the onset of a seizure is suspected. In 2004 the FDA expanded approval of this device to include use in people with severe depression that is resistant to drug treatment. The same year a related device, the SNS (Sacral Nerve Stimulator), was approved for use in people with lower spine impairment who experience lack of control over the urge or ability to urinate as a result of damage to the sacral nerve. The FDA approval specified the SNS as a "Humanitarian Use Device."[9]

What these three neural prosthetic systems share is their establishment of an aggressively intimate relationship between a digital technology and the nervous system of the user's body. They differ in their relationship, however, to both conscious control and the emotional and social meanings of the behaviors they regulate. The sacral nerve stimulator is used to regulate parasympathetic signals to urinate, returning control or relieving the user of an agonizingly persistent demand of a system gone awry that can

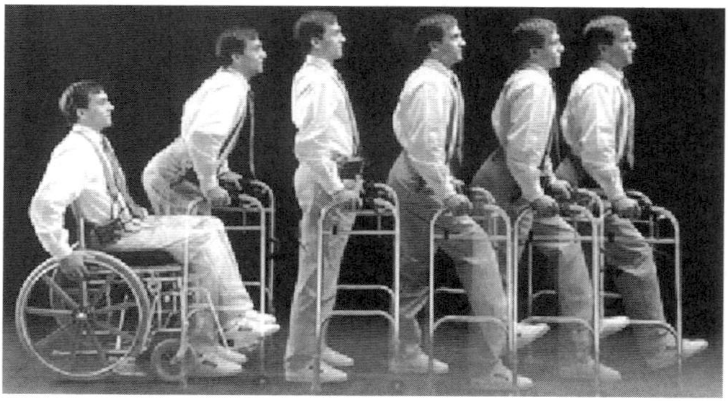

Figure 8.3 Multiple-exposure strobe image of user of Parastep FES system.
Source: Sigmedics, Inc. of Fairborn, Ohio.

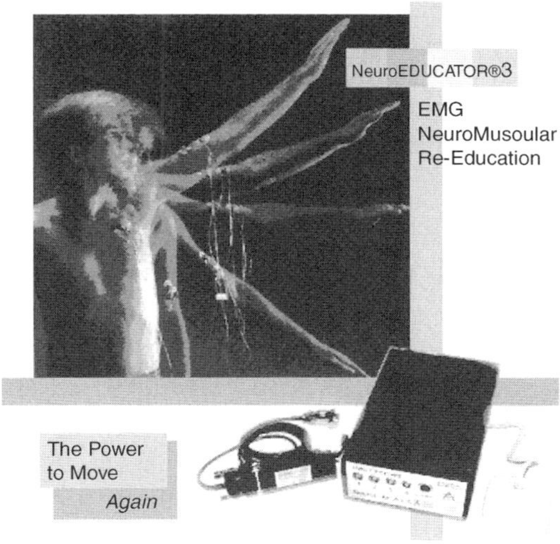

Figure 8.4 Therapeutic Alliances, Neuroeducator system.
Source: Therapeutic Alliances Inc. website.

interfere with work and deny the subject the ability to turn away from a persistently felt need. To prosthetically control that persistent hailing of the subject's conscious attention, whether it is the urge to urinate or somatic pain and disturbing thoughts of depression that hail us, even if control is not fully in the hands of the user, is to restore control to the subject. The Neurocybernetic Prosthesis similarly forestalls a loss of bodily control that is a seizure. As Jeanette Stirling (2001) explains in her rigorous reading of discourses of epilepsy, public display of loss of control in the form of a seizure is strongly symbolic to onlookers, invokes the potential for bodily damage, and is therefore highly regulated.

The FES system differs in that its function is to promote rather than to check motility or sensation. More importantly, the act prompted, walking, is one that typically involves conscious volition. A range of emotional and social meanings relative to agency and autonomy cling to the act of walking. Purposeful mobility, like speech and gesture, is a key signifier of human agency and personal expression. To inscribe a normative gait in the paralyzed subject, Shapiro implies, is to deny that subject the right to autonomous self-expression that is implicit in the willful act of walking or controlling a chair unassisted.

Here it might be productive to point out a linguistic hurdle in our analysis of these forms of prosthesis. The language of consciousness and subjectivity that we have been using is insufficient to describe differences in the somatic and autonomic nervous systems and their complex relationships to consciousness, control, and desire. Its terms are inadequate to describe the feeling, the fear and confusion, of losing control over functions that are autonomic and sympathetic [somatic]. We have yet to determine a schema for theorizing the relationship between consciousness, desire, and autonomic and sympathetic [somatic] functions, especially when the latter are subject to external facilitation and control. As the case of the Sense-of-Feel hand suggested, prosthetically facilitating the senses invokes a complex set of problems concerning the relationship between control and feeling in all of that term's intersecting meanings, between the nervous system and the emotions, and between the senses and desire.

Grosz notes that any willful plan of action requires a bodily schema (Grosz 1994). We might ask, what is the cognitive and material organization of that schema? Phillips (1991) notes that adjusting the body schema is an important part of learning to use a functional electronic stimulator to walk. Research in neural prosthetics since the 1990s has focused directly on analyzing the production of body schema in order to put it to use in the control of prostheses. At the Caltech Center for Neuromorphic Systems Engineering, researchers have been studying the parietal cortex where visual schemas are understood to form. It is there, they surmise, that arm movement is anticipated and planned, with the visual cortex playing a major role (Shenoy et al. 2001). Studying cortical activity through the interface of electrodes implanted in the parietal reach region of the brain, they have demonstrated that monkeys can directly control a robotic arm using brain signals. The researchers aim to replicate this system in a neural prosthetic system. This experimental project differs considerably from the myoelectric limb systems discussed previously because the source of control is the subject's own gaze. The goal is first to analyze the sensory-motor interface, where plans to move the limb first form, and to track and interpret neural signals expressing desired arm-reach locations specified by the gaze. The next step of this project is to replicate this series of events in a system that allows an individual control over the reach of their paralyzed arm (through muscle stimulation) or an attached prosthetic arm (through robotics) using brain signals, anticipated through visually based mental planning.

In this experimental model, *visual* awareness of the arm as an entity in space plays a major role in motor control of a virtual limb. The subject's relationship between internal body image and physical performance is made explicit and becomes a planned and controlled part of the system. In this schema, the idea of representational realism is very much in play: the external visual field must be matched with the "internal" body image. For researchers involved in this study, the boundary between mental image and the performance of body in space is an object of great importance brought into play as a pivotal feature of the prosthetic system.

Visual prostheses

It has been widely argued among theorists of the psyche that vision constitutes a special case among the senses, grounding ego formation and other developmental stages and organizing subsequent psychic processes (Schilder 1978 [1935]; Merleau-Ponty 1963; Spitz 1965; Lacan 1977; Grosz 1994). Prosthetic devices for replacing or enhancing vision have lagged far behind neural prostheses for hearing. Whereas the cochlear implant was first approved for use in adults by the FDA in 1985,[10] the ocular prosthesis has yet to gain regulatory agency approval in the US. Human experimentation has been limited to small numbers of test subjects, with some researchers setting up offices abroad, an arrangement that offers the potential to skirt US regulatory restrictions.[11]

The idea of plasticity and the rerouting of pathways extends to the senses. The majority of ocular prosthetic research and experimentation in the late 1990s and early 2000s has focused on the development of retinal implant systems designed to address forms of blindness in which the retina and other aspects of optical neurology are largely intact.[12] The various designs under development all involve the implantation of an electrode array on or below the surface of the retina.[13] These electrodes directly stimulate retinal cells sending signals to the optic nerve. In most cases these electrodes are activated by a signal transmitted from a miniature video camera outside the eye mounted to a pair of glasses, though initial testing and training of patients involves the direct transmission of image data from a computer display. Currently retinal implants comprise a relatively small grid of electrodes, providing resolution on par with a rudimentary LED display.

An exception to the focus on the eye as the locus of vision is the work of William Dobelle, a leading researcher in ocular prosthetics who has reported success with an artificial vision system using implants in the region of the visual cortex. Here we find, once again, an example of a system that locates the object lost, in this case sight, elsewhere than in the missing or damaged organ, in this case the eye. Rather than supplementing the eye (as the retinal implant does), the Dobelle device prosthetically impacts sight elsewhere, in the visual cortex. In this system, sight is explicitly understood as a translocational process that can be augmented or replaced elsewhere than in the organic region of the eye. One advantage of locating the prosthesis in the parietal region is that it can be used in individuals whose blindness involves physical loss of or severe physical trauma to the eye. But this process is highly experimental and controversial in its use on human subjects. Dobelle's research findings are based on implants in a small number of subjects (eight) across six countries. His design uses an external camera mounted on a pair of eyeglasses to communicate with an electrode array implanted in the visual cortex, the area of interest to the Caltech researchers studying virtual-arm control systems.[14]

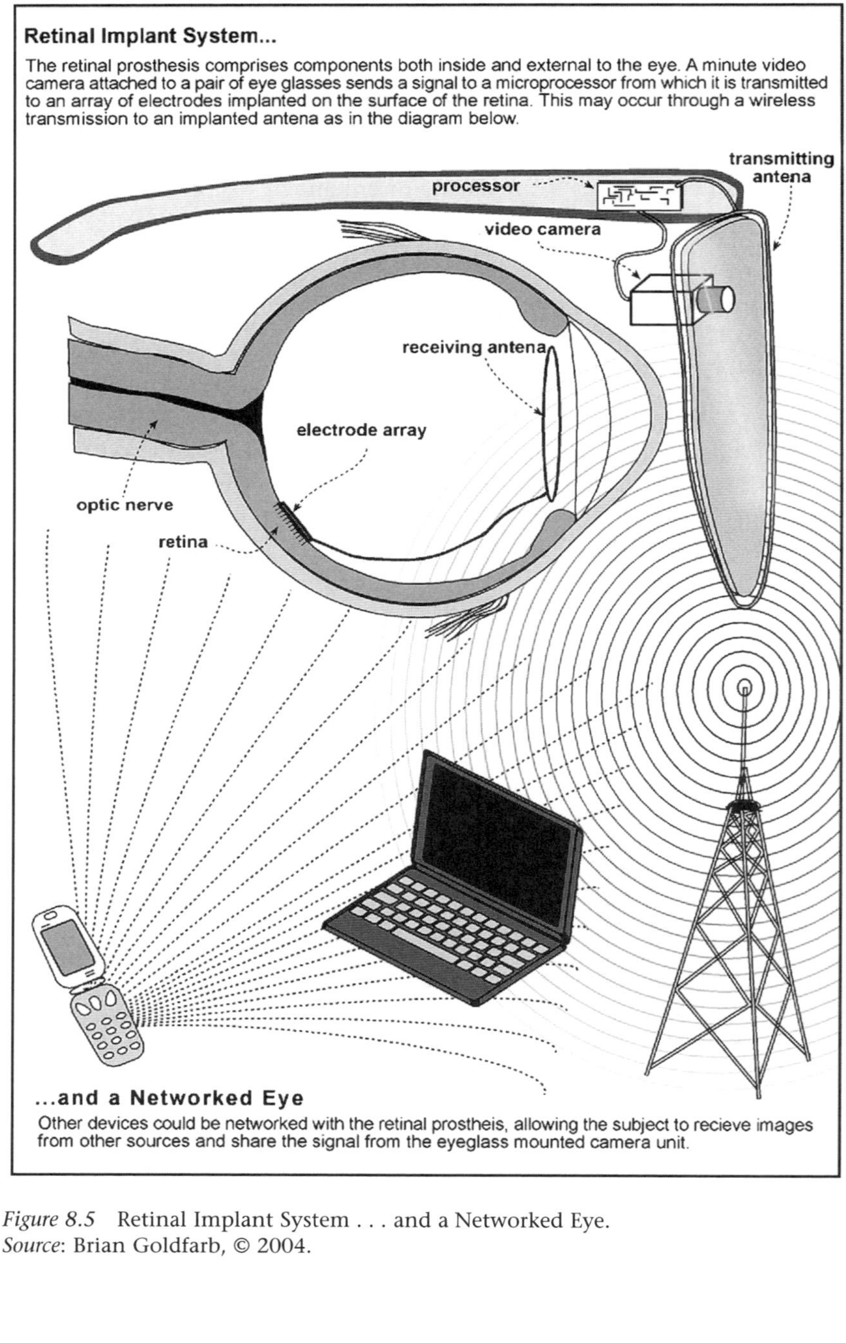

Figure 8.5 Retinal Implant System . . . and a Networked Eye.
Source: Brian Goldfarb, © 2004.

We are prompted to ask, is the latter also as much a visual prosthetic as a limb prosthetic system? The question of the physical location of "eyesight" comes to the fore in research that indicates that cortical implants fail to offer maximal results in image quality and detail because much of visual processing occurs between the eyes and the parietal cortex. With this concern, the significance of nerve conduction (the question of "image-data transmission," to use the computer-network metaphor) becomes an area of heightened attention in research into what we might have wanted to call "eye" prostheses.

What does it mean to recast what we call vision as a dispersed process? Transmission, in this case, is more than a supplement precipitated by the necessity of bridging the imagined center of subjectivity to peripheral senses. It is a material condition embedded in senses previously understood as more or less regionally situated and organ-based. One of the main concerns of ocular implant researchers in the early 2000s has been how to overcome the "transmission" problem intrasubjectively, between body and implant: the biocompatibility problem. This has led to the design of implants that share physical qualities with the nervous system. "Wet" chips, for example, are designed to more closely match the qualities of human neurology in their physical ability to conduct chemical charges. Increasing the resolution of implants addresses the problem of manufacturing instruments at the minute scale of the body's neural architecture. Researchers are attempting to create systems to reproduce the fine muscular ability and control over eye movement that plays a large role in the organic constitution of vision, a task that requires detailed understanding of nervous system functions relative to muscle control across the autonomic-sympathetic [somatic] divide.

Challenges like these, perceived by researchers as *technical* problems, have eclipsed concerns about the psychical and social aspects of making vision "work" for subjects who have spent the majority of their lives blind or visually impaired. When cochlear implants were introduced in the 1990s to profoundly deaf subjects with limited or only an early life history of hearing, broad discussion ensued about the need for these subjects to learn how to understand the abstract sounds they could newly hear. Memory could be invoked in the form of mental concepts of sound represented in other sensory forms (through signed descriptions for example—say, of the tone in which a word is uttered, and through mental impressions), but how might these memories be different from the memory of sound meanings as experienced through years of hearing? Only now is a scholarship emerging on the representational mapping of sound meaning into other registers, the idea of sound memory and experience in profoundly deaf subjects (acoustic memory and experience, for example) and the process of teaching and learning sound meaning for congenitally deaf subjects implanted with these neural-sensory prosthetic devices.

Similar questions will arise with the eventual introduction of ocular prostheses to the public market. What will the visual field mean for

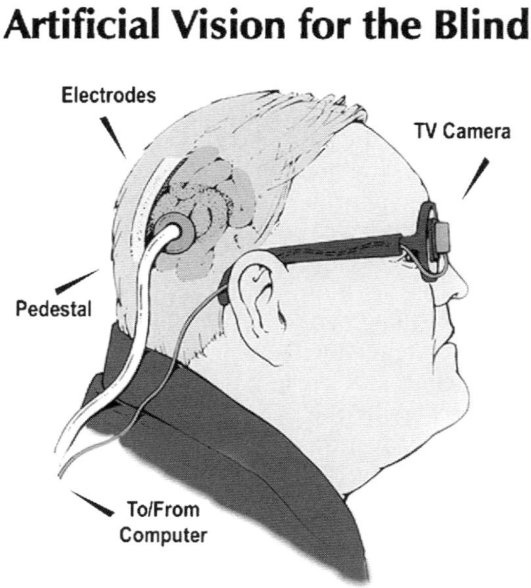

Figure 8.6 Diagram of the Dobelle artificial vision system.
Source: The Dobelle Institute's website.

formerly blind subjects? How will it be experienced? The eventual need for techniques of teaching and learning visual meaning in profoundly and congenitally blind subjects who acquire visual prostheses are barely hinted at in the professional literature of the late 1990s and early 2000s. The ideas laid out in the remainder of this chapter will suggest that in current research the question of visual meaning will not be viewed as clearly distinct from meaning in other sensory domains, as functional sensory areas are tapped for cross-sensory service, and as the synesthetic interrelationship of sensory domains imagined in the body without organs concept emerges as a concrete paradigm of lived body schema.

This prompts us to further consider the question of the location of organic and prosthetic sensory ability raised most clearly in the example of the Dobelle visual-cortex prosthesis. The Caltech remote arm control research showed that the visual field and body image are major factors in limb mobility (we add, for sighted subjects). Visual perception is linked so closely to motor control that paralysis of the muscles controlling the eyes would result in virtual blindness. Work on cortical and retinal implants has shown that bodily placement and movement in space are key factors in achieving visual ability. Control over bodily mobility, as psychoanalysis and cognitive psychology have shown, is a key factor in the formation of a sound body

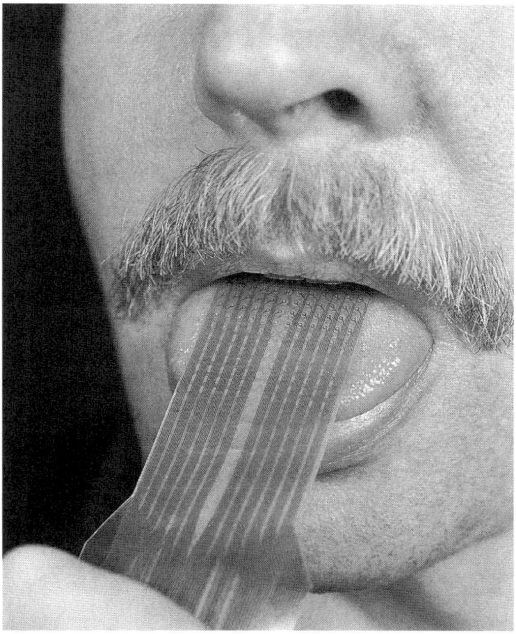

Figure 8.7 Tongue sensor developed by Paul Bach-y-Rita of the University of Wisconsin.
Source: Photo by Jeff Miller Courtesy of Wisconsin Madison.

schema. Physical orientation in space, body sense, is an important factor in the blind or sighted subject's negotiation of what is, curiously, called the body "image" and the "visual" field.[15]

The very definitions of the senses, their organs, and the abilities that cling to them change with the reorganization of the body on the neural model. Paul Bach-y-Rita of the University of Wisconsin at Madison provides a concrete realization of the idea of intersubjectively rerouting the senses by introducing an electrical sensor that interfaces with the tongue and can be used to transmit visual information (in firefighters or divers working in situations of low visibility, for example, or among gamers negotiating virtual worlds). "You don't see with your eyes," Bach-y-Rita asserts. "You see with your brain."[16] The tongue, "loaded with nerves and bathed in its own conducive saline solution," is put forth as the ideal "wet" medium for the array of 144 electrodes positioned there that can receive and send to the brain images from an external camera, computer, or other source. If we can tap nerves in the parietal cortex to produce vision, why not that nerve-laden organ conveniently situated at the junction of interior and exterior, the tongue? The arbitrary nature of the link between organ and function, like that between signifier and signified, knows no limits.

But is this linkage really arbitrary? Surely the sighted "taste" their food in part with the eyes. And surely we can rationalize the motivations of technology that offers us interaction with a stick or a tongue sensor as an alternative to a US $120,000 surgical brain implant as a means to gaining some of the abilities that the senses confer. It is good to remember that nerve plasticity is not dependent on direct surgical or electrical stimulation, and that for all of the advances in the art of surgery, bodies are not yet the assemblage imagined by Félix Guattari and Gilles Deleuze, the structure that "flattens all its dimensions on a single plane" (Deleuze and Guattari 1987, 90). Interventions in the body involving surgery continue to produce pain and risk of death. The body seen as nervous system is always already a body without organs, but it is nonetheless one in which differences between deep and superficial structure can be matters of life and death (Deleuze and Guattari 1987, 90).

What can we say of research that reorganizes the body to locate sight in the mouth, an erogenic orifice so strongly linked, like the bodily openings that are the eyes and the genitals, to sexual pleasure, consumption, and taste? We are tempted to remark about how we have overreached the limits of taste in this the literalization of the metaphor of visual consumption, this organization of the body that returns its subject to the oral phase. Deleuze and Guattari characterize the paradigm of the drugged body, the experimental schizo:

> The human body is scandalously inefficient. Instead of a mouth and an anus to get out of order, why not have one all-purpose hole to eat and eliminate?"
>
> (Deleuze and Guattari 1987, 150)

We might ask, in making the mouth a portal to sight, is Bach-y-Rita suggesting we might have one all-purpose hole out of which to see and to taste when vision fails us? Instead of pursuing what this construction might mean with reference to this research team's desires, the desperation for an erogenic location for sight ability that it suggests, we will return to Schilder on how such relationships between parts are psychically built in the abstract case. Schilder writes that it is as if energy "amasses" in zones invested with desire. "There will be lines of energy connecting the various erogenic points, and we shall have a variation in the structure of the body-image according to the psychosexual tendencies of the individual" (Schilder 1978 [1935], 126). The neural and sensory prostheses we have been discussing all contribute to an ongoing transformation of the body schema and the psychosocial subject that schema facilitates. But Schilder did not anticipate the degree of transformation possible that the Bach-y-Rita project suggests, and he did not anticipate the extent to which the researcher might determine the reorganization of the schema. This example of neural prosthesis recalls Deleuze and Guattari's questions: "Is it really so sad and dangerous to be fed up with seeing with your eyes, breathing with your lungs, swallowing with your mouth, talking with your tongue, thinking with your brain?" Surely it is "a question of life and death" to pursue the body without organs (Deleuze and Guattari 1987, 151). But it is not a matter of personally negotiated resistance.

We conclude by returning to the question of arbitrary signifers, and the experience of nerve degeneration and regeneration. Days after spinal neurosurgery I (Cartwright) began to experience the same muscle twitches and numbness that before the surgery had signaled the process of nerve death. This time I was prompted by my medical team to read these as ambiguous signifiers. Either my limbs were coming back to life, or further damage was caused by the surgery. I waited (am still waiting) to hear from my body whether my nerves are degenerating or regenerating, whether the regions damaged responded with the plasticity and regenerative ability so vaunted in the new century's hopeful discourse of the resourceful nervous system. I cannot fix a schema of this newly organized body yet; the body schema, I come to realize, is never fixed but, rather, is durational not only in the sense that it moves, but in that it ages, injures, heals. I dwell on the image of the lesion (dubbed a syrinx) that resides untouched inside the cord, because it is an entity about whose presence I can have some certainty. Neither I nor my surgeons have seen this lesion. It showed itself only under the most expensive and rarified of imaging processes (contrast MRI). Is the lesion still present? Am I willing it to grow, to swell with meaning through my investment of desire (for it to disappear)? Lines of energy connect this location in my spinal cord with my arms and feet, my trachea, and other zones noticeably impacted by my spinal injury. Living in this terrain of semiotic ambiguity, where life, death, and health hinge on such differences in signification, and where the body's functions are perversely organized around a schema that incorporates injury and damage as processes that are generative of changes in meaning that structure experience, leaves us hesitant to dismiss texts that consider prosthesis and the nervous system as tropes. Rather, we suggest that we investigate more deeply the semiotic meanings concretely incorporated into our bodies with each new trauma or loss, and with each new schema put forth in neural prosthetic research designed to work through that loss. If the body without organs is a neural and synesthetic body, we might ask cultural theorists to look even more closely at the particular material-semiotic practices that construct the subject of neural prosthetics.

Notes

*The authors and editors are grateful to MIT Press for allowing the chapter to be reprinted from *The Prosthetic Impulse*.

1. References in critical theory to this trope most often were to the work of Norbert Wiener (1948, 1954) and Marshall McLuhan (1962) to support claims concerning the emergence of enhanced interactivity not only among peoples and across geographic spaces (McLuhan's contribution), but also between biology and engineering, and between human cognition/biological life and machine intelligence/artificial life (Wiener's legacy). To supplement Wiener, we might add the following sources—works that took these conversations from neural network theory to neurocomputing and biological cybernetics: McCulloch and Pitts (1943), Hebb (1949), Widrow (1960a, 1960b, 1973), Kubrick (1968), Minsky and Papert (1969), and Hecht-Nielsen (1989). This sample of supplements began with artificial neural networks built on the computational model of hypothesized cognitive processes with electrical circuits. It ends with biological cybernetics, distributed intelligence, and shared cognition, as well as theories of intrasubjective nerve regeneration and neural plasticity. Research areas that emerged out of the intersections of these traditions include cyborg anthropology and the ethnographic, sociological, and cultural analysis of distributed cognition, ubiquitous computing, and science, technology, and information in medicine (STIM).
2. It is not in the purview of this chapter to review the historical literature on nerve plasticity and regeneration. We refer readers to the comprehensive bibliography in Nesmeyanova (1977) for the early history of these concepts, and to Vikhanski (2001) for an excellent historical overview.
3. Psychologist Elizabeth Wilson is one of the few critical theorists engaged in the conversation about technologies of the subject to have considered the role of the peripheral nervous system with respect to sensory experience and subject formation. See Wilson (2001) and Wilson (2004).
4. See the excellent collection edited by Ott, Serlin, and Mihm (2002) and Serlin (2004).
5. On somatization see Kurzman (2002) 244–5 and Kleinman (1986) 51.
6. Text is from the Novacare/Sabolich web site: www.novacaresabollich.com. Accessed in 1998 and no longer live. Sabolich, whose historic prosthetics company is now independent of NovaCare, maintains information on sensory prostheses at www.scottsabolich.com/about/research.
7. According to Willis and Coggleshall's 2004 textbook on the sensory mechanisms of the spinal cord, the term *sensory modality* was introduced by Helmholtz to describe the classification of the senses (Boring 1942). Müller's *Elements of Physiology* (1840–2 [1838]) classifies touch as a single modality, however the contemporary view is that touch, pain, cold, and warmth are discrete modalities (Willis and Coggleshall, 2004, 1).
8. On the matter of the norm of upright walking see Yuan (1997).
9. In Europe the device is CE Marked under the requirements of the Active Implantable Medical Device Directive 90/385/EEC (Corcos 2004, 537). Research into electrical stimulation of the micturition reflex and incontinence can be traced back to the early 1970s (Bradley 1971, Godec 1976).
10. http://www.earaces.com/cochlearimplant.htm.
11. For example, William Dobelle, whose research will be discussed shortly, maintained a laboratory in Switzerland between 1983 and 2001, and at Columbia-Presbyterian in New York from 1995 through 2001. Currently his laboratory is based in Lisbon, Portugal. See http://www.jwen.com/rp/articles/cortical.html, http://www.dobelle.com.

12. http://www.opticsreport.com/content/article.php?article_id=1007, http://www.stevemushero.com/retinal_implants.html.
13. Locations of ocular implant research include Second Sight, Los Angeles; Optobionics, Wheaton, Illinois http://www.optobionics.com/; Boston Retinal Implant Project (formerly the Harvard-MIT Retinal Implant group), http://www bostonretinalimplant.org/; Intraocular Retinal Prosthesis Group, The Wilmer Eye Institute, Johns Hopkins Hospital, http://www.irp.jhu.edu/.
14. http://www.jwen.com/rp/articles/cortical.html, http://www.dobelle.com. Also see http://insight.med.utah.edu/research/normann/normann.htm for University of Utah research on cortical implant chips designed for subjects with profound blindness.
15. Here we are referring to the dominant practice of taking the visual as both the metaphor and the primary sensory field in constructions of subject and world in philosophy, psychology, psychoanalysis, and myriad other fields concerned with body and life.
16. http://www.news.wisc.edu/story.php?get=5837.

References

Baudrillard, Jean. *The Ecstasy of Communication*. New York: Semiotext(e), 1988.

Beeker, T. W., J. During, and A. den Hertog. "Artifical Touch in a Hand Prosthesis," *Medical Biology and Engineering* 5 (1967): 47–9.

Bernard Widrow, Narendra K. Gupta and Sidhartha Maitra. "Punish/Reward: Learning with a Critic in Adaptive Threshold Systems," *IEEE Transactions on Systems, Man, and Cybernetics*, Vol. SMC-3, No. 5 (September 1973).

Boring, E. G. *Sensation and Perception in the History of Experimental Psychology*. New York: Appleton-Century-Crofts, 1942.

Bradley, W. E., G. W. Timm, and S. N. Chou. "A Decade of Experience with Electronic Stimulation of the Micturition Reflex," *Urology International* 26, 283–302, 1971.

Cartwright, Lisa and Brian Goldfarb. "Ocular Prosthetics and the Embodiment of Digital Visuality," *doKU: Wirklichkeit inszenieren im Dokumentarfilm*, ed. Christina Lammer. Vienna: Turia & Kant, 2002.

Corcos, Jacques and Eric Schick. *Textbook of the Neurogenic Bladder*. London and New York: Taylor and Francis, 2004.

Deleuze, Gilles and Félix Guattari. *A Thousand Plateaus: Capitalism and Schizophrenia*, trans. Brian Massumi. Minneapolis: University of Minnesota Press, 1987.

Foucault, Michel. *Power/Knowledge: Selected Interviews and Other Writings, 1972–1977*, ed. and trans. Colin Gordon. NY: Pantheon, 1980.

Godec, C., A. S. Cass and G. F. Ayala. "Electrical Stimulation for Incontinence," *Urology* 7 (1976): 388–97.

Grosz, Elizabeth. *Space, Time and Perversion: Essays on the Politics of Bodies*. New York: Routledge and Sydney: Allen and Unwin, 1995.

——. *Volatile Bodies: Toward a Corporeal Feminism*. Bloomington, IN: Indiana University Press, 1994.

Hebb, D. O., *The Organization of Behavior: A Neuropsychological Theory*. New York: Wiley, 1949.

Hecht-Nielsen, Robert. *Neurocomputing: Reading*. MA: Addision-Wesley, 1989.

Hurley, Susan and Alva Noë. "Neural Plasticity and Consciousness," *Biology and Philosophy* 18 (2003): 131–68.

Jain, Sarah. "The Prosthetic Imagination: Enabling and Disabling the Prosthesis Trope," *Science, Technology and Human Values* 24, no. 1, Winter (1999): 31–54.

Kleinman, Arthur. *Social Origins of Distress and Disease: Depression, Nurasthenia, and Pain in Modern China*, New Haven: Yale University Press, 1986.
Kubrick, Stanley. *2001: A Space Odyssey*. Warner Brothers, 1968.
Kurzman, Steve. "There's No Language for This: Communication and Alignment in Contemporary Prosthetics," in Ott et al. 2002.
Lacan, Jacques. *Ecrits: A Selection*, trans. Alan Sheridan. London: Tavistock, 1977.
McCulloch, Warren S. and Walter H. Pitts. "A Logical Calculus of the Ideas Immanent in Neural Nets," *Bulletin of Mathematical Biophysics*, Volume 5 (1943).
Merleau-Ponty, Maurice. *The Primacy of Perception*. Evanston: Northwestern University Press, 1963.
Minsky, Marvin L. and Seymour S. Papert. *Perceptrons: An Introduction to Computational Geometry*. Cambridge, MA: MIT Press, 1969.
Mitchell, David T. and Sharon L. Snyder. *The Body and Physical Difference Discourses of Disability*. Ann Arbor: University of Michigan Press, 1997.
Müller, J. [1838]. *Elements of Physiology, Second Edition*, trans. W Baly. London, 1940–2.
Nesmeyanova, Tat'yana N. *Experimental Studies in Regeneration of Spinal Neurons*. New York: John Wiley and Sons, 1977.
Ott, Katherine, David Serlin and Stephen Mihm (ed.). *Artificial Parts, Practical Lives*. New York: New York University Press, 2002.
Pfeiffer, E. A., C. M. Rhode, and S. I. Fabric. "An Experimental Device to Promote Substitute Tactile Sensation from the Anesthetic Hand," *Medical Biology and Engineering* 7 (1969): 191–9.
Phillips, Chandler Allen. *Functional Electrical Rehabilitation*. New York: Springer-Verlag, 1991.
Schilder, Paul. *Studien zur Psychologie und Symptomatologie der progressiven Paralyse*. Berlin: S. Karger, 1930.
——. 1978 [1935]. *The Image and Appearance of the Human Body: Studies in the Constructive Energies of the Psyche*. New York: International Universities Press. An expanded version in translation of *Das Körperschema*, Berlin: Springer, 1923.
Sears, Harold H., J. Thomas Andrew, and Stephen C. Jacobsen. "Experience with the Utah Arm, Hand, and Terminal Device," *Comprehensive Management of the Upper Limb Amputee* (eds) D. J. Atkins and R. H. Meier III. New York: Springer-Verlag, 1989.
Serlin, David. *Replaceable You: Engineering the Body in Postwar America*. Chicago: University of Chicago Press, 2004.
Shapiro, Joseph. *No Pity: People With Disabilities Forging a New Civil Rights Movement*. New York: Time Books, 1993.
Shenoy, Krishna, Sohaib Kureshi, Richard Andersen, Shiyan Cao, and Joel W. Burdick (2001). "Toward Prosthetic Systems Controlled by Parietal Cortex." Caltech. http://cnse.caltech.edu/Research01/reports/shenoy1full.html.
Sobchack, Vivian. "Beating the Meat/Surviving the Text, or how to Get out of this Century Alive," *Body and Society* 1 (1995): 205–14.
Spitz, René. *The First Year of Life: A Psychological Study of Normal and Deviant Development of Object Relations*. New York: International Universities Press, 1965.
Stirling, Jeanette. "The Subject of Epilepsy: A Textual Study," doctoral dissertation thesis, Department of English, University of Sidney, August 2001.
Vikhanski, Luba. *In Search of the Lost Cord: Solving the Mystery of Spinal Cord Regeneration*. Washington, DC: Joseph Henry Press, 2001.

Virilio, Paul. "The Third Interval: A Critical Transition," in *Rethinking Technologies*, ed. Verena Andermatt Conley. Minneapolis: University of Minnesota Press (1–12), 1993.

Widrow, Bernard and Marcian Hoff. "Adaptive Switching Circuits," *IRE WESCON Convention Record*, Part 4, August 1960a.

——. "An Adaptive 'Adaline' Neuron Using Chemical 'Memistors.'" Technical Report Number 1553–2, Stanford Electronics Laboratories, October 1960b.

Wiener, Norbert, *Cybernetics, or Control and Communication in the Animal and the Machine*. Cambridge, MA: The Technology Press; New York: John Wiley & Sons, Inc., 1948.

——. "Cybernetics in History," excerpted from *The Human Use of Humans*, New York: Houghton Mifflin Co, 1954.

Willis, William D., Jr., and Richard E. Coggleshall. *Sensory Mechanisms of the Spinal Cord, Third Edition*. New York: Kluwer/Plenum, 2004.

Wills, David. 1995. *Prosthesis*. Stanford: Stanford University Press, 1995.

Wilson, Elizabeth A. 2001. "Darwin's Nervous System: Investigating Critical and Physiological Psychologies," *Australian Psychologist* vol. 36, no. 1 (2001): 62–9.

——. *Psychosomatic: Feminism and the Neurological Body*. Durham, NC: Duke University Press, 2004.

Yuan, David. 1997. "Disfigurement and Reconstruction in Oliver Wendell Holmes's 'The Human Wheel, its Spokes and Felloes,'" in Mitchell and Snyder (ed.).

Part IV Art, Aesthetics, and Imaging

9

Final Fantasy, or the (Dis)Illusion of Computer Graphic Life*

Vivian Sobchack

The title of this chapter takes its point of departure from Alan Cholodenko's introduction to his edited anthology on animation, *The Illusion of Life*. There he tells us, "[a]nimation always has something of the inanimate about it . . . a certain inanimateness that both allows and disallows animation. . . . And this would suggest that in a sense the 'uncanny' is never not with us, for it would mark the always already returned of the ghost, the zombie, and the dead in us—lifedeath."[1] It is this "lifedeath" of animation I want to engage here, particularly, however, as it is foregrounded—but reversed—into the "deathlife" of animation (the "dis-illusion of life") in a film that is literally (or, should I say, computer-graphically) an overt exploration of "lifedeath" and "deathlife" and filled with ghosts, the dead, and zombies (the latter taking the form of "synthetic human actors"). That film is *Final Fantasy: The Spirits Within* (Columbia Pictures and Square USA, 2001), conceived and directed by Hironobu Sakaguchi who was also the originator, in 1987, of the "Final Fantasy" interactive video game franchise (that has had surprisingly little to do with the critique of the film and will not concern me here).

Touted as "the first major release to feature an entire cast of computer-generated human characters,"[2] *Final Fantasy* was also a major flop at the box office, its failure attributed primarily to its apparently confusing and emotionally unengaging narrative.[3] While its CGI attempts at "photorealism" were acknowledged as "ground-breaking," the general sense was that—for a variety of reasons, not all of them having to do with the convoluted story—the majority of reviewers and the audience found *Final Fantasy* an unsatisfying experiment that just "didn't work."[4] Why is this? One possible reason comes from a viewer posting on the Internet Movie Data Base. "Well," she writes, "it's quite simple in my opinion. Animation films are entertaining when we know that they are animation films. They are something different from reality, and all the imperfections we find in them don't count. All the holes we . . . find are filled with our imagination. . . . But, when the le of perfection of an animation film crosses the line between animation and reality, then we change our scale of values, and we judge the film by comparing it with non-animation film. [This] is when we notice . . . that there is still an abyss between a real and a virtual actor."

In what follows, then, I want to focus on "our scale of values" as it slides and shifts in our phenomenological relation to animated film in general and, as I go, to explore as well the computer-graphic conundrum—or "lifedeath"—of animation presented to us as "deathlife" by *Final Fantasy* in particular—but relevant also to several subsequent attempts to create "computer-graphic humans" such as appear in the "The Final Flight of the Osiris" episode of *The Matrix* DVD spin-off, *The Animatrix* (Warner Brothers, 2003),[5] and, most recently, in the animated feature, *The Polar Express* (Warner Brothers, 2004). Thus, although I will begin with rather grand theoretical questions about "the 'problem' of animation," most of my subsequent remarks will be of smaller scale and more piecemeal in approach—this elicited by the way in which *Final Fantasy* itself calls intense spectatorial attention to the smaller scale of the "detail" and, in piecemeal fashion, constitutes our epistemological engagement with the ontology of its CGI characters and world as a "patchwork" of heterogeneous, shifting, and incompatible values impossible to reconcile and resolve to either narrative or representational satisfaction.

The grand questions, then, are "What do we want from animation?" and "What does animation want from us?" In terms of the first question, as spectators, there are some of us—although we are in the minority—who prefer, from animation, its reflexive acknowledgment of its own (and our) "metaphysical effort": that is, the visible play or struggle of "becoming" we see in the work of the Brothers Fleischer on the one—and "lightning"—hand, and on the other, in the work of the Brothers Quay. What we privilege is animation always visibly marked by the threat of inertia and "lifedeath," animation filled with the "uncanny" gaps, starts, stops, and stutters that, as Cholodenko suggests, "simultaneous[ly] bring . . . death to life and life to death."[6] Nonetheless what a majority of spectators seem to want and value from animation is not a gloss on "metaphysical effort" but rather, as film theorist Noël Carroll has said of "trick films," "metaphysical release"[7]—that is, the vicarious playing out of the "plasmatic" possibilities for subverting and/or substituting the immutable laws of gravity and physics (and here I might add, the laws of mathematical calculation) with the contingent laws of imagination.[8] Furthermore, if one reads through reviews and IMDB viewer comments about *Final Fantasy*, it also seems that what most spectators want from animation is not abstraction, "sterility," or "coldness" but, rather, the "emotional substance," "flowing rhythm," and sense of "improvisation" of line, action, and world primarily associated with "auratic" hand-drawn pen-and-ink animation, if also apparently achieved by the "playful" CGI animations of the *Toy Story* films (1995/1998); *A Bug's Life* (Pixar, 1998); *Antz* (Dreamworks, 1998); *Shrek* (Dreamworks, 2001); *Monsters, Inc.* (Pixar, 2001); *Ice Age* (Blue Sky, 2002); and, most recently, *Finding Nemo* (Pixar, 2003) and *The Incredibles* (Pixar, 2004)—all films that, both in narrative and representation, have assiduously shied away from character "realism" of any kind but emotional.

The second question, "What does animation want from us?," seems, at least on the surface, of another order—one not only about aesthetics but also about the technology of the medium and its utopian dream of "overcoming" the inertia and stasis that would stall its freedom of expression. In this regard, film critic Anthony Lane's ironic comment about CGI and the recent *Finding Nemo* might well be said of the earlier *Final Fantasy*: "the ideal vehicle for the demonstration of an improving medium. To authenticity and beyond!"[9] And here, for its analogical—and, dare I say, analog—value, I would invoke André Bazin's famous essay, "The Myth of Total Cinema," which argues that cinema was "an idealistic phenomenon," and, very early on, was imagined by its various inventors as "a total and complete representation of reality . . . [a] reconstruction of a perfect illusion of the outside world in sound, color, and relief."[10] Thus, Bazin tells us, "[t]he guiding myth, then, inspiring the invention of cinema, is the accomplishment of . . . an *integral realism*, a recreation of the world in its own image."[11] What, we might ask, is the comparably idealistic "myth of total animation"? Is it identical to cinema's? Or, in its perceived place (rightful or not) as a "subset" of cinema, is animation guided by a similarly directed but, nonetheless, "second-order" idealism—that is, by a desire for the "perfect illusion" not of the "outside world" but of the perceptual correspondences that inform "photorealism," and, in particular, "*cinematic* photorealism," with "authenticity"?[12] But then, again, perhaps insistent on its own discrete metaphysics, animation might desire not "integral realism," but "integral *irrealism*"—and thus be guided by an alternative ideal founded on the "total" creation (not re-creation) of the world in its own image (the referent of "its" here *not* the "outside world" but, rather, "animation" itself).

The answer, then, to "What does animation want?" is certainly ambiguous—if not as downright ambivalent as it turns out to be in *Final Fantasy*. Here CGI animation wars with itself—attempting, on the one hand, to *re-create*, through the simulation of cinematic photo-realism, "perfect computer generated human beings" (this was the initial hype that later changed to "hyper-real" human beings) and, on the other, to *create* and animate an imaginative and irreal (if perceptually realistic) world in—and through—its own CGI imagination. Again, both analogically and in terms of the analog image, it is worth looking backward to make the "problem" of animation, particularly the CGI animation in *Final Fantasy*, even more problematic. Thus I will evoke *Pinocchio* (1940) and ask, again, "What does animation want?" Is it the fulfillment of Pinocchio's dream of "authenticity"—that is, to become a "real boy"? Ergo, to become "real cinema"? And, if so, "authentic" and "real" according to what criteria?—for, in the Disney feature, Pinocchio the wooden puppet and the "real boy" he is finally rendered (the word here no accident) both draw upon and are drawn through the *same* representational logic that would assert their ontological *difference*, each from the other. (That is, the puppet and the "real boy" he finally becomes emerge from the same hand-drawn line, the same "order of things," their ontological difference distinguished only through that line's angularity on the one side and its curvilinearity on the other.) The irony of this theoretical "il-logic" is that, in practice, it produces its own kind of "integral realism" from the coherence of the patently irreal, and thus a very emotionally satisfying phenomeno-logic—one not found, significantly, in what would seem to be the similar representational circumstances of *Final Fantasy*. Unlike in *Pinocchio*, as one reviewer notes, here the CGI "real boys" (and girls) give "emotionless performances" in a film with "a distinctly stiff, *wooden* flavor."[13]

The conundrum, then, is that while we never really question or find fault with either Pinocchio's "wooden-ness" or his later "fleshiness," in relation to CGI heroine, Dr. Aki Ross, and the "real boys" of *Final Fantasy*, we hear again and again—and down to the most minute details—litanies of disappointment at their rendering. This disappointment is as much about the characters' (and the film's) lack of affective charge as it is about their lack (or excess) of computer-graphic "life." Thus, a reviewer writes, "[t]he sight of these characters getting romantic is about as involving as watching two expensive mannequins kissing in a Macy's window."[14] This comment, however, might well evoke what seems a counterexample to the "wooden" characters of *Final Fantasy*—this from "The Final Flight of the Osiris" segment of the later (and more technologically advanced) *The Animatrix* DVD. The sight of two exquisitely rendered computer-graphic human characters "getting romantic" in the foreplay of a seductive martial arts engagement is quite erotically involving. However, it is precisely the rendering of the "expensive" mannequin-like qualities of these characters that touches us sensually. Eros emerges from a purposeful heightening of an "air-brushed" photo-realism that gives us extremely high resolution and a low level of detail. The bare architectural space, the ultra-sheen of light on perfectly smooth and pore-less skin, the intense saturation of red lip color and fabric, the reflexive sense of calculated "play"—all assert that what we are watching is not "really" real. This is not the assertion that *Final Fantasy* wants to make—indeed, insisting on and rendering a level of "realistic" detail in its human characters that is, unfortunately, "dead serious."

If we're willing to accept animation's "irreality" in its *representation* of human beings (a tradition that attaches not only to Disney but also to *Final Fantasy's anime* forbears and to the CGI "humans" of *Toy Story* and *Monsters Inc.*), why, then, is there such a problem with accepting *Final Fantasy*'s "hyper-real" *simulation* of human beings? Clearly, in this particular instance, the problem has something to do with a significant "incompatibility" that emerges between the different scale of values and expectations attached to representation and simulation when a particular kind of utopian "myth of total animation"—as the IMDB viewer wrote—"crosses the line between animation and reality." And, indeed, in *Final Fantasy*, this "incompatibility" can be found everywhere: not only in the differing epistemological demands made on the spectator by the contrary semiotic status of the animation but also in the irreconcilable rhetoric that surrounds the film and in the incompatible premises of the narrative that serves as their perverse allegory.

Let me touch first on the narrative. This focuses on "humanity's" need to defeat a highly successful "alien invasion" of a futuristic and decimated Earth by creatures (later found to be the "phantoms" of beings from a devastated planet) who, for their energy, "suck the life" out of (CGI) human beings. (In this regard, one reviewer notes both a basic ontological and epistemological incompatibility between the narrative and its CGI realization when he asks: "Is this some kind of sick joke, or at least an in-joke? A virtual human plays one of the few real humans left,"[15] while another highlights the irony underlying—or is it over-laying?—the fact that these CGI "human facsimiles have devoted every pixel of their beings to preserving life as we know it."[16]) One faction of the remaining are "mechanistically" inclined "humans" and wants to blast the red-orange and multi-formed aliens with a "Zeus cannon"—which may well end up also destroying the Earth. The other more "animist" faction, led by scientists Dr. Aki Ross and her mentor Dr. Sid, believes the invasion can be stopped without planetary threat if a certain small number of remaining and representative Earth "spirits" (eight to be exact) can be collected and amassed to form a "wave" of positive and life-giving energy that will overcome and transform the alien phantoms and their angry planetary spirit—or "Gaia"—into a more "pacific" (and literally blue) form.

In short, the narrative serves up a simplified version of James Lovelock's "Gaia theory"; that is, both the Earth and the ghostly remnants of the destroyed alien planet are regarded as an organic, animate, and purposeful totality. As noted, however, the use of CGI animation confuses and makes ironic this narrative opposition between "mechanists" and "animists," between the technological and the organic, between what it means to "suck the life" out of human beings and to "inspire" or "animate" them with an *élan vital*. Here, on the one hand, quite literally and computer-graphically, *in toto* and in the end, the "organic" phantom spirits of both alien and Earth "Gaia" co-constitute what, for Sergei Eisenstein, was the essence of film animation itself: "plasmaticness," "the potentiality of the primal plasma, from which everything can arise."[17] Nonetheless, on the other hand, the graphical "plasmaticness" of this Earth- and life-affirming "modulation, transitivity and the continuous coming into being of images" of which Eisenstein speaks, is, paradoxically, accomplished by—and allegorizes—the same CGI that would and does, in *Final Fantasy*, appear to "suck the life force" from organic human representations (and, more significantly, from the representation of organic humans)![18] Here, while CGI gives us the "infinite changeability, modulation, and transitivity" of the "plasmatic" in its passionate orange-red and pacific cool-blue visualizations of Gaia and its phantom spirits, in its human forms (however "diverse" in its multi-ethnicity), the "plasmatic" is bent on "being" a "fixed entity," and the rigidity of its CGI algorithms and geometric forms forestall any sense of organic or improvisational "becoming." As Keith Broadfoot and Rex Butler put it (years before the release of *Final Fantasy*), "[a]nimation corresponds, as Eisenstein implies, to a condition of organic production. Animation is not the result of the lifelessness of a mechanical reproduction but is itself life-giving. What, however is the form of life it creates? Is it a life that we know?"[19] And, in relation to *Final Fantasy*, the answer is that it is not "a life that we know"—or, at least, not quite—for, as one reviewer notes of his experience of the film: "You miss the unchoreographed wayward tilt of a head or an improvised double take; the unpredictable physical chemistry of actors that computer science hasn't mastered."[20] What emerges is not organicity but, quite visibly, "the lifelessness of mechanical reproduction."

We can see, then, that the foundational *visual* premises of *Final Fantasy*—that is, the computational graphic grounding of the film's "humans"—are semiotically incompatible with the film's *narrative* argument: that what animates, inspires, and gives life is organic, not technological, production.[21] Faced by this deep structural incompatibility—one that affects both the process of our "reading" and our emotional engagement with the film—we watch *Final Fantasy* in a state of what, in another context, Scott Bukatman has called "epistemological hesitation."[22] What might this hesitation mean in terms of cinematic effect?—particularly given that, at base, *all* cinematic imagery is technologically achieved? Again, why do we have problems with *Final Fantasy* and not with *Pinocchio*? Peter Wollen, in *Signs and Meaning in the Cinema* is helpful here. He tells us in relation to all cinema and its iconicity: "The iconic sign is the most labile; it observes neither the norms of convention [the symbol] nor the physical laws which govern the index." Thus, given its foundational iconicity, "[cinematic] depiction is pulled toward the antimonic poles of photography and emblematics. Both these undercurrents are co-present in the iconic sign; neither can be conclusively repressed."[23] If this antimonic tension is an undercurrent in *all* cinematic iconicity, nonetheless, *one* of its poles—the photograph or the emblem—usually tends to dominate in consciousness, regulating our phenomenological perception not only of genre but also of the representational "ontology" of what we see on screen. Here we might safely generalize that photographic iconicity (in which resemblance is informed by existential indexicality) is dominant in live-action, photo-realistic film whereas emblematic iconicity is dominant in animation (in which resemblance is condensed to basic structures and functions verging on the symbolic). Both, in their own semiotic modality, are able to create, respectively, an "integral realism" or "integral irrealism" that immerses the viewer in a coherent textual world. Thus, in *Pinocchio*, where both the wooden puppet and the "real boy" are hand-drawn and posited in the integrity of an irreal world, there is little antimonic tension between an iconic physical resemblance that is not photographically indexical and an iconic emblematics that is structurally symbolic. In *Final Fantasy*, however, this antimonic tension between photographic iconicity and emblematic iconicity is not only heightened but also ambivalently "double-valued." The film attempts to represent indexical, photo-realistic "human characters" who inhabit a world that is created as emblematic, symbolic, irreal—and thus it visibly *equivocates* within what, narratively and semiotically, is meant to achieve the integration attached to "perceptual realism." That is, without textual *purpose* for its double system of representation (as is the case in a film such as *Who Framed Roger Rabbit?* [1988]), *Final Fantasy* solicits from the viewer two different and incompatible modes of epistemological apprehension and aesthetic judgment. Finally, then, what we pay most attention to (or what most distracts us) is the oppositional—or antimonic—struggle between resemblance and emblematics, and this phenomenological attention (or distraction) reveals, even if only latently, the "illogical" or contradictory foundations of the sign system, and, quite blatantly, the "illogical" and contradictory premises of its narrative argument.[24]

The deep structural incompatibility that exists between *Final Fantasy's* narrative rhetoric and the semiotic mode of its visual representation also finds expression in the deep structure of the ambivalent rhetorical discourse that has surrounded not only the film's promotion and reception but also the perceived "success"—or not—of its then "state-of the-art" CGI renderings of human characters. In this regard, rhetorical theory (not often used in film or visual studies) is particularly illuminating about what constitutes the "double bind" of animation that sits on the fence between iconicity and emblematics, between representation and simulation. Chaim Perelman and L. Olbrechts-Tyteca's seminal *The New Rhetoric: A Treatise on Argumentation*, in a chapter aptly titled "Arguments Based on the Structure of Reality," provides us with a rhetorical—and structural—parallel to "the myth of total cinema" and "total animation" called the "argument of unlimited development."[25] Such an argument, we are told, "insist[s] on the possibility of always going further in a certain direction without being able to foresee a limit to this direction, and this progress is accomplished by a continuous increase in value."[26] This said, however, it bears emphasis that the argument of unlimited development can progress in *either* direction—that is, not just toward extremes of "enlargement" but also toward extremes of "diminution" (on the one hand, *Fast and Furious* [2001] and then *2 Fast 2 Furious* [2003] and, on the other, *Dumb and Dumber* [1994] and then *Dumb and Dumberer* [2003]).[27]

Unlimited development is, of course, the argument that accompanies the industrial (if not aesthetic) promise of CGI, particularly as it pursues what has been regarded (according to one IMDB viewer) as "the Holy Grail of CGI creators": "a realistic, living, breathing and acting human being." Hence statements from Sakaguchi and Square, made in 1999, about the then state-of-the-art CGI planned for *Final Fantasy*: "It is beyond anything that has ever been done before," and "We are creating 'perfect computer generated human beings.'"[28] (By 2001, this was changed to: "We can do anything we want now. In this type of filmmaking, there are no limitations." The "perfect computer-generated human beings" were now described, however, as "'hyper' real, not 'photo' real."[29])

Of particular relevance to *Final Fantasy* is the fact that the argument of unlimited development can be used not only to project "limitless" extension but also "to *depreciate* a state or situation which *could* have given satisfaction but to which a more favorable condition is supposed to be able to follow."[30] Hence, the following viewer comments on IMDB: "A level of CG detail that really has never been seen before, but it isn't completely perfect"; the terse dictum, "[d]on't ever make a movie about real people if you can't portray real people"; and "[There was a lack of] chemistry and warmth between the characters . . . [but] to dismiss computer animation as inferior to live action is to not give the spirit of technological advancement its due. I have no doubt in my mind that someday we will not be able to the tell the difference between computer-generated people and real people without serious evaluation."

Perelman and Olbrechts-Tyteca also tell us that, confronted with an argument of unlimited development (whether in the interest of an increase or decrease of value), listeners (in this case, viewers) "are often more interested in the value which such argumentation confers on *certain terms which fall short of the ultimate term*, but are really the *center* of the debate, than they are in the ultimate, always receding term in a given direction."[31] Throughout the discourse surrounding *Final Fantasy*, while the "ultimate, always receding term" may be an ideal "realism" indistinguishable from "cinematic photorealism," the term that "falls short" but is central to the debate is "*hyper-realism*"—a realism that is, at one and the same time, "*not real enough*" and "*too real*." Thus, on the one hand, IMDB viewers present extraordinarily detailed litanies of CGI failures: "errors" of movement, wooden facial expressions, poor lip-synching and the like. On the other, they also express disappointment in comments such as "[s]omething about the animation . . . makes the whole thing a bit too perfect." Particularly ironic in this context, then, is Lev Manovich's point in *The Language of New Media* that "although we normally think that synthetic photographs . . . are inferior to real photographs, in fact, they are *too perfect*. But beyond that we can also say that paradoxically, they are also *too real*. . . . From the point of view of human vision, [what we see] is hyperreal. And yet, it is completely realistic."[32]

Final Fantasy can't win for losing—and, given the semiotic and ontological incompatibilities and contradictions of its visual and narrative premises as well as the discursive field the film's CGI animation generates, this is no mere accident. What is perhaps most fascinating about the argument of unlimited development in the context of *Final Fantasy* is that, as Perelman and Olbrechts-Tyteca explain, the argument *contains the premises for its own rebuttal*. That is, the argument's "refutation . . . lies in the statement that it is impossible to go indefinitely in the direction indicated, either because one encounters an *absolute* or because one ends up with an *incompatibility*." Furthermore, the argument is always also in "danger of appearing *ridiculous* as a result of incompatibility with values one is loath to give up."[33] This awareness of the absolute and ridiculous extension of *Final Fantasy*'s visual argument and its two "incompatible" ontologies (or "myths of total animation") are rendered strikingly clear in two reviewer "rebuttals" of the film. One writes, "[t]he photo-realism of the humans doesn't always work, and frankly that is a relief. . . . If the illusion always worked, how would we know they weren't really human actors, and some joker just told us they were computer generated?,"[34] while the other notes: "If computer animation could ever get to the point of creating an entirely believable human representation (and that is a dubious possibility), then, as animation, it will have entirely missed the point. If you can create a computer being . . . unmistakable from a real human being, why not just use a real human being? Is it animation any longer, except in the most technical sense?"[35] In sum, the argument of unlimited development always provides the premises for its own rebuttal. In the case of *Final Fantasy*, CGI animators, the film itself, and the film's viewers are all caught up in a version of the argument of unlimited development—the misguided (and doomed) dream of a "totally" "photorealist" animation that constitutes, through its tiniest failures and greatest successes, the grounds for its own refutation and rebuttal.

And, here, in relation to the conflicting values entailed in the possibilities of both "enlargement" and "diminution" inherent in the argument of unlimited CGI development, we might turn again to the troublesome term in *Final Fantasy* "that falls short" of the "ultimate" term: "photorealism." As mentioned previously, that term of semiotic disappointment is "hyperrealism." Paradoxically, its disappointments have to do at one and the same time with both the enlargement and diminution of an ideal "photorealism." That is, the hyper-real is simultaneously "too much" and "not enough." To understand this doubled—and contradictory—semiotic (as well as rhetorical) effect, it is particularly useful to consider the "doubled" nature of the iconic sign—this dualism generally forgotten since the advent of photography and cinema, although, as I've indicated earlier, Peter Wollen points to it in his *Signs and Meaning in the Cinema* when he writes of the iconic sign in cinema as being pulled between photography and emblematics.

According to Charles Saunders Peirce, the iconic sign "represents its object mainly by its similarity to it; the relationship between signifier and signified [sign to referent] is not arbitrary but one of *resemblance* or *likeness*." Iconic signs, however, can be "divided into two sub-classes: *images* and *diagrams*. In the case of images, 'simple qualities' are alike; in the case of diagrams, the 'relation between the parts.'"[36] Until recently, however, because of the historical dominance of photography and cinema, the iconic sign generally has been seen only in terms of the "image," with its resemblance to its referent based on simple—that is, physically visible—correspondences between the two. The advent of computer-graphics has troubled this one-sided vision. In its attempts to achieve the "simple" resemblances of photo-realism, computer-graphic rendering, by its very digital and computational nature, cannot avoid basing its "imagistic" resemblance to its referent on the foundational *relation between its parts*—in other words, although it may aim for simple *physical* resemblance, computer-graphic representation always originates in and makes visible to some degree (either as an underlayer or overlayer) *structural* resemblance. Thus, unlike photography or cinema, when the computer-graphic attempts photo-realism, its iconic resemblance to its referent is "doubled": we are given both an *image* and a *diagram*. This is, in effect, the "too much" of the hyper-real. It stands as *both* the image of a referent and the diagram of that referent (the latter also revealing the structure—or emblematics—of the former).

Furthermore, as indicated earlier, this "doubling" also has the effect of giving us "too little." Against the normative and single-valued image of the "photoreal," "too much" sign can result in a "too little" illusion of life—that is, of live-action that does not always reflect back on its own structure, of human temporality that is not always experienced in a reflective mode. The "live-action" human simulations of *Final Fantasy*, however, seem embodied in a reflective mode even as their "motion capture" movement would suggest otherwise.[37] These algorithmic humans are, to quote Louis Menand (writing in another context), "always slightly overspecific." And, he continues, "[i]t's realism from an instruction manual: literal, thorough, determined to leave nothing out. But it has a vaguely irreal effect. . . . There is something animatronic about them. They are simulations of humanness figures engineered to pass as 'real.' What it means to be really human is always a problem for them. Can you just copy . . . people? Would that take care of it?"[38]

"Too much" sign puts a conceptual drag on improvisational live-action and physical materiality—even as that drag may be nearly imperceptible. As Akira Lippit writes, "[t]he intersection of a *material condition* with a *conceptual value* . . . makes hyperrealism possible only as an after-thought, an après-coup that flashes backward a representational structure . . . only possible as something that has (just) ended."[39] Our normative sense of the photo-real and live-action as "just right" becomes, in its CGI simulation, stalled in its diagrammatic grounding and reflexivity—the iconic sign as both physical likeness and conceptual diagram, "a refusal of the distance that separates *things* from *ideas*." Thus as Lippit says of the economy of the hyper-real: "*Synthesis* and *analysis* at once and one after the other." Too much here becomes too little. Enlargement and diminution. Something will always be wrong—if fascinating—with this picture.

Which, in conclusion, brings me to Dr. Aki Ross's hair. Writing about the "too real" nature of CGI animation although unaware of the argument of unlimited development, Manovich points to (among other things) CGI's capacity for "unlimited resolution and an unlimited level of detail."[40] On the one hand, this increase in resolution allows us to see more of less and, on the other, the increase in detail allows us to see less of more. In a sense, then, the extremity of CGI's "unlimited" resolution and detail constitutes a "hyperbolic" effect in relation to what we perceive (conventionally or otherwise) as "the structure of reality." Thus the "hyper-reality" of the human characters in *Final Fantasy* stands as a visual articulation of the major *rhetorical trope* that Perelman and Olbrechts-Tyteca identify as central to the argument of unlimited development: namely, *hyperbole*.[41] The hyperbolic nature of *Final Fantasy's* CGI's resolution and detail, in particular relation to its "hyper-real" human characters, demands from its viewers a heightened and hyperbolic form of judgmental attention. That is, whereas, in our normative mode of attention, we might have only looked at either cinematic photo-realism or irreal CGI animation "in general" (as we do with the non-human representations of the film), the CGI resolution and detail of *Final Fantasy's* human characters solicit a directed and intensified *scrutiny*. As one admiring IMDB viewer put it: "The detail to the faces and hair can't be denied. You could see every spot of stubble . . . every mole and freckle, wrinkles, eyelashes, even moisture in the eyes if you look closely enough. . . . If you pay attention, there's another world of detail to appreciate here, you just have to be watching for it." In either a live-action photo-realist film or in an animated fantasy that is metaphysically "released" from "the structure of reality," our scrutiny and judgment would not be activated at the small-scale level that they are in *Final Fantasy*. The film's own semiotic and rhetorical construction as well as our distancing "epistemological hesitation" at the film's dual ontology ask us to scrutinize—to either admire or judge wanting—*smaller and smaller semantic units of the image*: to wit, every one of the 60,000 hairs on Aki's head. Thus as another admiring IMDB viewer unselfconsciously writes: "Subtleties such as hair movement that are generally lost were not so with this film." It is not for nothing that, speaking of the CGI "ideal of photorealism" and the difficulty of "creating fully synthetic human actors," Manovich singles out a 1992 SIGGRAPH paper titled "A Simple Method for Extracting the Natural Beauty of Hair."[42]

Nonetheless not every viewer was pleased at having their attention directed to the micro-level of the putatively "natural" beauty of CGI hair. "When I read that animating Aki Ross's 60,000 strands of hair gobbled up one fifth of the entire budget for *Final Fantasy* I was just amazed," says one. "How can they spend so much money on what is really nothing more than an *irrelevant detail*, and in comparison give so little thought to things like dialogue and storyline?" Yet another viewer finds Aki's hair "really, really annoying, like someone must have followed her round and cyberbrushed it every five minutes. It actually detracted from the movie because it was *so irritatingly noticeable*." And yet another notes that, in rendering time, Aki had "something like $35 million of hair care," and that "not even Jennifer Aniston has worried that much about her hair." Furthermore, given the level of scrutiny demanded by the film's rhetorical and semiotic structure, even the hair is doomed to photo-realist failure: "There was overemphasis on the main character's hair, but at the same time," says one viewer, "not one strand goes out of place. How could someone with straight limp hair, when faced down, still have it fall toward the shoulder instead of covering the face?"[43]

The great irony here, of course, is that the detail is in the observer. Aki Ross's hair is *not* constantly foregrounded in the film, but that is what we look at. We look at the characters' skin, their liver spots, their freckles and wrinkles, and—given the epistemic call—watch the physics of clothing, movement, and (oh, yes) the way hair falls. In reverse—but in concert with the CGI animators—we, as viewers, also get caught up in the film's argument of unlimited (photo-realist) development. And thus we spend a large portion of our own time (dare I pun?) "rendering" judgment and "splitting" ontological hairs. Unfortunately, then, our attention—and that of the filmmakers—is greatly misdirected from a focus on "the illusion of life" to the "dis-illusion of life."

Certainly there are those who would argue with me and, in the name of unlimited computer-graphic development, point to *Final Fantasy's* attempts at simulating human beings as dated. However, one only need read through various reviews of the much more recent (and often quite exquisite) *The Polar Express* for evidence that—even with "performance capture" using 152 "reflective pellets, applied to each actor's face" to more closely simulate "the illusion of life"[44]—computer-graphic humans will always be closely scrutinized and ultimately found wanting.[45] One reviewer imagines that "most moviegoers will be . . . concerned by the eerie listlessness of [the] characters' faces";[46] another suggests the protagonist children "look slightly unnerving with their pod-people eyes";[47] another mentions "the waxy animated faces" as "off-putting" and "a bit like watching mannequins act";[48] and yet another points to the characters' ranging "from astonishingly lifelike to creepily waxwork-esque."[49] Indeed, it appears that "the largest intractable problem . . . [was] that the motion-capture technology used to create the human figures . . . resulted in a film filled with creepily unlifelike beings." Comparing these "beings" unfavorably with the non-human Gollum in *The Lord of the Rings* [2001], a reviewer writes: "With their denatured physiognomy, the human characters in 'Polar Express' don't just look less alive than Gollum; they look less alive than the cartoon family in Brad

Bird's 'Incredibles.' It's baffling that Mr. Zemeckis . . . didn't see how dead these animated characters look."[50]

Thus watching the computer-graphically simulated humans in *Final Fantasy* (as well as those in *The Polar Express*), we are most certainly more in the realm of animation's "deathlife" than its "lifedeath." Certainly, this inflection might well be recognized, used explicitly, and successfully exploited for aesthetic and narrative effect—that is, for its particular "disillusion of life." However, insofar as that all computer-graphic animation wants is to create "real boys," it has reached the semiotic and rhetorical limits of its supposedly "unlimited" development, and its effects will continue to disappoint—serving, indeed, as their own rebuttal to the impossible "final fantasy" of realizing a "perfect" simulation of cinematic photo-realism.

Notes

*The author and editors are grateful to John Libbey for allowing the chapter to be reprinted from *Animated Worlds*.

1. Alan Cholodenko, *The Illusion of Life: Essays on Animation* (Sydney: Power, 1999), 28–9.
2. *Variety*, July 9, 2001: 20.
3. Other reasons that might underlie the film's failure to engage its mainstream audiences (particularly in the United States) are tied to genre. In the first instance, at the time of the film's release, most of that audience was not familiar with either the narrative or visual conventions of Japanese *anime*. (The film was extremely successful in Japan where its ecological concerns coupled with apocalyptic science fiction and a female protagonist were hardly novel.) In the second instance, as David Surman has pointed out, "the CGI feature film as a genre is highly culturally specific, albeit specific to the hegemony of the North American animated cartoon tradition" (dominated by a Disney "realism" that was not photoreal). See David Surman, "CGI Animation: Pseudorealism, Perception and Possible Worlds," unpublished thesis (Autumn 2003).
4. The film cost $145 million to make and lost the new motion picture division of Square $20 million. (That division was subsequently closed down.)
5. Warner Brothers Home Video is the DVD distributor. The production company for "The Final Flight of the Osiris" segment was Square, USA.
6. Cholodenko, op cit., 29.
7. Noël Carroll, "Notes on the Sight Gag," in *Comedy/Cinema/Theory*, ed. Andrew Horton (Berkeley: University of California Press, 1991), 25.
8. It is worth noting here that narrative animated films provide us "metaphysical relief by doing both"—that is, *subverting* physical laws within a world perceived by the viewer as physically akin to our own (hence the delight when a character walks off a recognizable representation of a cliff and doesn't fall); and *substituting* physical laws with created ones (the film creates a consistent physics of its own).
9. Anthony Lane, Review of *Finding Nemo*, *The New Yorker*, June 9, 2003: 109.
10. André Bazin, "The Myth of Total Cinema," in *What is Cinema*?, trans. Hugh Gray (Berkeley: University of California Press, 1967), 20.
11. Ibid., 21.

12. On "perceptual realism" and "correspondence," particularly in relation to photorealism and CGI, see Stephen Prince's seminal essay (originally published in 1996), "True Lies: Perceptual Realism, Digital Images, and Film Theory," in *Film Quarterly: Forty Years—A Selection*, ed. Brian Henderson and Ann Martin with Lee Amazonas (Berkeley: University of California Press, 1999), 193–411.
13. Sean Axmaker, "Review of *Final Fantasy: The Spirits Within*," *Seattle Post-Intelligencer*, July 11, 2001.
14. Kenneth Turan, "'Final Fantasy' Sticks to Game Plan," *Los Angeles Times*, July 11, 2001.
15. Bob Graham, "Review of *Final Fantasy: The Spirits Within*," *The San Francisco Chronicle*, July 11, 2001.
16. Stephanie Zacharek, "Review of *Final Fantasy: The Spirits Within*," Salon.com.
17. Sergei M. Eisenstein, *Eisenstein on Disney*, ed. Jay Leyda and trans. Alan Upchurch (London: Methuen, 1988), 45.
18. There has been another allegorical reading of the film which sees the opposition set up between Earth's Gaia and the Phantom Gaia "as an allegory for the rivalry between analog cinema/analog media and digital cinema/new media . . . as well as for the digital's obligatory detour through the analog" (Livia Monnet, "Invasion of the Movie Snatchers: Mimesis and the New Uncanny in *Final Fantasy: The Spirits Within*," unpublished manuscript).
19. Keith Broadfoot and Rex Butler, "The Illusion of Illusion," in *The Illusion of Life*, 273.
20. Elvis Mitchell, *The New York Times*, July 11, 2001. *Final Fantasy* did use "motion capture" meant precisely to inform the CGI simulations with the "organic" quality we sense in "the unchoreographed wayward tilt of a head" or "improvised" movement—and yet the movement still seems "wooden" to most. This is a function of the disjunctive relation between the plasmaticness of movement and the algorithmic rigidity of simulation which, it seems to me, is phenomenologically "felt" by the spectator.
21. Of interest here is the relation between semiotics and rhetoric, particularly in relation to the iconic sign. Umberto Eco views a variety of codes as operative in the iconic sign, including "perceptual codes (the domain of the psychology of perception)" and "iconic codes proper, subdividable into (a) figures, (b) visual premises, and (c) visual arguments" (Robert Stam, Robert Burgoyne, and Sandy Flitterman-Lewis, *New Vocabularies in Film Semiotics: Structuralism, Post-Structuralism and Beyond* [New York: Routledge, 1992], 31).
22. Scott Bukatman, "Online Comics & the Reframing of the Moving Image," unpublished manuscript, forthcoming in *Matters of Gravity* (Durham, NC: Duke University Press, 2003).
23. Peter Wollen, *Signs and Meaning in the Cinema, New and Enlarged* (Bloomington: Indiana University Press, 1972), 152.
24. This tension and felt contradiction between resemblance and emblematics also emerges in a tension between the use of indexical "motion capture" to figure human movements and the algorithmic emblematics the simulated human bodies that move.
25. Ch. Perelman and L. Olbrechts-Tyteca, *The New Rhetoric: A Treatise on Argumentation*, trans. John Wilkinson and Purcell Weaver (Notre Dame: University of Notre Dame, 1971), 287–292.
26. Ibid., 187.

27. References are to *Dumb and Dumber* (Farrelly Bros., 1994) and *Dumb and Dumberer* (Troy Miller, 2003), and to *The Fast and the Furious* (Rob Cohen, 2001) and *2 Fast 2 Furious* (John Singleton, 2003).
28. Tim Ryan, "It's reel Fantasy," *Honolulu Star Bulletin*, January 15, 1999.
29. Barbara Robertson, "Reality Check," *Computer Graphics World* (August 2001), 24.
30. Perelman and Olbrechts-Tyteca, 288.
31. Perelman and Olbrechts-Tyteca, 290.
32. Lev Manovich, *The Language of New Media* (Cambridge: MIT Press, 2000), 202.
33. Perelman and Olbrechts-Tyteca, 289.
34. Bob Graham, Review of *Final Fantasy: The Spirits Within, San Francisco Chronicle*, July 22, 2001.
35. David Luty, "Review of *Final Fantasy: The Spirits Within*," *Film Journal International* (August 2001), 74.
36. Wollen, 123.
37. Given this sense of reflection underlying unreflective "live-action" movement, there is one moment in *Final Fantasy* that seems to me semiotically coherent. It occurs when, after a group of characters strip down to decontaminate after a battle, one of the women is shown in the corner of the frame slightly flexing one of her biceps for her own admiring gaze. This somewhat "privatized" reflective moment enacted through bodily reflexivity is, perhaps, the most "authentic" and thus seemingly "spontaneous" moment in the film insofar as it recognizes its own constructive premises. (It is also the only witty moment in the film and, not foregrounded, seems a private "in-joke" for the CGI animators.
38. Louis Menand, "Something about Kathy," *The New Yorker*, March 28, 2005, 79. Menand is writing about several novels by Kazuo Ishiguro.
39. Akira Mizuta Lippit, "Extimacy: Chronography and Hypperrealist Films (1963–71)," unpublished translation of "Extimité: Chronographie et Cinéma Hyperréaliste (1963–71)," in *Hyperréalism, USA: 1968–1975*, ed. Jean-Claude Lebensztejn (Paris: Les Musées de Strausborg, 2003), 79–97.
40. Manovich, 202.
41. Perelman and Olbrechts-Tyteca, 291.
42. Manovich, 194. Reference is to K. Najyo, Y. Usami, and T. Kurihara, "A Simple Method for Extracting the Natural Beauty of Hair," *Computer Graphics* 26.2 (1992): 111–20.
43. In terms of different cultural response to the film in the United States and Japan, it is worth noting here that, as Gilles Poitras writes, "Hair is often an important stylistic element in anime. . . . A character's hair is one way of making him or her instantly recognizable. . . . The movement of a character's hair is quite noticeable in anime." See Gilles Poitras, *The Anime Companion: What's Japanese in Japanese Animation*? (Berkeley: CA: Stone Bridge Press, 1999), 62. (I thank David Surman for this reference.)
44. Ron Magid, "'Polar', Bared," *Premiere* (November 2004), 54.
45. The film was initially disappointing at the box office, taking in only $23 million dollars when it opened theatrically. However, a 3-D IMAX version released somewhat later picked up the slack—but it also changed spectatorial focus from the details of waxwork faces to immersion in dimensional kinetic effects. See Raymond Finn, "Big Picture Thinking," *Entertainment Weekly*, January 14, 2005, 22.
46. Manohla Dargis, "Do You Hear Sleigh Bells? Nah, Just Tom Hanks and Some Train," *New York Times*, November 10, 2004.

47. Lisa Schwarzbaum, "Review of *The Polar Express*," *Entertainment Weekly*, November 19, 2004, 63.
48. B. D. , "Film Notes: *The Polar Express*," *The New Yorker*, November 22, 2004, 32.
49. Steve Daly, "Claus and F/X," *Entertainment Weekly*, November 12, 2004, 44.
50. Dargis.

10
A Scene in a Digital Library: Imaging Literature

Chris Burnett

Introduction—Fox Talbot's 'A Scene in a Library'

What is it for a book to be part of the electromagnetic spectrum? My inquiry into the odd relations between imaging technologies and textuality begins with this kind of question as I look at an illustration of the entire electromagnetic spectrum from an old manual on color photography. Regions of the spectrum are intersected with exploded images of tangible artifacts and experiences of modern life. An atomic bomb blast, an X-rayed bone, the colors of visible light, an infrared light warming a tiled bath room, a vintage 1970s color TV set, FM radio, electric towers marching through an industrial landscape. Most items suggest the various registers, media, and scales of communication in tangible terms and hint at the twin duties of imaging technologies: first, to extend vision, the eye of knowledge, beyond light into the hither and yon reaches of the electromagnetic spectrum; and, second, to make sensible and more concrete the numeric abstractions and measurements of engineering and science.

But the illustration lacks reference to literature, and any account of imaging technology concerned with the modern conditions of knowledge would be inadequate without it. For centuries, of course, literature and books have formed the primary vehicle of knowing and the storehouse of learning. As an "image" of knowledge, visual allusions to literature in the graphic form of books have figured in the pictorial field of art at many periods for a variety of purposes. Meyer Shapiro used three medieval mosaics to summarize the different ways of inserting written, scriptural words into pictures.[1] Figures of saints may hold out their open books directly for the literate eyes of an external reader. Alternatively, the figure holding the book might invert the writing to accord with his own viewpoint as an internal viewer. Or finally, the inset books might bear a set of mysterious marks, viewable, but legible to no one. The alternatives suggest the odd accordance and disparity of graphic manifestations of books at play within and around images and their viewpoints. Perhaps my original question on seeing literature in terms of imaging and the electromagnetic spectrum bears little sense given the logical disparity between literature, composed of arbitrary signs, and the analogical sweep of the spectrum. The discrete marks of writing do indeed exist and work at many points along the electromagnetic spectrum, even as there are wormholes of intersecting, conflicting meanings cutting across any point within that range. Having their own prickly logic, words just do not fit easily into the continuous mold of the spectrum.

In a similar combination of fitting and not fitting, the graphic reality of literature has its "show down" with emerging imaging technologies. In parallel, both literature and imaging may be taken as extensions of sensible experience, knowledge, and inscriptive practices. Our use of media and information technologies in everyday life is cross-threaded with the imaginative extensions of literature and the literary inclusion of imaging. In the opening sequence of Disney's classic animation *Pinocchio*, an ornately bound book opens to become a window through which the planographic camera takes our vision on a literary roller-coaster ride. Using technologies of simulation, we are now seeing writing on multi-windowed computer screens interwoven as a tapestry of juxtaposed frames and narrative threads. After centuries of the image locked within and controlled by literature's conditions of textuality, the image is breaking out, and imaging technologies have in part exploded the hole in text for their escape. After all, imaging was first a matter of the library as we will now see.

To encounter this first scene, we need to go back to the seminal work of the invention of photography, *The Pencil of Nature*, by William Henry Fox Talbot.[2]

This serially produced publication explains and demonstrates by example the variety of uses of photography, uses that were entirely new to the world in 1844. In twenty four calotype plates with commentary, Talbot details the peculiarities and possibilities of photography. Many aspects of the photographic image, so commonplace that we now take them for granted, are wonderfully eccentric and fraught with potential: an artist can draw a picture of several figures as easily as he might portray one; a "negative" might serve as well as a positive image for purposes of representing pattern, such as lace; a variety of different aesthetic effects may be derived from diverse angles and points of view; a photograph represents a specific time and place; a chimney pot may be as important as the *Apollo Belvedere*; the eye may take an imaginative journey out of the frame down a road not actually shown in the picture. But, Talbot's most enigmatic and imaginative journey occurs in Figure 10.1, "A Scene in a Library," in which he prophetically speculates on photography as an imaging technology (Figure 10.1). Mysteriously, he does so as an encounter with a library. So here is a literature of the electromagnetic spectrum. Significantly, we find imaging

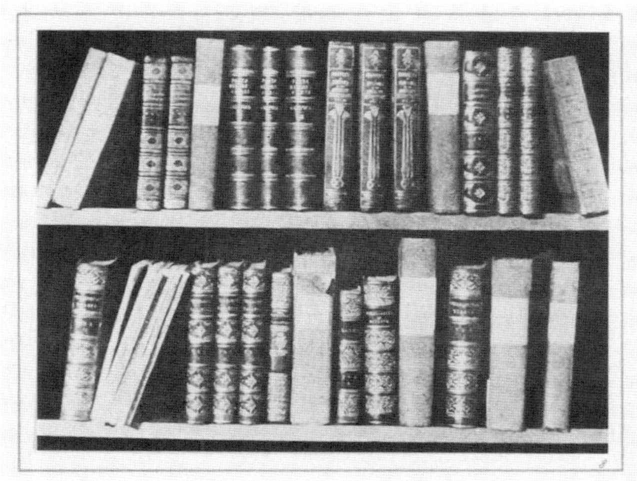

Figure 10.1 "Plate VIII. A Scene in a Library" in *The Pencil of Nature* by William Henry Fox Talbot. Courtesy George Eastman House.

and literature together at the threshold of speculation about photography's impact on knowledge and its use as a poetic medium.

Talbot opens by framing his commentary, separate from the others, as a "rather curious experiment or speculation." He points out the existence of the spectrum and that experiments have shown that the violet end produces the principle effect on light-sensitive paper. Moreover, he claims that certain "*invisible rays* which lie beyond the violet, and beyond the limit of the spectrum" can be revealed. He conjures for his readers an apartment filled only with these invisible rays. "If there were a number of persons in the room, no one would see the other: and yet nevertheless if a *camera* were so placed as to point in the direction in which any one were standing, it would take his portrait, and reveal his actions." His thought experiment becomes distinctly allegorical with the following line that hints at photography's use as an investigative tool to extend knowledge. "For to use a metaphor we have already employed, the eye of the camera would see plainly where the human eye would find nothing but darkness." And if that is not enough, he adventures further into layers of speculation by imagining that "this peculiarity is somewhat too refined to be introduced with effect into a modern novel or romance; for what a *denouement* we should have, if we could suppose the secrets of the darkened chamber to be revealed by the testimony of the imprinted paper." Talbot's thought experiment, as a metaphoric imagining, poses photography as an imaging technology, which is itself likened to a literary climax and unraveling of a plot line. Imaging reveals secrets of darkened chambers, and photography imprints "the testimony" as a new kind of literature.

The crosscurrents between the imaginative text printed on the recto of the book with the calotype tipped on the verso is telling. The picture is a direct rendition of the bookends of two shelves of books. This juxtaposition marks a departure from the usual straightforward use of words to comment on, describe, and elucidate the image. Nowhere in the body of the commentary does Talbot mention a library. "A Scene in a Library" results from the imaginative stitching of commentary and illustration, word and image. Parallel to and interlinked with each other we find literature at the technical discovery of the electromagnetic spectrum. Even more prophetically, this is a new hybrid literature ready to blow images loose; not encase them in worded texts. Talbot places his speculations on photography as a novel form of imaging, an imaging technology, in the domain of literature. The encounter raises a host of questions and suggests links to the contemporary proliferation of imagetexts.

To approach these questions, I draw on Talbot's scene to explore how imaging technologies encounter (both parallel and intersect) the textual conditions of digital libraries. The "scene" of this inquiry is full of suggestions such as: Talbot's prototypical imaging, understood as an expansion of the visual spectrum, inspired a vision of visual literature and a poetics of photography; the inventor found a new kind of literature in the imaging possibilities of photography; photography as an imaging technology opened up a new mode of the visualization of literature (just as digitization is introducing a new mode of literary temporality); Talbot's scene foreshadows "a writing that is not derived from speech," as explored later by Jacques Derrida and others; later, imaging as expansion of the spectrum shifts to an imaging of algorithmic processing (spectrum to coded signal) and inspires new media literature, a virtual literature. Taken together, these propositions hint at current attempts to find in digital poetics a "writing of images." The scene ultimately proposes that imaging technology has at heart a skewed literary impulse that leads to the unstable textual conditions of our emerging imagetexts and virtual poetics.

X-rayed literature

Of course, imaging technologies still had a long way to go from Talbot's time with the invention of photography to the digital signal processing and visual poetics of today. There were decades of progress still necessary to arrive at what now has become a paradigm of sorts for imaging technology: X-ray imaging. From the beginning in 1895, when Wilhelm Roentgen discovered what he called "a new kind of light," X-ray technology has captured the technological imagination with roads leading to both past and future.

Already, Talbot's thought experiment with photographic "invisible rays" was on the way, past the blue point of the spectrum, to the near-ultraviolet and the X-ray regions beyond. The experiments that Talbot alluded to were probably those of Johann Wilhelm Ritter who discovered near-UV light in 1801 using silver chloride crystals as a detector. Ritter was a contemporary of Sir William Herschel who also examined both near-UV and infrared parts of the spectrum. (Interestingly, his son Sir John Herschel contributed much to the development of photography and astronomical photography.) Out of these experiments, Talbot set the stage for the actual and metaphoric extensions of sight by describing photographic imaging as a kind of "penetrative seeing" of figures in space. His staging added narrative twists of deception and ephemerality. X-rays inflected these implied penetrations from the library scene to the actual inner space of the body.

The mysterious X-ray image of Bertha Roentgen's hand was the first to circulate both in the popular press as well as the medical community. The skeletal hand is eerily limned in shadowy forms that seem willowy, vulnerable, and brutally exposed. As Lisa Cartwright comments, the hand suggested an ambiguous display of death in life.[3] The ambiguity straddles prior suggestions, foreshadowed by Talbot, that "light no longer simply reflects off the surface of an object of knowledge, having shone upon it so that we may better observe it. Rather, light becomes a brutal force that physically penetrates its object, stripping away its concealing surface to lay its structure bare." Living flesh penetrated by a deathly light, Frau Roentgen's hand also becomes a fulcrum of the fundamental modern ambiguity between personal subjectivity and the anonymous gaze of technology. A ring on her finger, a token of personal intimacies and social decoration, becomes a flattened orb, a black cipher of a stripped down, medicalized existence.

Many artists and writers have used X-ray images as icons and switching points between the ambiguities of modern life. Elaine Defibaugh is a contemporary artist based in Rochester, NY, who works in this tradition of artists and writers using X-ray images as icons that switch between the personal and anonymous (Figure 10.2).

The point of intensity on Frau Roentgen's hand, her flattened ring as a hole in the image, becomes for Defibaugh the actual hole in her heart, the discovered atrial septal defect that she had from birth. Following her surgery to close off the hole, she collected all sorts of scientific data—her X-rays, EKGs, cardiac cauterization—pertaining to her case. She then began incorporating the data into her paintings and drawings, which are large, X-ray-like constructions consisting of transparent layers that can be viewed, like X-rays, either by reflected light or light-box. The works become both a reflection of medicalized regulation and a concrete process of self-healing. She has used the drawings to bring her accelerated heart rate down before her medication "kicks in."

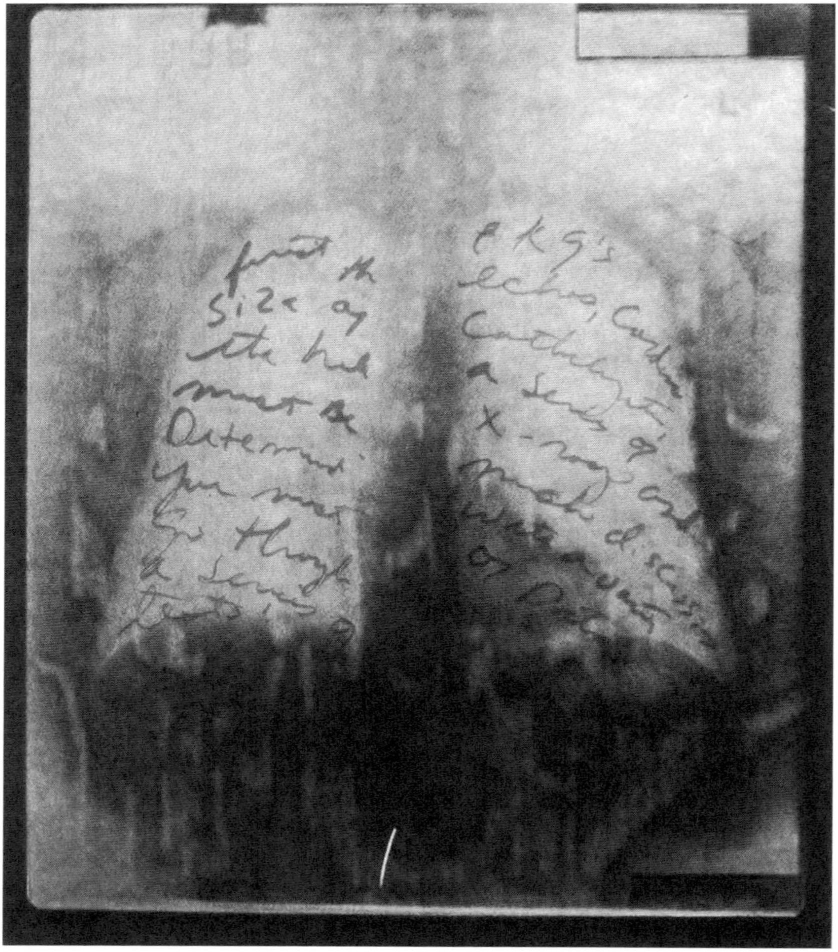

Figure 10.2 Elaine Defibaugh, "Blue Chest," 2001, from the series *Book of Hearts,* mixed-media on mylar, polyester sheet, and blueprint of X-ray, 11 × 8.5 in.

The dual use of the drawings as reflective image and self-healing tool relates to the uncanny identity of X-rays as cultural objects shifting between poles of personal and impersonal meanings. We often don't know how to place X-ray images in relation to personal identity: either as impersonal medical data or revelations of the self. Unorthodox uses of X-rays have highlighted the ambiguity: an illustration in a 1945 book, *Mass-radiography,* includes an ordinary snapshot of an ordinary woman, "Mary," juxtaposed with her unearthly X-ray. Both are incongruously signed with the strange effect of depersonalizing the self and personalizing the objective. In a more recent case, a radiologist decorated his home with rows of old X-rays and was sued by previous patients for an invasion of privacy. In these cases, we simply do not know how to situate such objects. Defibaugh's "X-rays" are further problematized by being signed as appropriated impersonal record, as self-reflection, and as work of art. In any case, no process of signing can fully personalize, or depersonalize, the death-like embrace of imaging technologies.

Apart from the variety of "signature effects," Defibaugh's extended use of language connects X-ray conceits to Talbot's interest in a novel form of visual literature. Defibaugh writes that she "perceived the new language suggested in the data as a challenge." Drawing on the challenge, she has used the X-ray forms as guidelines for extensive diaristic writing. The eerie penetrative space of the X-ray becomes a writing space and form of visual literature. There seems to be no "writing surface" per se in the images, and the words seem themselves to stand exposed in a penetrative light. It is significant that Defibaugh feels that her writing and drawing bridge a gap between technical data and more "imagistic data." She also sees her artwork as involving acts that "translate the language systems of the medical profession into an artistic language." This translation is related to an imaging of literature as first displayed in Talbot's "A Scene in a Library."

The literal translation of artwork into scientific data is familiar to us with the use of X-rays and other imaging technologies in art conservation laboratories. The title of a classic work in the field, "Art Criticism from a Laboratory," suggests that the literature of art criticism was itself translated into the imaging technologies of the laboratory. And it is not so strange, since criticism is often taken to mean "seeing deeply." In any case, X-rays used in the art conservation laboratory link imaging technology with interpretation. They penetrate murky details and layers of a painting for hidden motifs, telling brushstrokes, and the underlying effects of material decay. Frequently, the X-rays reveal or disturb attributions of authorship. Early in the development of X-ray technology a student of Roentgen used the process to discover the signature of Albrecht Dürer beneath the surface of a disputed canvas. Such forensic uses of imaging accords with Talbot's detective-like imagination—an inchoate literary genre at that time germinated and later developed with the penetrative seeing of X-rays. (Crime Scene Investigator or CSI is the current rage in this line.) And, like paintings, actual works of literature are subjected to the laboratory rays of this penetrative kind of forensic work. The Dead Sea scrolls are famous as an example of the use of infrared photography to bring out the lost writing obscured by centuries of accumulated staining of the parchment. It is here that we can take Talbot's experiment as a way of literally penetrating books and the material objects of literature. "A Scene in a Library" is simply another way to see books in terms of their materiality, beneath the surface.

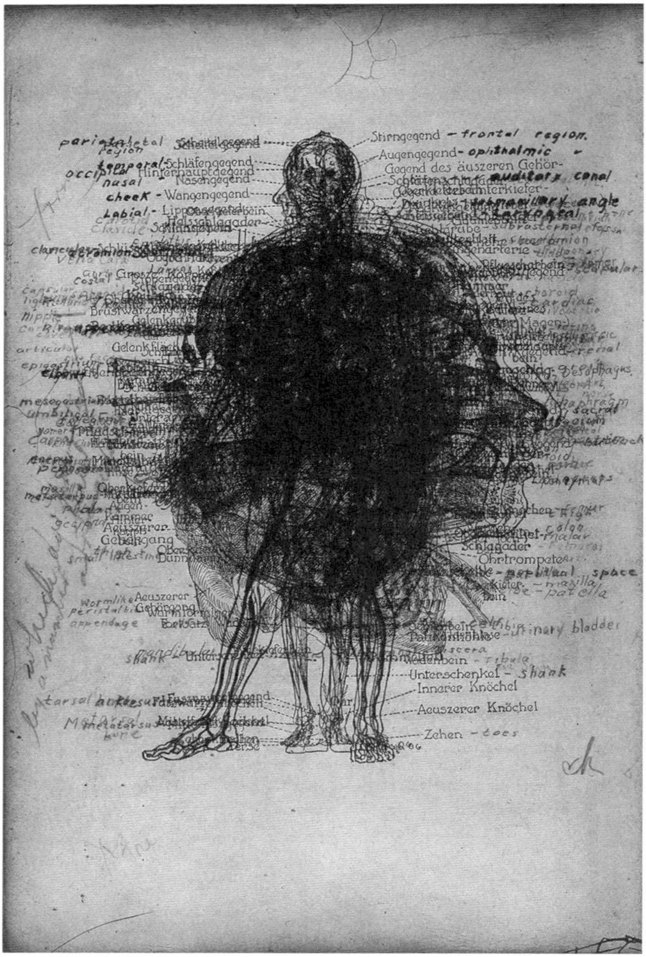

Figure 10.3 Doug Manchee, "Medical German," from *Concretions*, 2003–6.

Looking deeper, Doug Manchee is another contemporary artist who skewers literature with his imaging technique. He looks so deeply in fact that he seemingly flattens literary texts in a deconstructive visual process. He digitally scans each and every page of a book, usually a thick canonical work of literature, and layers the transparent pages in precise registration as one composite image (Figure 10.3).

The visual effect is an X-raying of the book where the lines of sentences pile on top of each other to reveal a textual skeleton. Around the edges of the text's skeletal features are minute details of features that stand out from the averaging effect of line superimposed upon line: superscript, subscript, accents. In Manchee's precisionistic X-ray, a swarm of marks and marginal details coexist with averages as bits of interlineal noise accent the regular, uniform pace of the gridded blocks of text. The book is rendered transparent as a totality and flattened in space, available to the gaze all-at-one. The images are a shadow and projection of marks in space, but they also stand as a residue of time. They resemble in this way composite portraits (such as those of Francis Galton in the 1890s) where individual faces specific to a self-identical point in time become condensed or "averaged" as a common form. Like X-rays, Manchee's textual composites penetrate and reveal the common visual surface of literature.

The penetration of surface effects bears on traditional processes of reading and writing. The matter is similar, if in reverse, to Mieke Bal's supposition that when images can be written and read, an intricate "flatness" ensues.[4] For Bal, this flatness is a matter of visual tropes and finely developed optical figuration in the novels of Proust. But for Manchee, the X-ray flatness situates the visual process of reading and writing on the margins. The total transparency of the book (as if the pages were glass) paradoxically renders the main line of text illegible as one sentence piles on top of and cancels the other. Only the exceptions remain visible, and Manchee's images are designed to reveal and make writing on the margins stand out. In an essay that influenced Manchee, Odell Sheppard concludes that marginalia is writing addressed to no one: an active gesture of reading that forms another kind of writing (one that Sheppard claims forms the origin of the essay). A piece of writing without a clear sense of address is for Sheppard "a dark matter to be explored." Marginalia extends literature in the same suggestive way that Talbot's near-ultraviolet photography moved imaging and literary life into darker regions of the library. Manchee's X-rayed literature severs the works from their usual kind of address, seeing them in a new condition of light where marginal marks and meanings stand out. Writing and reading become fractured and flattened in an extended process of looking. Visual appearance itself becomes textual marginalia in Manchee's new type of imaging essay.

A favorite piece of Manchee's work for me is his rendition of Thomas Carlyle's *Sartor Resartus* (first published in 1834, approximately ten years prior to Talbot's *Pencil of Nature*)[5] (Figure 10.4).

Most of Manchee's pieces carry a subtle, droll interplay of theme and image, the sense that the X-ray-like image parrots the book's subject matter in a peculiar way. The "Eclipse of God" is seen as eclipsed. *Sartor Resartus* is even more pronounced in its potential interplay of literary meaning and materially imaged literary work. Carlyle's original work is almost impossible to classify, and Manchee's scanning of it is entirely a part of the book's enigmatic cloth. Ambiguously, it is fiction posing as a factual work. In its

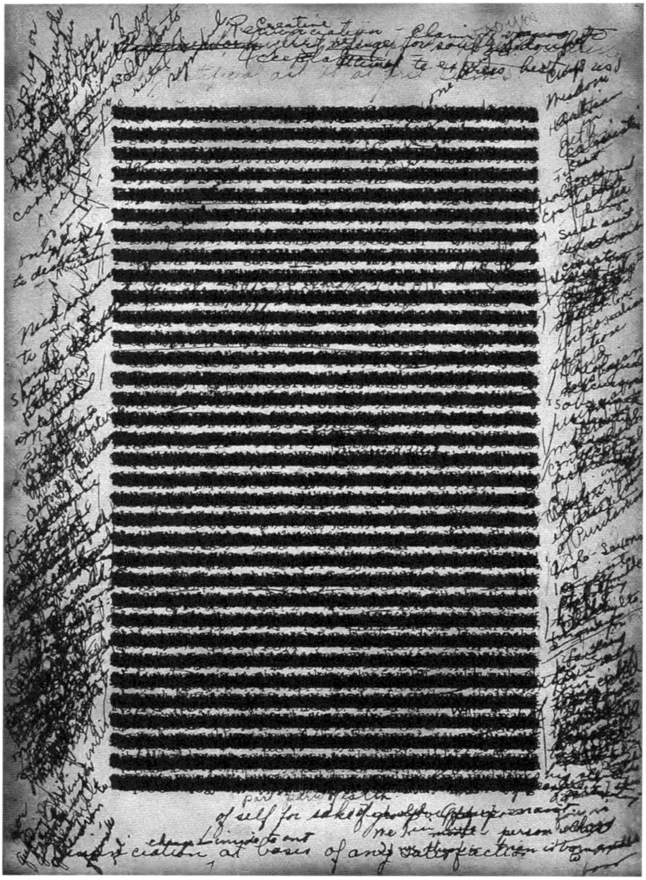

Figure 10.4 Doug Manchee, "Sartor Resartus," from *Concretions,* 2005, 17 × 10.75 in. pigment print.

premise, the book claims to be a biography of Diogene Teufelsdröckh, the author of *Clothes, Their Origin and Influence*. This inserted work, too, is difficult to explain but nevertheless holds a clear, insistent grip on the imagination. *Sartor Resartus* (the tailor retailored) is an exposition not simply of the idiosyncratic book of philosophy embedded within itself but of the narrator's efforts to understand the work and its strange fascination. The fascination with the inset book equals the fascination with the nominal subject of the book, clothes. Clothing works as a metaphor for presentation of self to society and the outer world. Carlyle and Teufelsdröckh imply that all social beings (people and things) communicate indirectly; words and images are merely external signs of inner states, always shrouded. Philosophy accordingly is a matter of negotiating clothes: stacking, folding, cutting through garments, and revealing the invisible flesh behind.

Manchee's metaphorical X-raying of Carlyle's "Philosophy of Clothes" seems to be saying with Carlyle that books, too, are clothes, difficult to classify, and very much a matter of appearances. *Sartor Resartus* situates interpretation within a hall of mirrors, while Manchee's imaging flattens the work to a textual X-ray: a sartorial roadkill (clothes run over on the road).

Like Talbot's "A Scene in a Library," *Sartor Resartus's* complications (and Manchee's intricate flattening of them) are early textual emanations and disruptions of the electromagnetic spectrum that are familiar today, along the roadway, as the printed page. As a final point on the X-ray imaging of literature, it should be noted that X-rays can also refer to the more outward disruptions of a landscape where words coexist with invisible rays and atmospheres. The rays of science fiction—the glow of atomic blasts, gamma rays, and so on—cut through the fog of signs in an era of media saturation. An iconic example is the 1957 film, *The Amazing Colossal Man*, where the explosion of a plutonium bomb causes a man who is exposed accidentally to the radiation to grow big and out of control.[6] Consequently, the man loses his bearings in a strange landscape of texts and signs. His effort to find comfort in a book proves futile because he has literally grown too big to hold and see the book; he has physically outgrown the bookish medium. Breaking away, he careens through Las Vegas where, on the other end of the scale, he stands up to the gigantic signs only to dash them to the ground in frustration. Out of scale, the signs become disjointed and illegible in the new light of the vast west. Another more recent example of X-rayed text in the new landscape is the notorious graphic design of David Carson in *Ray Gun* magazine.[7] Text and image alike are chewed away and decayed by a hard, corrosive light. The imagetext aesthetics of *Ray Gun* points to a futuristic literature where text may be illegible by traditional standards, but still meaningful according to the effects of light. Imaging technologies and X-rays extend and transform our sense of the legibility of graphic figures and cause us to question the scale of our lives. They also expose graphic codes operating at registers of which we are not always aware.

Hidden words

Talbot's thought experiment about photography as an imaging technology was also a detective story: an unknotting of a situation under another condition of visibility (near-ultraviolet light and the camera). Talbot introduces "this peculiarity" into a "modern novel," which at that time would have been the emerging detective stories of Edgar Allen Poe which involved the intriguing character, Monsieur Dupin. It is questionable whether Talbot knew of Poe's work, but he was a worldly man sensitive to the imaginative directions of his time. For the detective, a scene of a crime (or library) is a labyrinth of details and clues that indicate an event beneath the surface. Similarly, readers are common detectives who take the meaning of the story as the revelation of something hidden within the text. This is most often a matter of engaging with metaphor in a process of combing figures and tropes from the surface language. But for pictures, detecting hidden figures from the surface image can be quite literal. The art historian James Elkins has studied the identification of hidden images as one of the defining strategies of late twentieth-century art, and he has looked for its sources in early puzzle pictures, anamorphs, cryptomorphs, and aleomorphs. In these forms, the viewer finds hidden images either planted by the painter or occurring naturally, like seeing a face in a cloud.

Seeing hidden words or patterns within texts is not as common as it is within these pictorial examples, but it is no less significant. It may indeed figure as another defining feature of our time to take texts quite literally and to examine them materially for "secrets of the darkened chamber." Material literature has an unconscious analog to the human psyche, in which imaging techniques can assemble and reveal. Books, as Tom Conley phrases it, have a "graphic unconscious."

The experience of books and libraries as graphic scenes for unconscious processes is wonderfully enacted by Peter Greenaway's film *Prospero's Books*. Sections of Shakespeare's *Tempest* are interlaced with sumptuous depictions of 24 wonder-working books taken by the deposed duke, Prospero, to his haunted island kingdom. The books seem all encompassing. There's *A Primer of the Small Starts, A Harsh Book of Geometry, An Atlas Belonging to Orpheus*. Their powers of wonderment emanate from their visual capacities: especially from such volumes as the *Book of Colours* and the *Book of Mirrors*. As a spectrum of colors the book becomes a prismatic imaging device. As a sequence of mirrors, the book reflects its reader in an eerie sequence. One page as a mirror shows one's appearance a year's time hence; another as he was three minutes ago. Prospero's books are spectral entities whose texts reconfigure the relationships between light, form, and time.

This imaging of literature is of course an effect in the medium of film, special effects made possible by digital technology. The representations of books float image layer upon image layer. Classic renaissance books at the dawn of print technology anticipate the digital future in their power as screen image. Through the use of paint box manipulations, digital compositing, and computer animation, Greenaway suggests that such codex books will be translated and reborn in digital space. A scene in a Library will be a matter of future imaging technologies.

Similar to Greenaway, Douglas Holleley visually enters the past of classical print as a revolving doorway to the digital future. He has used digital imaging technology as a way of "reading" late renaissance texts in order to intervene in a classic era of print through modern imaging processes. His digital photography of books operates in a way that is reminiscent of Talbot in "revealing the secrets of the darkened chamber." And like Greenaway, these revelations create an awareness of the book as susceptible to the actions of the spectrum and capable of unleashing hidden forces within and beyond light.

Like Greenaway, the prismatic pages of special books may bear specifically on geometry as a mediator of light and sight in renaissance perspective. Greenaway's Prospero had in his *A Harsh Book of Geometry* a place of knowledge that Holleley fills with Albrecht Dürer's treatise on theories of proportion, *Hierin sind begriffen vier Bücher von menshlicher Proportion* (from the Cary Collection at Rochester Institute of Technology (RIT) (Figure 10.5).

If the former book reveals geometry as "harsh," it is because it tends to be associated with architecture as containing structures to define human existence as abstract, plotted, and preordained. Dürer's geometry however relates to proportion, especially as it is related to the human form. Here the lines of geometry emanate from the body rather than being set upon and surrounded by them. Holleley's reading by "photographic analysis" supports a way of understanding geometry as adding to the body in addition to dismembering it. His digital processes, colorizing, tone manipulation, and so on, "give flesh to the outlines" and become an active manner of digitally supplementing Dürer's proportion studies. The figures consequently become more animated, and often seem to act in dialog with other shadowy forms next to them. Given a life of their own, Holleley's sense of geometry and imaging analysis is

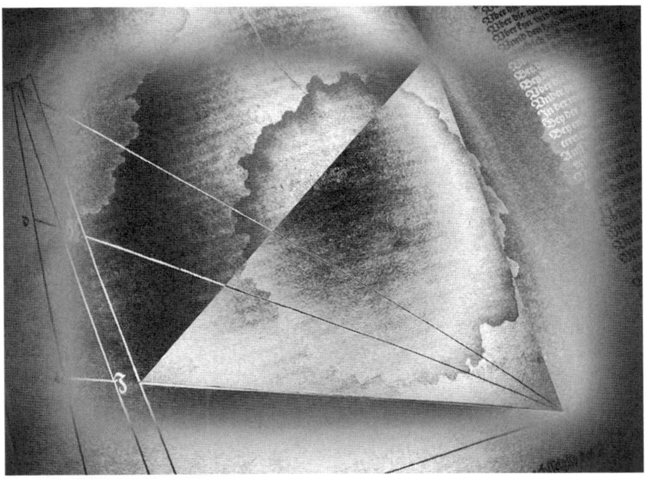

Figure 10.5 Douglas Holleley, from the portfolio "Re-reading the Book, A Photographic Analysis of Albrecht Dürer's *Hierin sind begriffen vier Bücher von menschlicher Proportion*," November 2000.

animistic and very much aligned with the overarching perspective expressed in *Prospero's Books* that the library is a wild place. Books run wild in relation to geometry as ambiguously hidden schemas are set free by imaging.

The capacity to set geometry free stems from Holleley's self-defined photographic procedures, in effect his "angled" reading of the text. He demonstrates that different angles of camera viewing establish incongruous geometries and reveal different imaginative relations to the book. From a different angle, Dürer's book is not about "flat" geometry. Holleley's oblique views bring out the spatial contours of the page. In his statement, Holleley explicitly differentiates his angles of imaging from more normative, squared-off geometries that align the film-back of the camera in a strict parallel relationship to the object plane (in this case, the book page). He calls this arrangement the "perpendicularly possessive perspective," a practice familiar to any common library copy stand (a crushed book in compliance with the superior camera on the fixed rail above). The book is literally pressed, if not subjectively oppressed in this visual relationship. Plate IX in the *Pencil of Nature*, immediately following "A Scene in a Library," demonstrates with photography this "perpendicularly possessive perspective" (PPP). Talbot titles the plate, "Fac-simile of an Old Printed Page." He uses it to illustrate photography's ability to copy a page from a book or manuscript, to show that "[t]o the Antiquarian this application of the photographic art seems destined to be of great advantage." In some respects it represents historically the first Xerox-like photocopy in that the fac-simile is, in Talbot's words, "copied the size of the original by the method of superposition." The last word indicates the ideal practice of Holleley's PPP, which is a marked departure from the subtle insinuations of library, text, and image in the previous plate, "A Scene in a Library." In one consecutive plate to the next, Talbot switches from using imaging technology as a poetic medium of metaphor to its practical use as a clear, flat-footed process of copying and distribution. He moves in the step of one page from poetics to simple exposition.

Holleley's angled readings and copying resist this simple transition and division between poetics and exposition that might be involved in imaging and copying. In a signature image used as an icon for Holleley's book publishing enterprise, a digital scanner cuts laterally across the open pages of the book. Angled scanning digitally flays the book and combines the clarity of incisive detail with a defamiliarization of the book's structure. The combining of these effects of scanning and skewed angles leads to fundamental questions about how types of imaging processes determine strategies of reading. How does a version of imaging as facsimile coexist with a more deconstructive variant where metaphors of imaging extend the spectrum and reveal hidden words? Holleley shows us that meanings can differ as imaging penetrates the graphic nature of text from varying angles.

Moreover, these meanings play on the suggestions of surface and depth and implicate the unconscious. Holleley claims that his altered images and unusual viewpoints are a conscious metaphor for the act of reading itself. Yet this consciousness of multiform reading extends into deeper regions of the text, regions that may have eluded Dürer and his publishers. The reading by imaging reveals, according to Holleley, personal aspects of Dürer himself, his sexual desire and ambiguity, through "cutting into" the woodcut "Melancholia." These unconscious elements are highlighted by Holleley's "small reading light" which intensifies vision, illuminates, "inquires," but does not possess. The process of revealing the unconscious of text is explicitly associated with imaging technology when he claims that the method is "not too dissimilar from how distant planets are imaged and how barely legible ancient texts are revealed by infra-red photography." And, as we know from Talbot's "A Scene in a Library," "data . . . is only part of the story."

A discourse of parts, Holleley's version of the unconscious is a matter of details. Fragments have long been associated with the unconscious ever since Freud deployed his archaeological motifs and images for describing the inner workings of the psyche. However, Holleley works with details, not fragments, which are distinctive in being isolated or bracketed parts retaining an imaginative relation to the whole that still exists (Talbot's first plate is a detail of a whole scene, not a fragment, since we can still take an imaginative detour off the frame down High Street). Moreover, details are more readily sequenced to form alternative chains of meaning. Holleley exploits sequencing and the juxtaposition of details to support the reading of unconscious elements in the graphic image. The sequencing of a book's graphic details suggests even more elaborate strategies for conjuring unconscious levels of textual meaning using the book as a reference system.

Tom Conley demonstrates this method in his innovative, scholarly work *The Graphic Unconscious*.[8] This book studies in the light of psychoanalytic theory, the patterns of typographic elements in French Renaissance literature, especially the poetry of Clémont Marot, Rabelais's *Gargantua*, Ronsard's sonnets, and the essays of Michel de Montaigne. Conley shows that the graphic material of books, printed letters and word arrangements, both reinforces and undermines the linguistic meaning of the texts. The literature displays revealing composition patterns, graphic slips of tongue that do not appear to be under the control of the author. For example, Ronsard's sonnets evoke graphic constellations of word forms that seem to mimic the arabesque and spirals of baroque pictorial art and architecture. Ronsard's text seems to have unconsciously assimilated in print the formal motives of his time. Montaigne's motives appear more politically based, with letters floating in a visual field from which a diligent reader can parse hidden writing "admonishing the way censors . . . go about destroying the planet."

What is striking is how Conley shifts investigative methods to literature and text that are far more common in art history. Again, Elkins looks at intriguing marginal traditions of hiding figures and images within pictures: nineteenth century puzzle-pictures and cryptomorphs. He locates in these arcane practices more fundamental tendencies of modern interpretation to identify hidden images and to delineate pictorial ambiguities. Iconology generally operates according to an investigative method, and the same may be said of rhetoric according to Conley's principle of the graphic unconscious. The key figure of rhetoric at work here is *ekphrasis*, the verbal description of pictures or works of art in literature. However, Conley approaches an inverted, image-rebellious movement of ekphrasis. In such literary works, images instead describe texts and influence unconscious drifts of meaning between word and image. In classical ekphrasis, the text controls pictures, while in the late Renaissance we find that texts in print lose control of the image in novel ways.

Again, we find an imaginative view of the dawn of print as an entry point into the textual conditions of the electronic future. The frantic transmission of signs in the information society creates a confusion of imaged words and worded images where, according to Conley, "[t]he image is detached from but conjoined to a shard of writing in order . . . to show us that an image is *not* writing, while, on the other, the placement begs us to transgress the opposition."[9] Whatever the confusion of imagetexts, they beg for the investigative method offered by imaging technology. Conley's method of discerning unconscious graphic patterns in literature implies the astronomical process of linking random star configurations into coherent clusters and visually mapping them. Imaging literature sets the stage for graphical procedures and psychoanalytic paradigms of interpreting texts.

The logic of these procedures rests not only on their visuality but also on their technicity. Imaging literature implies that there are mechanisms of meaning at work in texts. Writing is a technology certainly, and the possibility follows that text can be considered as a machine. The impact of this possibility on questions of consciousness arises with great force in the existence of early writing machines, especially the eighteenth century writing and draughtsmen automata of Pierre Jaquet-Droz. The most perplexing instance is the clever little figure who could compose the enigmatic passage, "Je ne pense pas . . . ne serais-je donc point?"—roughly translated as, "I don't think . . . therefore I don't exist?"[10] In this convolution of the Cartesian cogito, the writing machine in the form of a little boy twists the existential relationships of writing, meaning, and consciousness. It undermines the normative view of text as a conduit of meaning from one conscious mind to another (reflective writer to silent, solitary reader). The automaton, who cannot think, can only simulate writing by being programmed to draw pictures or graphical inscriptions that legibly resemble text. The difference between the writer and draughtsman automata lies only in our eyes. Just as there is no consciousness behind the little boy's negative-cogito, there is no text in that text (the machine's written sentence). The auto-writing is just an image. All writing potentially invokes this paradox since writing by definition is the graphic rendition of language. Imaging technology like writing machines participates in the elusive play of image and textual meaning. This play raises questions beyond the spectrum leading to code.

From spectrum to code

So far, this inquiry into the literary impulses underlying imaging technology has been regulated by actions and figures belonging to the spectrum. Talbot prognosticated photography's use as an imaging technology because he saw that it could extend the reach of visual representation beyond the visible regions of the spectrum. Somewhat mysteriously, in the same breath, he filled this region of photography's possibility with literary tropes and a library setting. Interestingly though, we find some 50 years prior to Talbot's discovery of photography an automaton which, like the little writing boy, foretells the potential breakdown of the spectrum as a force governing relationships between imaging and literature. The boy's twist of the cogito proleptically shatters the meaning and role of the line as the supreme graphic union of text and image.[11] The line in this instance operates in the analog mode, like the spectrum. The shattered line has now yielded to the digital bits that constitute an information society; it becomes the meeting point between visual and textual systems.

This meeting is obvious in the new media blending of aural/visual and textual/graphic signs churning through the twenty-first century channels of communication and computer interfaces. But more to the point here are the less obvious, behind the scenes work of computer software and the mutual "coding" of image and word. Imaging technology likewise has become an operation of code and mathematical processing rather than of extended spectral sensitivity, as it was in Talbot's day. *The Pencil of Nature* construed image and word by the actions of the "line," and its speculative extensions of the image worked according to its linear, analogic nature. Talbot extrapolated the line of photography (the photographic pencil) into the unseen regions of the spectrum and metaphorically into literature. He could not see the latent mathematization of the image as a basis for imaging literature. However, in this era of computers, it is impossible not to discern the mathematics and code that drive the engines of image processing.

Current brain imaging, such as PET scans (positron emission tomographs), for example, allow the visualization of non-visible phenomena, like "A Scene in a Library." But, unlike Talbot's scene, the PET image is dependent on mathematization. These functional maps of brains are not simple pictures, but are the result of statistical processing embedded at several levels (see Megan Delehanty in this volume). The image depends on the electromagnetic emissions of oxygenated blood, but that is only part of it. The question remains how the imaging work of embedded codes changes the literary tropes at the heart of imaging technology and the literal "imaging of literature." How do we "read" images and texts (imagetexts) consisting of mathematics and code? The answer follows from looking at the computers that supply both the framework and underlying bit: the front end and back end of image-laden literature.

Imaging books with computers makes reading a two-ended matter of decryption at the back end and interface at the front end. Even if Talbot made his fac-simile according to Holleley's "perpendicularly possessive perspective," he augured the unleashing of the book with the pervasive use of imaging in the digital storage and transmission of bookish text. Moreover, in cybertexts, the digital imaging of books is not limited to possession or capture by facsimile reproduction. The book image operates freely as a design metaphor for building graphical interfaces in order to access texts. Design metaphors, as any interface designer knows, deploy icons of familiar, known things in order to make the use of the computers tacit, "natural," and transparent. Texts accessed on the computer become readily legible according to the known characteristics of chapters, pages, indexes, and so on. Even nonfunctional details like hard cover, paper texture, dog-eared pages become emotionally significant. They may seem gratuitous, but these effects and icons of the book make us feel at home as we read the digital text.

The computer game and work of interactive fiction *Myst* (Robyn and Rand Miller, 1994) serves as a classic example. The image of vintage books of the heroic age of print culture again serves as the gateway to the digital future as they did in Greenaway's film and Holleley's "reading" of Dürer. But here they are deflected away from spectrum and gravitate more toward hidden codes and machines at work within an island of bookish images. *Myst's* staging is a remote isle spotted with fabulous machines, garden follies, and whimsical ruins that come straight out of romantic literature. Among these fabulous points of attraction, the island features a library as the most central scene of the island for both orientation and control. In this library, images of spectral transmission familiar to Talbot, give way to figures of decoding and decipherment. Knowledge is gained less by penetrating the unseen than by finding and arranging clues in order to unravel an ambient mystery. *Myst* replaces the mysterious "denouement" of Talbot's scene with decoding and decipherment. Books offer the iconic interface and literary images for unraveling a code rather than simply presenting a knotted fictional narrative.

A fundamental ambiguity slides through bookish text and computer codes, mediated by imaging that makes us wonder what are we looking at (or reading for that matter). For example, in the 1987 exhibition *Information Art*, the Museum of Modern Art mounted gigantic photomicrographs of computer microprocessors.[12] Imaged and presented on an abstract expressionist scale, gallery goers stood in awe before the enlarged chips as formalist works of art, like rectilinear Mondrian paintings. The ambiguous power of the microprocessor as image is not so bizarre given that in some respects they owe their existence to an image-making process with intricate patterns burned and etched into silicon wafers. Norbert Wiener called such cybernetic patterns "operative images" in that they not only bear relationships of resemblance but also have, at the core of their existence, behaviors.[13] The media theorist Friedrich Kittler goes further into the ontology and ambiguous presence of chips by reading them as hardwired text.[14] Paradoxically, what he reads in the etched iconic surface of the microprocessor is the hieroglyphic message that "[w]e do not write anymore." He continues that "[t]he last historical act of writing may well have been the moment when, in the early seventies, the Intel engineers laid our some dozen square meters of blueprint paper (64 square meters in the case of the later 8086) in order to design the hardware architecture of their first integrated microprocessor."[15] In Intel's revolutionary layout, the architectural representation of information has interrupted the normal course of literature in somewhat the same way that photography in Talbot's day broke into and altered the sense and scene of the library.

Kittler goes even further in saying that "[p]rogramming languages have eroded the monopoly of ordinary language and grown into a new hierarchy of their own." He touches here on the ambiguous status of code as an operation of machines and a special script and determinative writing of programmers. Code is both machine-readable and human-readable depending on the level of processing and display. Fundamentally, code acts as writing when we can see it, and many ironies of code grow out of its relation to visuality and imaging. An amusing example is the T-shirt protesting the DVD Copy Control Association. DVD CCA is a not-for-profit corporation that licenses CSS (Content Scramble System) to manufacturers of DVD hardware, discs, and related products. The T-shirt emblazons on its front the "ban" sign crossing out "DVD CCA" while the back carries the full-text of the algorithm for "ripping" CSS protected discs. This renegade code-as-software, the executable program, is prohibited by copyright protection laws. However, the code-as-algorithm, the script of the programming language, is a matter of free speech. Any citizen has the right to read and write the code and perhaps compile it at their leisure as their own executable program for private use. Reading, information property and viewing pleasures are oddly aligned on this iconoclastic T-shirt.

Codeworks

This whirl of code, text, and imaging has entered imaginative literature and digital poetics with a perplexing force. Poets are having to deal with Kittler's challenge that in an information society, "[w]e do not write anymore." To the extent that his pronouncement applies to software, I might add that we also do not always know what our writing does, or for that matter, how it may look. These questions about the conditions of the visibility of text and poetics are not limited to the usual avant-garde experiments with typography, scale, and page layout but extend to logical procedures and text processing. Digital poetry questions "how code *looks*?" and its influence on the imaginative use of language. Strategies of questioning include medial self-reference, text-processing, interactivity, hypermedia, and networking. I would add imaging literature (imaging technology applied to literature) in the sense of working with textual codes in the attempt to find and explore a pattern in the text. A range of poets are deploying these strategies. I will sample two related to "A Scene in a [digital] Library."

In his networked meditations on issues of interiority, subjectivity, body, and language, Alan Sondheim has shown how optical motifs give way to the dispersal of code. His writing—snippets, extended monologues, alter egos, diary entries—permeate a multitude of sites with a sense of maniacal proliferation. He infuses computational linguist Terri Winograd's statement with an unbounded passion: "the computer is not a writing machine, it is writing." A key node of his writing called the "Philosophy and Psychology of the Internet" indexes hundreds of sections written over a period of ten years. Using metaphors of the spectrum, he divides subject matter into so called "short-waves" and "long-waves." Short waves are the individual sections, written in various styles. They harbor interrelated "characters" and play with narrative structures. The long waves, according to Sondheim, "are fuzzy topoi on such issues as death, love, virtual embodiment, the granularity of the real, and physical reality, which crisscross the texts. The resulting fragments and coagulations emerge from code and codework, surface and substrate." In emergent forms, Sondheim does not seem to write so much as deploy particles of texts that operate within quantum patterns, scattered and able to be in two places at the same time. They do not form coherent pictures, and yet they take significant shape on the screen.

Forming loose graphic shapes, Sondheim's text sometimes resembles ASCII art, the primitive form of computer graphics using letters as picture elements. But again, the result is not so much a picture as a visual "resonance" between elusive graphic pattern and textual meaning. From a Google search for Sondheim, I found this list entry that sums up the effect vividly.

> Regarding Alan Sondheim's [*sic*] net art titled "text on the missing." I can see rows and rows of picket fences in an extreme long shot. Is the text missing or am I missing the text?—Shohini Ghosh (Figure 10.6).

The comment brings out a language of film in the graphic text ("extreme long shot"), but the commentator implies something else closer to the iconic nature of imaging technology. Imaging produces pictures with missing standard picture elements: they cannot simply be looked at or looked through for meaning but must be read and decoded according to their own logic (PET scans of brains for example are not ordinary pictures of brains). Likewise, Sondheim's "waves" of writing are not to be read by plain poetic conventions but decoded as technical imaging on the surface

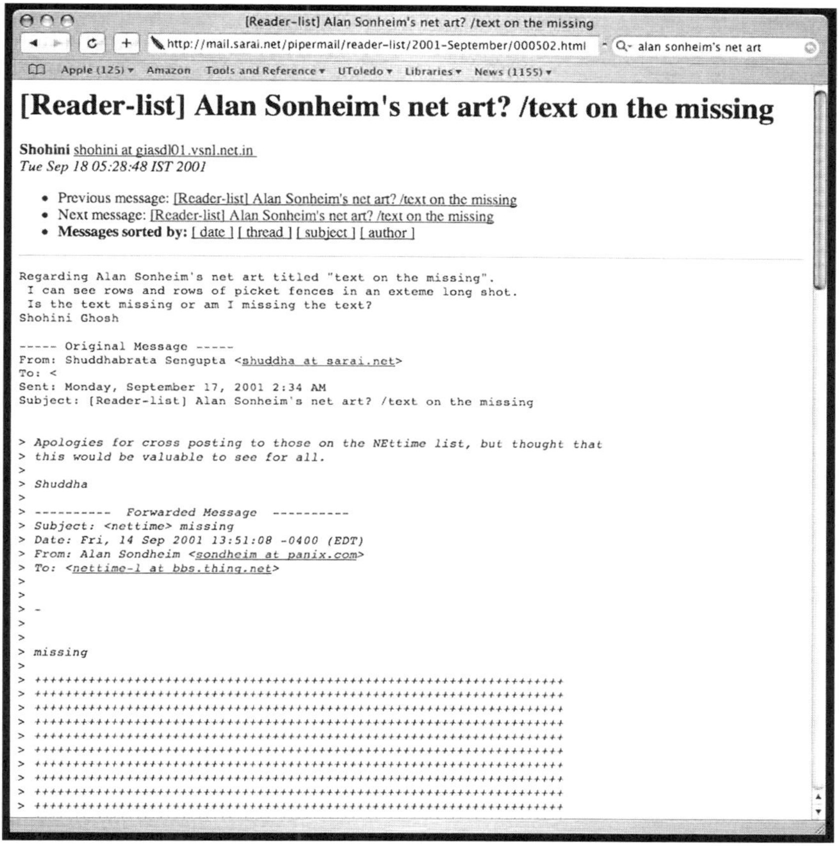

Figure 10.6 A post with a query about Alan Sondheim's imagistic net art, "Is the text missing, or am I missing the text?" Screen shot by Chris Burnett, January 10, 2005.

of writing. For Sondheim's project, the Internet is one vast textual scan of our virtual subjectivity.

This extended virtuality of scanned text is visualized from an entirely different angle in the poetic *Crystallography* of Christian Bök.[16] For Sondheim text is viral, it multiplies by digital forms of infection, while Bök presents text as crystal formations that have their own laws of reproduction and generation. The results are very different but both apply metaphors drawn from science and biology to the operations of digital poetics. And each engages with their images of language at the level of writing itself. Bök's particular move is to endow language with crystal form; it breaks down into planes, facets, and luminous transparencies. The sense of line, the graphical substance of writing, has a crystal presence, too, working as structural division and refractor rather than a continuous interface between image and text. If Bök has invented another *Pencil of Nature*, it is decidedly more chiseled. In one plate, he displays a bizarre looking alphabet for "plotting the transmission of light through a transparent medium." His poetry, and the alphabet underlying it, is both a paperless and screenless medium, providing a sense of transparency.

As a transparent medium, it is shot through with the condition of imaging technology. Imaging takes root in Bök's poetry through the very visualness of crystals. As a substance, they appear aligned with a visual nature and an unusual perceptual process, like a naturally designed lens. Crystal substance also suggests a higher order of ideas and form linked to vision and light. In one section on "Euclid and His Modern Rivals," Bök writes through inventors of X-ray technologies Max von Laue and W. L. Bragg:

> X-rays passing through the atomic scaffolding of a crystal cast the pointallistic shadows of its form across the page.
>
> Writing represents an obsolete machine for catscanning drops of frost that glitter in a spiderweb of dreams.
>
> X-ray machines illuminate a secret with a light that no one sees, the sound of the secret recognizable in the shape of its silence.
>
> Each one of us emits an infra-red aura as unique as a signature.

These passages bring us back to "A Scene in a Library" where an early model of imaging technology extended vision to make perceptible a new form of literature. But, in Bök, we also enter into a new kind of literary place (a library) informed by crystals. Works of literature, like crystals, can function like a microworld. In this way, *Crystallography* stands in the tradition of allegorical tales of geometry beginning with Edwin Abbot's *Flatland* (1884). In such tales, we imaginatively enter into a compressed space and see it expanded as a world that works according to its own laws. Narrative vies with geometry and geography as an ordering principle. Like Talbot's "A Scene in a Library" the parables stem from a thought experiment: testing a premise of what we might see with imaging technology and then expressing a hidden literary impulse. Talbot framed his scene as a mystery. The world that *Crystallography* opens is more like science fiction, an alien place formed by the cruel beauty of digital poetics.

The work manifests its digital nature as much out of this sense of a cool, alien terrain and its hybridity than from any overt reference to digital culture or new media. The subtitle *Book I of Information Theory* establishes information as a premise, but this topic is never overtly theorized in the text. The text does not employ the usual strategies and conditions of digital communication: interactivity, processing, multimedia, hypertext, and networking. The relatively staid design does not display the layered, digital pizzazz of Photoshop "comps." Rather, the digital condition comes out indirectly as an ambient feature of the book. Page after page points to the new medial conditions of writing. Phrase structure and its iconic distribution on the page hints at the work of underlying algorithms in its formation. Fractal geometry informs the layout and content of one poem: each component is progressively split according to "the ideal of redundancy." The book itself appears to be the result of hidden abstract procedures allied to the logic of crystals. Both creature of print and number, the book is a digital hybrid.

As a hybrid, this work of digital poetry plays out the ambiguities of crystals as both geometric and organic; static and generative in their growth. They fascinate the mind and the eye as an encounter with another order of being and substance; another approach to the nexus of the ideal and the concrete. As both material and abstraction, a poetics of digital crystals combines with imaging technology as an informational hybrid. Imaging technology presents a radically new way of approaching images and making pictures. As Peter Galison points out, imaging involves different mixes of making and observing and drawing on principles from previously distinct areas. Compound imaging systems mediate reality through mixed procedures of recording, simulating, and representing. Galison goes further in looking at how disciplines themselves have become hybrid in organizing themes around these mixed images. In a similar way, Bök's text finally reveals how a literary text can organize itself around a hybrid image, a crystallization of imaging technology.

Toward a media literature

The computer therefore has formed a specular kind of black box, and within it we find a hybrid framework of digitized scenes that crystallize the hidden literary impulses behind imaging technologies. We can sense, with the conjoining of imaging technology and literature, insights about the interdynamics of word and image that go beyond avant-garde experiments with "visual literature." As visually interesting as dynamic typography, nonlinear graphics, and intriguing comics may be, they rarely penetrate deeply enough into the possibilities of imaging technology to disclose its literary values. It could be (as Fox Talbot discovered long ago) that visuality itself does not go far enough into this expanding field of image and information that is often dark to the eyes. In truth, the particular examples of visual literature that accomplish this might be better termed a "media literature."

But confabulations of identity, or a newly labeled field of endeavor, do not resolve the final question of what these odd textual/imaging contraptions are as kinds of media or kinds of literature. The question spills over into matters of control and context such as we have already encountered with the rhetoric of ekphrasis. As mentioned, ekphrasis is an ancient means of controlling images in literary text that has never lost its currency. Age after age, writers use ekphrasis to control a circuit of connections between words and images with words playing the traffic cop of their comings and goings. This relationship of control has changed as images are now channeled by electronic circuits and software algorithms, as is language itself. Ekphrasis has perhaps become even more prevalent in computer culture by imagistically controlling patterns of information with imaging technology as a prime pattern maker. Ekphrastic imaging features patterns within a whole environment of control; it draws its subjects relationally in a dark room.

This preference of perceiving media pluralistically and environmentally stands out in current scenario planning as applied to the dizzying world of independent media today. In the Spring of 2005, the *Global Business Network*, NAMAC and a consortium of media arts organizations from the San Francisco Bay Area released their major study, *Deep Focus: A Report on the Future of Independent Media*.[17] The report captivates the imagination because it unabashedly favors a messy environment of swirling options and confusions about communications rather than trying to make sense through identity assignments. It doesn't try to identify and foretell the future so much as deploy competing perspectives of this world to see which one might win the imagination. In doing so, the report uses scenario planning as a kind of ekphrasis that recalls once again Fox Talbot's idiosyncratic thought experiment about the possibilities of photography as an imaging technology. Scenarios generate imaginative stories in the same way that the visualization of imaging technology might generate an entirely different perspective on knowledge and material reality. A visualization formed by imaging technology may offer a scenario or kind of story created by extending the means of sight. If visualizations are stories to be read, then a PET scan is more like a scenario (a story in pictorial form) than a photograph or common camera picture. Moreover, visualization might yet produce an extended form of scenario making that in turn would yield unthought-of stories and media literatures. The title "Deep Focus" is a metaphor drawn from filmmaking that works equally well for imaging technology in its literary aspirations. Just as deep focus in film renders close up and distant views with equal clarity, imaging technology models obscure subjects in depth, from foreground to background, with surprising lucidity.

The "deep focus" of this chapter has been between the obscurities and lucidities that hang off the two-sided interchange between our capacity to visualize and narrate. Both near and far, media and imaging technologies now form and inform our experience of literature, and they have etched seemingly boundless features into a growing information landscape and media ecology. Fox Talbot's "A Scene in a Library" was the initial encounter with this landscape, though it was scarcely narratable in Talbot's day because it could not be seen. Now code as a digital pencil of nature inscribes what it is possible to see and write and experience as literature. All of this bids us to look within the intricate artifacts of imaging technology (or what can be experienced with extended sight) and find literature—not just the words of literature—but literature within the extended, coded registers of visual experience.

Notes

1. Schapiro, M., *Words, Scripts, and Pictures: Semiotics of Visual Language*. New York: G. Braziller, 1996.
2. Talbot, W. H. F., *The Pencil of Nature*. New York: Da Capo Press, 1969 (1844).
3. Cartwright, L., *Screening the Body: Tracing Medicine's Visual Culture*. Minneapolis: University of Minnesota Press, 1995.
4. Bal, Mieke, *The Mottled Screen: Reading Proust Visually*. Stanford, California: Stanford University Press, 1997.
5. Carlyle, T., *Sartor Resartus*. Everyman's Library. Vol. 278. London: J. M. Dent & Co. 1908.
6. Gordon, B. I., *The Amazing Colossal Man*. Dir.: Gordan, Bert I, 1957.
7. Carson, D., *The End of Print: The Graphic Design of David Carson*. San Francisco: Chronicle Books, 1995.
8. Conley, Tom, *The Graphic Unconscious in Early Modern French Writing*. Cambridge; New York: Cambridge University Press, 1992.
9. Ibid., p. 166.
10. Chapuis, A. and E. Droz, *Automata: A Historical and Technological Study*. Neuchatel, Switzerland: Editions du Griffon, 1958.
11. Curtis, G., *Visual Words: Art and the Material Book in Victorian England*. Brookfield: Ashgate, 2002.
12. McCarty, C. (ed.), *Information Art: Diagramming Microchips*. MOMA: New York, 1987.
13. Wiener, N., *God and Golem, Inc.—A Comment on Certain Points where Cybernetics Impinges on Religion*. Cambridge: MIT Press, 1964.
14. Kittler, F. A., *Literature, Media, Information Systems*. London: Routledge, 1997.
15. Ibid., p. 147.
16. Bök, C., *Crystallography: Book 1 of Information Theory*. Toronto: Coach House Press, 1994.
17. Blau, A., *Deep Focus: A Report on the Future of Independent Media* (ed.) H. D. Michiel. San Francisco: NAMAC, 2004.

11
William Kentridge: Moving Pictures

Cyril Reade

The animated films of the South African artist William Kentridge (Johannesburg, 1955–) are moving, absorbing the viewer and eliciting marked affective responses. It is not only the subject of the films that stirs the film's public; it is also the artist's attention to the medium and his manipulation of its capacity for movement and its potential to have that motion represented, the passage of time being one of its prominent figures. This also characterizes the work that Kentridge effects in other mediums, prints, drawings, sculpture and collage.

I first experienced Kentridge's animated films at a solo exhibition at the New Museum of Contemporary Art in New York City in the fall of 2001 that introduced the artist to American audiences.[1] As is the practice in the art world, film, video and DVDs are projected in makeshift screening rooms, darkened spaces, as in conventional theatres, but that often do not offer seating and where the public wanders in and out following their own rhythm; even in this makeshift environment, the films exerted a riveting fascination. The New Museum exhibition introduced its public to the artist's working method as well: some of the 30-odd charcoal drawings that are the basis for each of his animated films were exhibited as well. These drawings constitute the field on which Kentridge pursues a continuous process of drawing and erasure and redrawing, the paper surface worked and reworked as the characters, objects and interiors, landscapes and cityscapes play out their evolving roles as each new drawing is filmed and the narrative advances. Since that exhibition I have seen his artwork presented in other venues including the screening of *Tide Table* (2003) at New York City's Marian Goodman Gallery where all the drawings that were the basis of that film were shown.[2] Etchings, drawings and collages unrelated to the film work were exhibited more recently at the Metropolitan Museum of Art that had acquired *Tide Table* and the other artworks shown.[3] A lone sculptural work, *Procession* (2000), was presented in the context of a group exhibition, *The Body Transformed*, focusing on the presence of the human figure in modern and contemporary art organized by the National Gallery of Canada.[4]

In other words, my experience of Kentridge's work has incrementally expanded to encompass a wide range of mediums.[5] Each subsequent viewing of Kentridge's production has confirmed and strengthened the initial impression of mobility that is evident in his films in a structural sense—that is, is bound up with the movement of the filmic image—but these subsequent viewings also confirm that mobility is a trait accompanying the transposition and translation of Kentridge's concerns to other mediums, even the static ones that I have just enumerated, drawing, etching, collage and sculpture.

If I have started this chapter with a first-person account of my viewing experience of an artist's production in the context of his exhibitions, it is because I want to emphasize here the experiential nature of apprehending an image, and specifically the capacity of an image to move a viewer. This point of view acknowledges the role of the artist as a maker of images, the artwork as a means of communicating and the viewer whose reading of the image gives it meaning. Beyond the fact that artists' images are what I know best, I also believe that visual artists are the least constrained of today's image makers, not constrained by the commercial imperatives of advertising, the informational imperative of photojournalism, or the instrumentality of scientific and medical imaging, for example. The visual artist is the freest to explore the potential of the image. While the art market, whatever its form, is not entirely absent in the production of images or the formation of taste, a contemporary artist exercises virtual freedom in forging a body of work—a vision—whose singularity is a mark of the artist's maturity. The art image does not have to yield its meaning easily, does not have to sell a product, does not have to recount a recognizable narrative; it can propose other ways of seeing and other ways of understanding the visible. The viewer in turn is rewarded for cultivating a world of images with broad historical, geographic and cultural boundaries. William Kentridge's work, then, is exemplary of the artist's ability to move the viewer through images, and allows us here to examine the inherent complexities and capacities of the image. Although Kentridge is identified with his animated films, the discussion here will open on his work in other mediums as well and the movement from one medium to another performed as much by the artist as by the viewer, a movement that locates the power of the image outside a specific medium.

I have been suggesting that the image has a powerful capacity to move. For me the pleasure of Kentridge's oeuvre is the complexity that it lends to the idea of movement whose contours I will suggest here. It can be a quality of the image and is evidently one of the primary ones of moving pictures—of film. It is perhaps redundant to reiterate that it is also a formal quality of static images—drawings, paintings, prints, photographs—where the orchestration of the formal qualities of an image—color, line, light, shape—moves the viewer's eye across, around, through, or into an image. Accompanying this formal movement and indeed, inextricably bound up with it, is the movement associated with the development of narrative. Beyond the confines of an individual work, Kentridge transposes his subjects to different mediums; the medium reshapes the subject and the viewer is invited to follow yet another movement of translation. As a South African, Kentridge's work reflects where he works, yet his work also makes reference to other geographic and historical realities that the viewer is invited to visit. Finally, beyond the breadth of these trajectories is the tone of Kentridge's work that strikes a human chord and moves the viewer. Kentridge plays with the fecund connotative potential of the image; indeed his work makes it clear that it is impossible to escape that potential of the image.

Let me now turn to each of these aspects of the idea of movement in Kentridge's work beginning with the physical displacement, for example, of a line on the film screen. This filmic mobility is the cumulative effect of Kentridge's filming the evolving images frame by frame as the charcoal drawing is continuously modified. The clearest illustration of this movement is the use of a blue chalk line—standing out from the tonal effects of Kentridge's usual black charcoal—emanating from the switchboard of a Johannesburg telephone exchange as it wakes up to the city's business and creates a web that envelops the exchange and entangles the city in the film *Stereoscope*.[6] In one image of the exterior of the telephone exchange housed in a four-storey modernist building, these blue lines echo those of the telephone and electric lines that create an urban grid framing the building. The blue lines also signal difference between registers, suggesting that the communication taking place along these wires is not identical with the instruments that carry those messages.

Kentridge's working method distinguishes his animated films from conventional animation where each cell is generated individually, or in computer-generated images, where the human hand has been removed from direct intervention. As an object moves across the screen in a Kentridge sequence, a trace of its former position remains visible, both because it is difficult to completely lift the charcoal dust from the paper, and the draftsman's desire to leave the trace. The movement of this object across the screen, as a piece of paper blowing across the street, leaves a ghostly trail in its wake, suggesting present and past simultaneously. This process bestows a palimpsestic aura to the work, without, I believe, the artist wanting to create the effect of either the palimpsest or the aura, but this effect is one that suggests movement in time and opens up the vista of memory. The drawing literally holds the memory of what preceded it, creating material traces or indexes of physical activities that emerge as surrogates of the functioning of the body and, by extension, of memory.

Even while using the mechanical means of an analog camera to capture the progress of the work and a digital format to project it, Kentridge conserves a trace of his drawing activity that I have suggested is a cipher of the function of memory, a human faculty that is otherwise not evident. Kentridge's commitment to the importance of memory of the individual or the collective arises out of a humanistic tradition forged during apartheid and that continues to express itself in the post-apartheid period when the past is continually and painfully being revealed and assimilated. In other words memory is not only a constituent part of individual bourgeois self-identity, but participates in a larger social and historical process that the individual and society are responsible to recall. Kentridge's work resists what Fredric Jameson had identified in the 1980s as postmodernism's growing amnesia, "the disappearance of a sense of history, the way in which our contemporary social system has little by little begun to lose its capacity to retain its own past."[7] Kentridge's work reminds us that history is not always relegated to cinematic or novelistic reenactment, as it tends to in consumer societies.

Beyond recalling memory, these imperfectly erased images lend an atmosphere to the scene described. Interior and exterior scenes are saturated with grays: landscapes are grisly, an industrial site is a wasteland, an interior is increasingly burdened with the dust and debris of comings and goings. These scenes are suggested impressionistically, a mode that the charcoal medium privileges, in particular when working with a thick blunt stick. A charcoal drawing thus elaborated is unable to render its subject realistically or precisely except through laborious effort. The rough drawings of Kentridge's animation, and all his other drawing-related activities, eschew what might be identified as virtuosity; or suggests, at least, that virtuosity is disguised in other less immediately visible and recognizable forms. Each drawing when filmed is rough and unfinished, as is the narrative in which it participates. The contingency of the drawing translates the contingency of the vision from which it emerges. The suggestion of the lingering presence of the hand despite Kentridge's use of technology belies the seamless surfaces of the simulacra that according to Jean Baudrillard animate screens and navigate cybernetic networks that "by dint of being more real than the real itself" destroy reality.[8]

By marrying his loyalty to a traditional way of making images—drawing, charcoal—with a contemporary technique—film projection—Kentridge provides the viewer a current point of view with an historical perspective, or at least with a view of history, or again, even more modestly, a view of the past that is not explicitly articulated by the artist, nor is it necessarily recognized by the viewer during the screening. That is, the drawing technique itself is a trace of the past, a relic of art history that conjures the illustrious history of western art without doing so in a self-conscious way. The qualities of the charcoal drawings also skirt the slickness of recent monumental photographic work and film, video and DVD projections, most often in color, that emulate the attention drawing posturing of contemporary commercial media. Kentridge's images are comfortable in their unfinished and

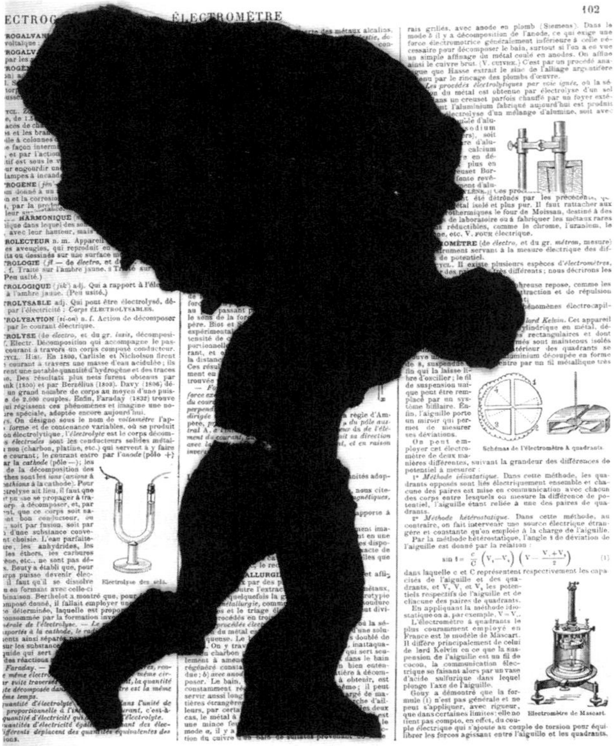

Figure 11.1 William Kentridge, *Portage*, 2000, collage.

incomplete state, as drawings, that in the history of western art, are perceived as preparatory and provisional—despite twentieth-century claims to a higher status—in their grayness and lack of clarity or sharpness, in their sense of muddling through Kentridge's static images—prints, collages, drawings—also imply movement by compositionally stimulating the eye of the beholder in a traditional visual way, as do the 17 pages of the reproduced collages titled *Portage* that constitute an extended frontispiece for the exhibition catalog *William Kentridge*.[9] Roughly torn black construction paper silhouettes—replicating Kentridge's desire to not create a precious drawing—of Africans, many burdened with large objects and packs, trudge heavily or scamper individually or in groups across the page/s that on closer inspection have been appropriated from a French language illustrated encyclopedia. The reproduced page also reveals a barely visible mysterious fault line, as if the encyclopedia sheet had been torn, and then recombined. As in *Portage* (Figure 11.1), the silhouette occupies the field proposed by the print, the stooped figure dragging his heavy pack along the bottom edge of the printed area that figures as the road being trod upon. But even as this edge offers stability and orientation to the cutout, Kentridge's figures challenge the sanctity of this frame by engaging the outer border and sometimes spilling over into the subsequent frame, as the bundle punches at the top of the page here. The grossly articulated figure and its jagged outline move dynamically if slowly across the page and engagingly shape the ground, creating jabbing negative spaces against which the figure struggles or which it leaves in its wake. The ground's more subtle gradations and textures of print and images invite perusal. The vintage French encyclopedia, an artifact of European civilization that has "burdened" these rough-hewn un-individuated figures, acts as negative space, but one that is meaningfully charged. Here, on this page, where the figure acts as a beast of burden, are entries from *electrogalvanic* to *electromètre*, instruments of civilized societies not made available to this act of unassisted manual labor. As the pages are turned, a physical and visual movement is initiated that accompanies the procession of ambulatory figures. The catalog for which this image acts as frontispiece is introduced foregrounding the dialectical movement of Kentridge's work. And just as this image preserves the image/ground relationship and its political dimensions, Kentridge's work invites the establishment of a similar relationship between an image and the political and historical ground out of which it emerges.

The willing participant can also walk the length of the scaffolding-like table of *Procession* (Figure 11.2), taking in and accompanying the 26 small walking bronze figures on their trajectory. The movement acted out is multilayered. In the first instance it consists of the perambulation of the length of or around the piece. This walk can be animated by scrutinizing the individual components and assigning the figures a place in a narrative that is both related to the specific configuration of this work, and for those who have prior experience of Kentridge's work, of the evocation of the history of those figures. Here one both zooms in and zooms out. The individual figures, while textured, are essentially flat silhouettes—like the

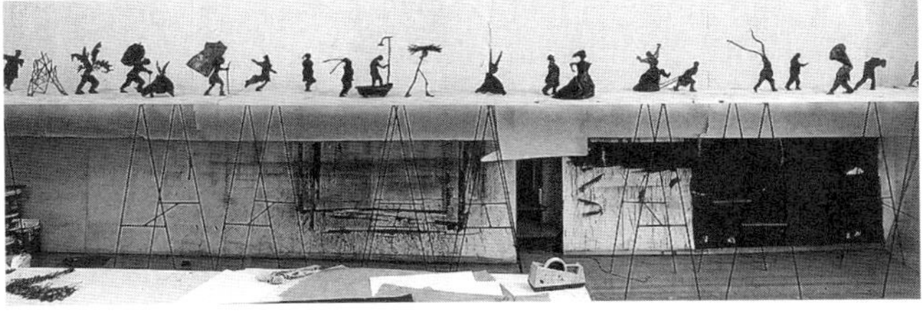

Figure 11.2 William Kentridge, *Procession*, 2000, bronze figures on wooden table with iron trestles, 1.3 × 11 × .6 m.

paper cutouts just discussed—but seemingly made of assembled planar and linear elements, and exhibit the expressionist style of the drawings. But it is in zooming out that the magical screen of Kentridge's work appears. In focusing on the negative space or the ground of the figures, the magical screen of Kentridge's work can appear; there, for example, the memory of the collages just discussed or the great procession of anonymous workers trudging through a desolate landscape as they leave Johannesburg, the opening scene of the animated film *Johannesburg, 2nd Greatest City after Paris* can be evoked.

In the drawing *Procession of the Dispossessed* (Figure 11.3), Kentridge inverts the usual relationship between text and image described by Roland Barthes in his essay *"Rhetoric of the image."*[10] While the inscribed JOHANNESBURG does serve to name the place of the scene described, the stability of the term is shaken by the desolate landscape in which its diminished presence is only one of many graphic elements, and in which no other component reinforces its referent. There are no signs of the prosperous South African city that the name normally conjures. As a sign it is overwhelmed by the desolation of the landscape and the evident misery of what the title suggests are the dispossessed. That is, the image reinforces the contradictory messages that the title and the inscribed text communicate. The image describes, yet the description is open ended, raising more questions than it answers. In this act Kentridge asserts the primacy of the image over the caption, or the visual over text, opening visuality's polysemic possibility to the viewer's interpretative capacity that in more instrumentalized uses of the visual are precisely diminished by the directive function of accompanying textual information.

In each of these three works featuring bodies in motion discussed above, realized in different mediums, Kentridge allows the specificity of each medium to inform its subject differently. My discussion of variations in movement, both actual and virtual, of the work and, complementarily, in the eye and the body of the beholder, only outlines the mechanisms that structure movement within a medium, and ignores the physiological aspect of perception of such movements. Rather, I have addressed the apperception of a work of art at the level of language where it is possible to talk about an image/object and an individual/body. The viewer's appreciation is inextricably entwined with the pleasurable apprehension of the surface of the work and the artist's manipulation of the medium. The viewer's engagement plays out through the senses, recognition, memory and affect, giving rise to a dialectical give and take between the artist, mediated through images and objects, and the reception experienced by the viewer. While this constitutes the basis of the way in which Western art normally functions, what Kentridge's work adds is the movement in register, or the movement in meaning that, to reopen a modernist binary, is part of the internal structure of the work. This creates the conditions for the opening that Roland Barthes,

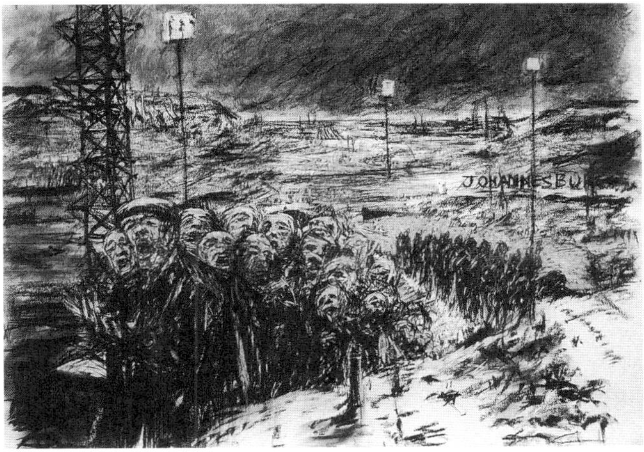

Figure 11.3 William Kentridge, *Procession of the Dispossessed* (detail) 1989, drawing for the film *Johannesburg, 2nd Greatest City after Paris*, charcoal on paper, 92 × 130 cm.

in discussing the affect of photography, identified as the *punctum*, an opening in which the viewer is moved by meaning.[11]

The complexion of movement on the surface of Kentridge's work is complemented by the complexity of meaning that its images generate. Meaning is generated by association between one image and a successive filmic one, by historical reference to an earlier aesthetic style, by metonymic reference to social and political realities, by the moods suggested by color and musical atmosphere, and by the reference to the realm of the individual or the realm of the collective. This wide range of possible meanings in Kentridge's work echoes the interpretive possibilities investigated by the humanist art historian Erwin Panofsky in formulating his study for iconology. In *Meaning in the Visual Arts* Panofsky foregrounds the idea that "the concept to be expressed" is separated "from the means of expression," and is investigated by a person interested in the human condition.[12] Certainly the investigation of meaning of images has taken many turns since Gombrich's foundational study, but what I want to preserve here is its link to a sense of human responsibility evident in Kentridge's production. Kentridge refuses the flattening of meaning that postmodern pastiche—or what Jameson described as blank parody[13]—has embodied and that has, according to Bill Nichols, subsequently been transformed into limitlessly interconnected cybernetic simulacra.[14]

Perhaps I am experiencing a nostalgia for meaning. Kentridge and his commentators have identified his drawing as executed in a German Expressionist style most closely identified with the work of the painter

Max Beckmann and by extension with the free wheeling cultural moment of the Weimar Republic (1919–33), a moment of progressive resistance before its extinction with the seizure of power by the National Socialists beginning in 1933. These expressionistic animated lines of Kentridge's drawings—transposed onto film, and translated into prints and even sculpture—transmit a dynamic energy immediately apprehended, without distorting the recognizability of the image's subject. Indeed, as expressionist drawings do, they translate the emotional field that the subject occupies. In addition though, the style itself conjures a past, even more specific than the history of drawing evoked above, one whose history is colored by emblazoned hope and bitter disappointment; one in which successive generations of intellectual and cultural historians have invested utopian aspirations and perfidious drama.

How appropriate it is then that this expressionist style resurfaces in South Africa where for Kentridge it underwent its gestation during apartheid and emerged on the international scene in the post-apartheid period. That is, Kentridge's work emerged in an historical context where the protagonists were easily identifiable, as were their relative moral positions, although in reality the positions of those inside and outside of the situation were more ambiguous. Indeed, it is this ambiguity that lends Kentridge's work some of its weight in what might be called a post-postmodern oeuvre, one that works through and complicates the easy polarization of identity work of the 1980s and 1990s that was (another) one of the hallmarks of postmodernism in the visual arts.

Perhaps part of the ambiguity of Kentridge's work lies with his being a member of the minority of South African whites, and one who lives in a prosperous Johannesburg suburb that has as yet to appear, to my knowledge, in Kentridge's work. His commentators though seem to be at pains to identify his belonging to a family of lawyers, his own father a lawyer prominently engaged in the anti-apartheid movement.[15] South Africa and its political struggles and tensions are the backdrop for first, the drawings, and second, the worlds of Kentridge's principal protagonists, Soho Eckstein, a businessman, developer and industrialist who according to the artist owned half of Johannesburg and Eckstein's alter ego Felix Teitlebaum. Soho Eckstein and Felix Teitlebaum are both absorbed with their daily preoccupations, and not, as a more radical position would have it, directly involved with the oppression of the majority, on the one hand, or in armed resistance and struggle, on the other. They are, rather, middle class, even if Eckstein, the businessman is the more successful, but nonetheless, identifiable both with Kentridge's social position, and with that of a European and North American art audience. But this middle class position is not innocent and flirts with stereotyping; Kentridge's names for these characters further identifies them as Jewish, as is his own family, extending his self-identification with a group that, in other contexts were the object of persecution, but within the South African context, benefited from apartheid. Kentridge's self-identification subtly underlines the complicity of the heterogeneously composed middle classes in supporting particular regimes.

What I am saying here is tinged with a touch of cynicism regarding the workings of the international art world. In its push in the 1990s to internationalize itself, it picked up a South African artist, but a white one whose work was identifiable stylistically and politically, palatable for consumption by the metropolitan art market (both critical and commercial) that could assuage its perennial guilt about its marginal impact on *Realpolitik* and its lip service to progressive politics. That in no way diminishes my enjoyment of Kentridge's work, and I use enjoyment advisedly to describe the pleasure of viewing, and viewing again, Kentridge's animated films, drawings, prints and sculpture, whose multilayered density intensifies with every new work viewed and re-viewed. That is, the richness of Kentridge's work occupies both axes of external contiguity and internal depth. Indeed the ambiguity that I have just discussed is one of the pleasures gleaned in experiencing Kentridge's work, and is not separate from a discussion of the image. Ambiguity creates a mobile space that I, as a viewer, identify with, aware of the complexity of one's social position that allows one to create or to write or to generate images in a middle class consumer society, aware that this privilege depends on the exclusion, if not the oppression, of others both within the society and beyond. As Kentridge acknowledges in discussing his position, "[a]t the edge of huge social upheavals, yet also removed from them. Not able to be part of these upheavals, nor to work as if they did not exist."[16] Thus the image is replete with nostalgia, ambiguity, and even melancholy, if not despondency, that the grays and blacks of the charcoal drawings—punctuated occasionally with blue or red chalk lines—communicate so effectively, echoed, if not nurtured by the accompanying audio tracks, often melancholic, of the animated films. Kentridge's images, borrowing Martin Jay's summary of Jean-François Lyotard's notion of figurality, "inject opacity into the discursive realm" working against the self-sufficiency of linguistic meaning, and I would add cybernetic image streaming, "introducing an unassimilable heterogeneity into putatively homogenous discourse" and networks. The figural, Jay continues, "is the principle of disruption that prevents any order from crystallizing into full coherence."[17]

The delectation of the object's qualities is bound up with the recognition of the body and memory as reservoirs of the traces of the past, as I suggested earlier. While the social position of Eckstein and Teitlebaum is determined by such factors as geography and class and education and race, Kentridge enunciates these factors, but does not dwell upon them. Kentridge does not reduce his characters to only these external factors, no matter how determining they are. Kentridge, as I argued earlier, shows himself in many ways to be a humanist—and perhaps is such because of the marginality of South Africa, engaged with its human drama and basic social needs, to post-humanist discourses associated with what Timothy Druckery identifies as the "virtualization" of culture[18]—evoking an intellectual past, associated with the expressionist moment, that acknowledges a familiar conception of the western individual. That is, Kentridge reconstitutes for us our interiority analogously, the paper on which he draws the receptacle for marking, as our skins are the surfaces of our encounters with the world whose traces are, for Kentridge's middle class audiences, engraved in our memories. The imprecisely drawn figures that I described above implicated in the indefinite narratives played out in the animations correspond to a humanist vision of the individual. Or, more accurately, a desire to humanize the consciousness-centered individual of that worldview by acknowledging the imperfections and shortcomings of the individual, embodied in a balding nondescript middle age body of an Eckstein or a Teitlebaum, and his daily unheroic undertakings. In other words, Kentridge proposes an image of ourselves that we can live with. It embodies a position that the historian Saul Friedlander locates in artwork that constitutes another body of memory work, that addressing the Holocaust. James Young encapsulates Friedlander's reflections: this artwork "remarks its own limitations, its inability to provide eternal answers and stable meanings. Works in this vein acknowledge both the moral obligation to remember and the ethical hazards of doing so." This artwork devotes itself "to the dilemmas of representation, an antiredemptory history of the Holocaust that resists closure, sustains uncertainty, and allows us to live without fully understanding."[19] Kentridge's self-conscious figures muddling through open-ended narratives whose background is a politically charged past and present, embodies the individual's dilemma in dealing with memory.

The emotional complexion of the images is, as I have suggested, composed of many elements, including the gray of the charcoal that has worked itself into the pores of the paper, an appropriate image in relation to one of the subjects of the film *Mine* (1991), that as the title suggests includes drawings of Africans working deep in the mines that were and are an important source of South Africa's wealth. One of the most compelling transformations of image effected by Kentridge is that involving a *cafetière* from which Soho Eckstein pours his morning coffee into his cup on a breakfast table resting on his bed. The coffee plunger begins to revolve and bores through the table, the bed and the floor and plunges us deeper and deeper into the bowels of the earth, that below Johannesburg are burrowed with mine tunnels. The coffee plunger takes on the function of a pneumatic drill, as it tunnels through the somber layers of the mine the crowded bunk beds of the dormitories, the shower rooms that lack privacy, dark faces further darkened by the pallid artificial subterranean light and the ever present dust. This transformation is as much metonymic as it is metaphoric as it creates a relationship between the instruments of leisure and the tools of labor. Visually linking the two realms, the plunger/drill creates a contrast between the other conditions that obtain for Eckstein, the white entrepreneur and the black mass of laborers. Eckstein for one has an identity and enjoys life's pleasures in a comfortable solitude. This realm rules over the layers below, revealed like a descent into Dante's inferno, but where the indignities and tortures are not the imagined ones of the afterlife but those imposed by the necessity of work. The Africans are nameless, without identity, and fulfill their daily functions in cramped quarters and inescapably in the company of others. And so the two realms exist simultaneously, one dependent on the other, the conditions of one shaping the conditions of the other. Thus the burrowing movement effected by the plunger transformed into a drill links not only two distinct spaces—above and below—but qualifies and relates these distinct but related realms. The physical displacement that is both graphically illustrated and figuratively enacted involves a temporal dimension that creates a memory, as one reality is substituted for another. The film's physical movement and transformations provoke a moving play of memory and association.

The way in which Kentridge's pictures move us as viewers signals the artist's respect for the power of the image, a respect that allows the image to communicate without foreclosing meaning in advance of either a passing glance or a sustained viewing. And while a sustained examination forges associations with other images in the oeuvre of the artist, or with the wider world of images, enriching the possibility of meaning, this mute world hovering around the signification of language continuously defies language to circumscribe the affect of moving pictures.

Notes

1. *William Kentridge*, New Museum of Contemporary Art, New York, June 3–September 16, 2001, was an exhibition initiated by the New Museum and the Museum of Contemporary Art, Chicago, curated by Neal Benezra, Staci Boris and Dan Cameron. The show toured to the Hirshhorn Museum and Sculpture Garden in Washington, Houston's Contemporary Arts Museum, the Los Angeles County Museum of Art and the South African National Gallery, Cape Town. An exhibition catalog *William Kentridge* was published by Harry N. Abrams in 2001. This was not the first North American viewing of his work: he first submitted work to New York's American Film Festival in 1982 and subsequently participated in group festivals and projects.
2. *Tide Table*, Marian Goodman Gallery, March 4–April 10, 2004.
3. *William Kentridge*, Metropolitan Museum of Art, November 12, 2004–April 10, 2005.
4. *The Body Transformed*, National Gallery of Canada, Cité de l'Energie, Shawinigan, June 15–October 5, 2003.
5. I have not had the opportunity to witness Kentridge's work in theatre where he has been very active and that has influenced his pictorial, graphic, sculptural and filmic approaches. Kentridge has been an actor, playwright, director, set designer and graphic designer with various companies and productions, including the Handspring Puppet Company that integrated puppets, actors and animation. His collaboration with the latter is included in a 1999 documentary titled *William Kentridge, Drawing the Passing* by Maria Anna Tappeiner and Reinhard Wulf.

6. A partial list of William Kentridge's animations includes: *Johannesburg, 2nd Greatest City after Paris*, 1989, 8'2", Music: Duke Ellington, choral music; *Monument*, 1990, 3'11", Music: Edward Jordan; *Mine*, 1991, 16mm., 5'50, Music: Dvorak's Cello Concerto in B minor, opus 104; *Sobriety, Obesity, and Growing Old*, 1991, 8'22, Music: Dvorak String Quartet in F, opus 96, South African choral music, Friedrich von Flotow "M'apparia" from *Martha* sung by Enrico Caruso; *Felix in Exile*, 1994, 8'43, Music: Philip Miller, composition for String Trio, Motsumi Makheme, *Go Tlapsha Dadiba*; *History of the Main Complaint*, 1996, 5'50, Music: Claudio Monteverdi, *Ardo* madrigal; *Ubu Tells the Truth*, 1997, 8'; *WEIGHING . . . and Wanting*, 1998, 6'20", Music: Philip Miller; *Stereoscope*, 1999, 8'22, Music: Philip Miller; *Tide Table*, 2003.
7. Fredric Jameson, "Postmodernism and Consumer Society," in *The Anti-Aesthetic: Essays on Postmodern Culture* (ed.) Hal Foster (Port Townnsend, WA: Bay Press, 1983) 125.
8. Cited in Bill Nichols, "The Work of Culture in the Age of Cybernetic Culture," in *Electronic Culture: Technology and Visual Representation* (ed.) Timothy Druckery (New York: Aperture, 1996) 133.
9. *William Kentridge* i–xvii.
10. Roland Barthes, "Rhetoric of the image," in *Image, Music, Text*, ed. and trans. Stephen Heath (New York: Hill and Wang, 1977) 32–51.
11. Roland Barthes, *La chamber claire: Notes sur la photographie* (Paris: Gallimard Seuil, 1980) 49.
12. Erwin Panofsky, *Meaning in the Visual Arts* (New York: Doubleday, 1955) 5–6.
13. Jameson, *Postmodernism*, 114.
14. Nichols, *Work of Culture*, 134.
15. For example, Dan Cameron's essay, "A Procession of the Dispossessed," in *William Kentridge* (London: Phaidon, 1999), 38–81.
16. Quoted by Lynne Cooke, "Mundus Inversus, Mundus Perversus," *William Kentridge* 41.
17. Martin Jay, *Downcast Eyes: The Denigration of Vision in Twentieth Century French Thought* (Berkeley: University of California Press, 1994) 564.
18. Timothy Druckery (ed.), *Electronic Culture: Technology and Visual Representation* (New York: Aperture, 1996) 13.
19. James E. Young, *At Memory's Edge* (New Haven and London: Yale University Press, 2000) 6.

Part V Ethics, Politics, and Imaging

12

Publicity and Indifference: Media, Surveillance, "Humanitarian Intervention"*

Thomas Keenan

> The price of eternal vigilance is indifference.
>
> Marshall McLuhan, *Understanding Media*

In his too-hasty indictment of the 1999 NATO air campaign over Kosovo, *Stratégie de la déception*, Paul Virilio suggests that there was a determined relation between "the 'humanitarian' dimension of this very first 'human rights conflict'" (p. 30/p. 21) and the "truly panoptical vision" (p. 28/p. 19) which NATO brought to bear on the battlefield.[1]

> After the eye of God pursuing Cain all the way into the tomb, we now have *the eye of Humanity* skimming over the oceans and continents in search of criminals. One gets an idea, then, of the ethical dimension of the Global Information Dominance programme, the attributes of which are indeed those of the divine, opening up the possibility of *ethical cleansings*, capable of usefully replacing the *ethnic cleansing* of undesirable or supernumerary populations. After oral informing, rumor, agents of influence and traditional spying, comes the age of optical informing: this "real time" of a large-scale optical panoptics, capable of surveilling not just enemy, but friendly, movements thanks to the control of public opinion.
>
> (pp. 31–2/pp. 21–2)

This "global telesurveillance" (p. 32/p. 22) is for Virilio the signature of the "globalist putsch" he denounces, "a seizure of power by an anational armed group (NATO), evading the political control of the democratic nations (the UN), evading the prudence of their diplomacy and their specific jurisdictions" (p. 74).

The tropes are all-too-familiar, and not just to readers of Virilio. Democracy sacrificed to speed, accountability to total visibility. As if surveillance were just one thing. As if the images produced by the global panoptics were self-evident in their meaning or effect. And as if every project taken on by the

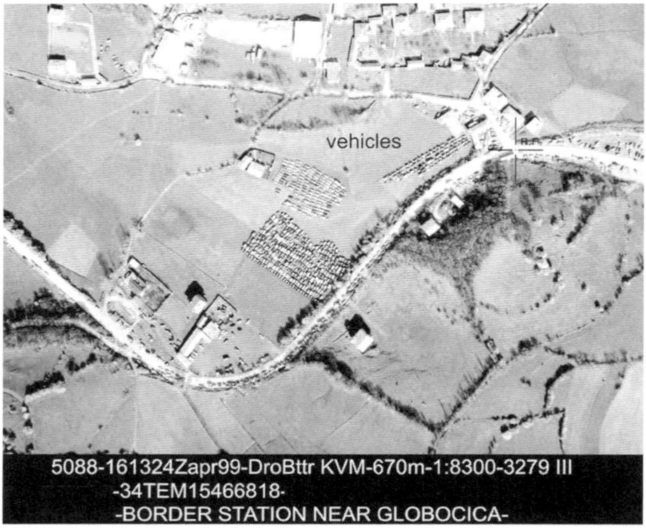

Figure 12.1 Border station near Globocica, Kosovo, April 1999; image from Bundeswehr CL 289 UAV (drone). *Source*: Bundeswehr.

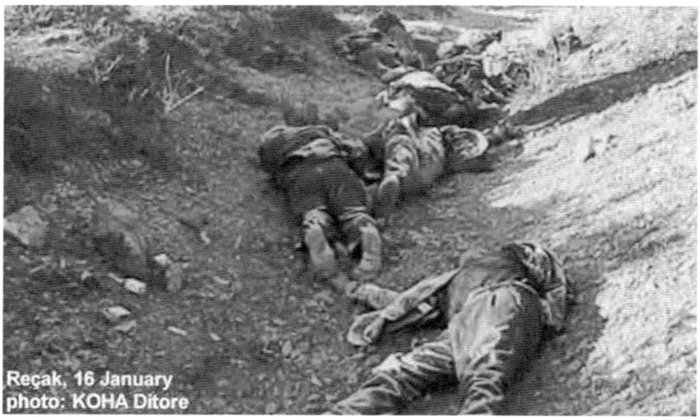

Figure 12.2 Bodies at Racak, Kosovo, January 16, 1999; news photograph from *Koha Ditore* (Pristina). *Source*: Koha Ditore.

Western military alliance, or what his Foreign Minister Vedrine memorably nicknamed the hyperpower, was irremediably contaminated. But those are obvious commonplaces. What is interesting is the question of betrayal, denunciation, of this "informing [*délation*]." What difference does all the watching make? Especially where 'ethnic cleansing' is at stake—or to call it

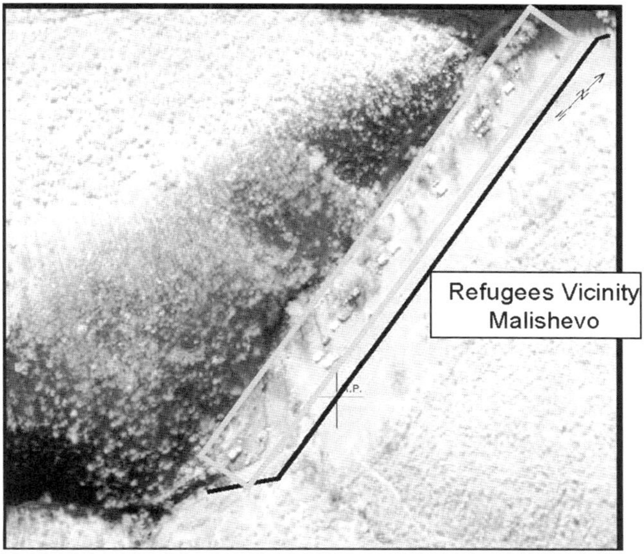

Figure 12.3 Refugees, vicinity of Malisevo, Kosovo, April 1999; overhead imagery released at NATO briefing, April 10, 1999. *Source*: NATO.

by its legal name, where genocide is underway? Virilio's dissident position, that what is truly to be feared and resisted is less the killing itself than the practices of global control it alibis or sets in motion, is at once unjustified and deeply flawed from a political standpoint.

But, interestingly, it also runs counter to the most cherished axioms of the international human rights and humanitarian movements. Since the end of the Second World War, indeed, the nongovernmental movement has looked forward to the prospect of up-to-date information about crimes in progress, coupled with access to the public opinion that might enable them to be interrupted. With the creation of a rich and increasingly robust global network of human rights monitors, and the ability to relay acts of witness and evidence around the world in near-real-time, something like this transparent world is increasingly real. "The media will carry the demand for action to the world's leaders; they in turn must decide carefully and positively what that action is to be," runs the axiom in its clearest formulation.[2] But what of the reaction, the action, and the public? Kosovo—where a limited military intervention probably prevented a genocide, protected a terribly endangered civilian population, and finally stopped a military and paramilitary apparatus that had terrorized primarily Muslim civilian populations in southern Europe for most of the 1990s—was rather the exception than the rule. Global telesurveillance and human rights monitors did not help much at Vukovar, Omarska, or Srebrenica. Nor did these names confirm the omnipotence of NATO, or the unaccountable power of the transnational human rights movement. After a decade of genocide, famine, and concentration camps, the very value of publicity—whether that affirmed by the movements or condemned by Virilio—seemed questionable.

* * *

Visiting Sarajevo at Christmas in 1993, less than a year into its suffering, the Archbishop of Paris Cardinal Jean-Marie Lustiger noted the strikingly public or visible character of the carnage there.[3] In an interview with Zlatko Dizdarevic of Sarajevo's *Oslobodjenje*, he compared the siege of the city to the horrors of the Second World War, but with a significant difference.

> Here, however, there are no secrets. There are journalists here, from here pictures are transmitted, there are satellite communications, all of this is known. In this city there are soldiers of the United Nations, well armed, and nonetheless it all continues to happen. This is unbelievable; this is overwhelming. One man yesterday told me that everyone here feels like they are animals in a zoo that others come to look at, to take pictures of, and to be amazed. And then, those up in the mountains also treat them like animals, killing them and "culling" them.

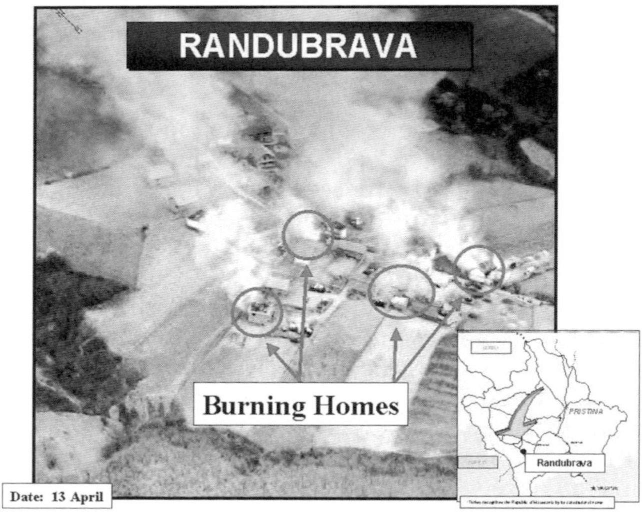

Figure 12.4 Burning homes in Randubrava, Kosovo, April 13, 1999; overhead imagery released at NATO briefing, April 14, 1999. *Source*: NATO.

Dizdarevic asked how it was possible that, "all of this goes on without any end in sight, in spite of the fact that we are surrounded by hundreds of cameras [. . . and] that everyone knows everything and sees everything"? Lustiger responded: "There is no answer for that—I really do not have an answer. However, that means that it is always possible to get worse and worse."

Lustiger's bold and uncompromising position, as rare as it was at the time, has now achieved the status of common sense. Among the too many would-be "lessons of Bosnia," this one stands out for its frequent citation: that a country was destroyed and a genocide happened, in the heart of Europe, on television, and what is known as the world or the West simply looked on and did nothing. "While America Watched," as the title of a documentary on the genocide in Bosnia broadcast by ABC Television in 1994 already put it.[4]

The surveillance was as complete as the abandonment. Bosnians, said one to the American journalist David Rieff, "felt as you would feel if you were mugged in full view of a policeman and he did nothing to rescue you."[5] Or, as Rieff himself put it, "200,000 Bosnian Muslims died, in full view of the world's television cameras, and more than two million other people were forcibly displaced. A state formally recognized by the European Community and the United States [. . .] and the United Nations [. . .] was allowed to be destroyed. While it was being destroyed, UN military forces and officials looked on, offering 'humanitarian' assistance and protesting [. . .] that there was no will in the international community to do anything more" (p. 23).

But what does "in full view" mean, and what is the particular ethico-political force of this condemnation: not just genocide, but genocide in the open, transparent mass murder?

There is no denying the simultaneity of this watching and that destruction. They happened together—and what happened should not have happened. But what did the surveillance and the watching have to do with what happened? What links the thing we so loosely call "the media" and its images with action or inaction? Or more precisely, when something happens "in full view," why do we expect that action will be taken commensurate with what (we have seen) is happening? And what about that humanitarian assistance: what sort of "action" is it?

This trajectory of this program—from the camera to a response, but maybe nothing "more" than a humanitarian one—appears everywhere today, in military and political and historical discussions of so-called postmodern wars or humanitarian crises, in legal or ethical commentaries on genocide and catastrophe, and in critical media-studies analyses of what has been called the CNN effect or the role of the media in contemporary conflict. And what seems to concern us the most, for better and for worse, are the media. It seems as if we cannot talk about what happened in Bosnia or Somalia or Rwanda without talking about the media.[6]

Consider, for (and unfortunately only for) example, the brilliant series of articles in the *New York Review of Books* in which Mark Danner chronicled the high and low points of the battles over Bosnia in the United States and Europe. He was in Sarajevo for much of it, but his articles insistently began with watching television.[7] "To the hundreds of millions who first beheld them on their television screens that August day in 1992, the faces staring out from behind barbed wire seemed painfully familiar," begins his December 4, 1997 report on the camps of Western Bosnia. The opening sentences of his November 20, article tells a similar story about Srebrenica:

> Scarcely two years ago, during the sweltering days of July 1995, any citizen of our civilized land could have pressed a button on a remote control and idly gazed, for an instant or an hour, into the jaws of a contemporary Hell. Taking shape upon the little screen, in that concurrent universe dubbed "real time," was a motley, seemingly endless caravan, bus after battered bus rolling to a stop and disgorging scores of exhausted, disheveled people. [. . .] every last one a woman or a child. The men of Srebrenica had somehow disappeared. Videotaped images, though, persist: on the footage shot the day before, the men can be seen among the roiling mob, together with their women and children, pushing up against the fence of the United Nations compound, pleading for protection from the conquering Serbs.

From 1992 to 1995, says Danner, we watched, and what we did and didn't do with what we saw was all the less forgivable, because we could see.[8] Many other versions of this protest could be enumerated, but the precise formulations of and differences among them are less interesting than their ubiquity. The recurrence of the gesture (we watched "all that" but we did not act as we should have), across so many different accounts and styles and methodological predispositions, mirrors somehow the phenomenon it describes: the omnipresence of the gesture is the very ubiquity of the camera, the image or specter of the camera that now seems to haunt our consciousness, and indeed, the in-full-view-of-the-camera seems now to have become the most privileged figure of our ethical consciousness, our conscience, our responsibility itself.

This was not always a rebuke. Television, publicity, surveillance of the affirmative sort, was supposed to help. This was the situation Michael Ignatieff described some years ago—before Bosnia and Rwanda, when the crises were those of starvation and Cold War proxies—in an essay on "the ethics of television," now the first chapter of *The Warrior's Honor*.[9]

> Television is also the instrument of a new kind of politics. Since 1945, affluence and idealism have made possible the emergence of a host of nongovernmental private charities and pressure groups—Amnesty

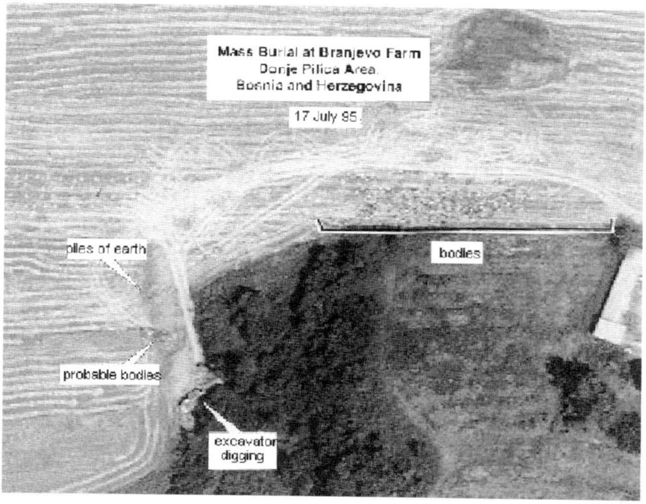

Figure 12.5 Mass burial at Branjevo Farm, Donje Pilica Area, Bosnia-Herzegovina, July 17, 1995; image from United States U-2 aerial reconnaissance aircraft, released by US officials in Bosnia, March 22, 1996. *Source*: US Department of State.

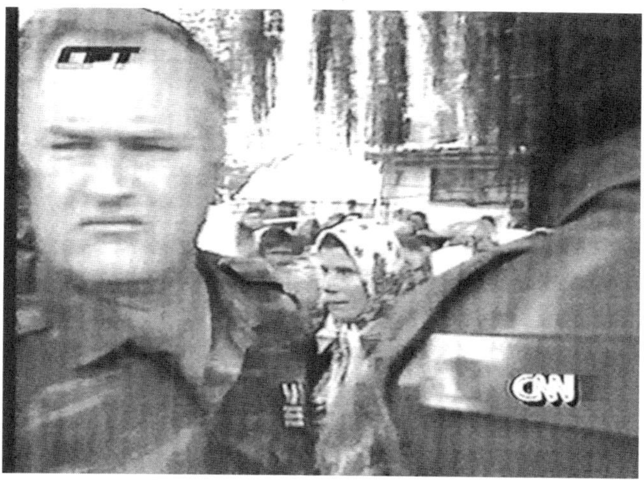

Figure 12.6 General Ratko Mladic, Bosnian Serb Army, Potocari, Bosnia-Herzegovina, July 12, 1995; video still of RTS image on CNN, date unknown. *Source*: Radio Television Serbia via CNN.

International, [. . .] Medecins sans frontières, and others—that use television as a central part of their campaigns to mobilize conscience and money on behalf of endangered humans and their habitats around the world. It is a politics that takes the world rather than the nation as its political space and that takes the human species itself rather than specific citizenship, racial, religious, or ethnic groups as its object. [. . .] Whether it wishes or not, television has become the principal mediation between the suffering of strangers and the consciences of those in the world's few remaining zones of safety. [. . .] It has become not merely the means through which we see each other, but the means by which we shoulder each other's fate. (pp. 21, 33)

Ignatieff allows us to orient this inquiry toward the special relationship between television and humanitarianism. International humanitarian action of the sans-frontières variety is unthinkable except in the age of more-or-less instant information. As Rony Brauman has underlined, the founding of the International Committee of the Red Cross in 1864 is linked noncoincidentally to the possibility of high-speed transmission by telegraph, and contemporary relief operations since Biafra and Ethiopia have been born and bathed in the light of the television camera and the speed of the satellite uplink.[10] Humanitarian action seems not simply to take advantage of the media, but indeed to depend on them, and on a fairly limited set of presuppositions about the link between knowledge and action, between public information or opinion and response. In some cases, like that of the international human

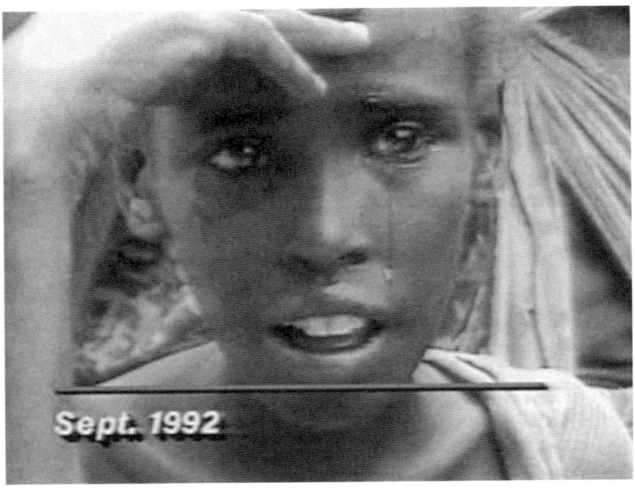

Figure 12.7 Somalia, September 1992; video still from CNN, date unknown. *Source*: CNN.

rights movement, as Alex de Waal has argued, the conditions of action rest all too heavily on the concept of "mobilizing shame."[11]

In the humanitarian arena proper, the pioneering French activist turned politician Bernard Kouchner put the coordination between media and intervention in a simple epigram: "sans médias, pas d'action humanitaire importante, et celle-ci, en retour, nourrit les gazettes [without the media, there is no important humanitarian action, and this, in turn, feeds the papers]." Kouchner calls this "la loi du tapage," the law of noise.[12] And among military thinkers, practitioners, and diplomats, the sense that television imagery or news dispatches "drive" decisions about intervention has by now gained a name of its own—"the CNN effect"—and is the topic of vigorous debates.[13]

What does it mean? Thanks to what is loosely termed "public opinion" in the media age, which displaces or warps state institutions and power through emergent alternative centers of power like the media and non-governmental organizations, the so-called famine movement (or what Alex de Waal has nicknamed the "Humanitarian International") has emerged as a political actor, and of a new sort: apparently unlimited by traditional notions of sovereignty, accountability, borders, interest, and the rest.[14]

We need to understand the "humanitarian action" which triumphed in Bosnia as something different from either of the two obvious options: it was neither inaction (a passive acquiescence or a cover-up, a fig leaf that disguises the actual doing-of-nothing), nor a heroic new non-state politics of the sort anticipated by many of the founders of the movement. It was an action that—precisely because it offered the possibility of a reference not to national interest or the defense of the state but to what it called, alternatively, "human beings," "victims," "misfortune" or "suffering," and did so by way of public opinion and the image, which is to say by reference to the order of the ethical—opened the possibility of a political discourse that, for better or more often for worse, did not have to justify itself in political terms. In Bosnia, humanitarian action was action indeed, action that threatened to totalize the field of all possible action: not simply to hide inaction or offer alibis for not doing other things, but more radically to interrupt, to render impossible, to actively block or prevent those actions.

And this action had as its field or condition the image, sometimes precisely the image and sometimes more generally what we nickname "the media" or "real time." Recall that for Walter Benjamin, in the Artwork essay at least, the invention of the motion picture introduced nothing less than a temporal explosion, "the dynamite of the tenth of a second," such that in what remained, the "far-flung ruins and debris" of our daily lives or our familiar terrain, would open up "an immense and unexpected field of action."[15] Film and today television do not only collapse and annihilate, as is so often said, time and distance—they also make unprecedented times and spaces available for action, real virtualities that are marked by the affirmation of possibilities of engagement, "action," as well as by the negativity of this "dynamite." Field of action, yes, but what kind of action? The answer is also Benjaminian, though this time in a different way. The privileged example at the close of the Artwork essay is war, what he labels the aesthetics of mechanized warfare, which he says is discerned more clearly or best "captured by camera and sound recording" and not the naked eye (pp. 242, 251). Today cameras don't simply represent conflicts but take part in them, shape not only our understanding of them but their very conduct. We need to attend to these sounds and images not just as accounts of war but as actions and weapons in that war, as operations in the public field which today constitutes an immense field of opportunity for doing battle, as weapons in what we too easily call "image contests" or "publicity battles."

* * *

"There was a cameraman there"—this is a fragment from a news report about a man shot by a sniper in Sarajevo.[16]

> Mr. Sabanovic got in the way at a particularly dangerous Sarajevo crossroads. That is why there was a cameraman there to film his near death. Because the spot is treacherous, the chances are good that a few hours of patience by a cameraman will be rewarded with compelling images of a life being extinguished or incapacitated.
>
> (p. 12)

What difference does it make that a cameraman is there, as he or she so often is? No matter where, it seems, a camera regularly happens to be there, when

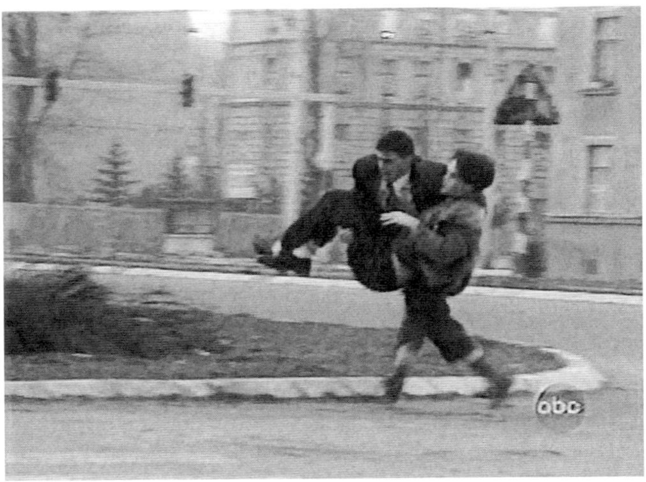

Figure 12.8 Faruk Sabanovich, outside the Holiday Inn, Sarajevo, Bosnia-Herzgovina, March 3, 1995; video still from ABC, March 3, 1995. *Source*: RTV/WTN pool material via ABC.

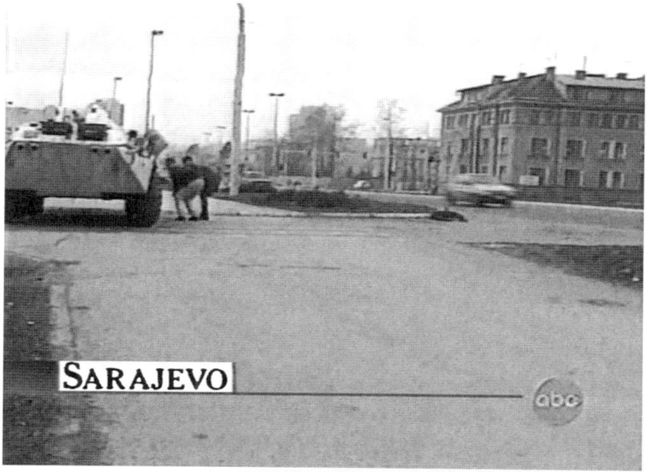

Figure 12.9 Faruk Sabanovich, outside the Holiday Inn, Sarajevo, Bosnia-Herzgovina, March 3, 1995; video still from ABC, March 3, 1995. *Source*: RTV/WTN pool material via ABC.

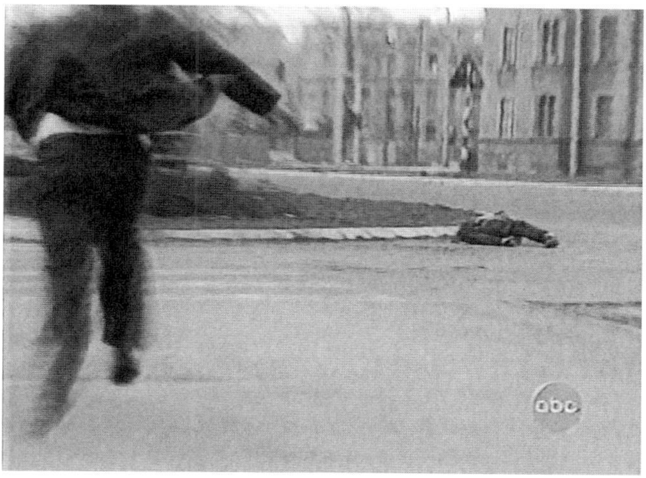

Figure 12.10 Faruk Sabanovich, outside the Holiday Inn, Sarajevo, Bosnia-Herzgovina, March 3, 1995; video still from ABC, March 3, 1995. *Source*: RTV/WTN pool material via ABC.

something happens to happen. So much so that it has become a cliché, a veritable commonplace, to say that today things don't happen *unless* a camera is there. Of course, it takes not just a camera, but an entire network of editing, transmitting, distributing, and viewing technologies—and agents—that extend out from the camera, to make what McLuhan so famously and confusingly called a global village.[17] But it begins with the camera and its operator, with their already having been there.

What the journalist here wants us to understand is the complex structure of that "there"—was it a place where cameras waited patiently for things to happen (a particularly dangerous crossroads), or a place where things happened because cameras waited patiently (compelling images of lives extinguished)? The camera is there because of the danger, but its silent witness transforms the event and its "there"—that is what matters here. Thanks to the camera, what it means for the event to occur, its taking-place, undergoes a mutation. The crossroads so precisely targeted in the sniper's gunsight is also the blurred intersection of what our impoverished theoretical vocabulary allows us to call only event and representation, occurrence and image. This confusion cannot be written off as one more version of a timeless ontological conundrum (which comes first?), nor caricatured as a postmodern prejudice for the discursive over the real, nor simply eliminated with a declaration of the moral superiority of the things themselves. The confusion itself is all too real and—especially in the case of events like those at this crossroads—it constitutes something like an exemplary ethico-political difficulty and opportunity for us.

What is at stake, finally, in this confusion is a certain experience and definition of public space and time, of publicity and of a crisis in our sense of public information and exposure today. The corollary, of course, of the cameraman's being there is that, in some sense, we are too. The camera metaphorizes the becoming-public of the event, because we who watch and listen are also caught in the intersection of the sniper's and the cameraman's viewfinders—not as potential victims exactly, but in some other sense as targets of those vectors (borrowing this sense of the word from McKenzie Wark in *Virtual Geography*).[18] What do we do in watching and listening? When I say "we," I mean that hazy thing called the public, a rich concept sent to us by the Enlightenment and the French Revolution and in need of extensive rethinking. If the public means us, us in our exposure to others, then today "we" cannot be something given in advance, not the sum total of all of us somewhere or sometime, not a community or a people but rather something that comes after the image, a possibility of response to an open address. The public, we could say in shorthand, is what is hailed or addressed by messages that might not reach their destination. Thinking about the images at hand, we could even say that what makes something public is precisely the possibility of being a target and of being missed.

So the television image constitutes a field of action—not just a representation of actions elsewhere but a field in or on which actions occur—a public field, we could say, but only if we're willing to part with some of the cherished predicates of that concept.

We can begin with this snapshot, or live feed. Somalia, December 1992. The first American soldiers of Operation Restore Hope land on an Indian

Ocean beach at Mogadishu, met not by clan fighters or starving children but by hundreds of reporters, camera people, technicians, whom, as it turns out, the American military had informed in advance of the time and place of the operation. Kouchner's claim that without television, there is no

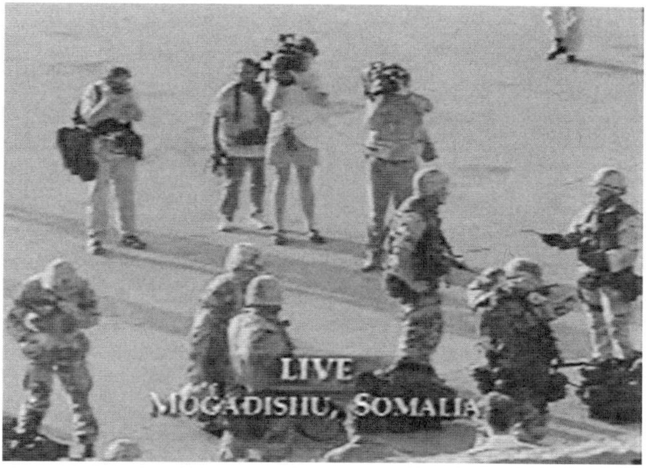

Figure 12.11 Mogadishu airport, Somalia, December 9, 1992; live video still from NBC. *Source*: NBC.

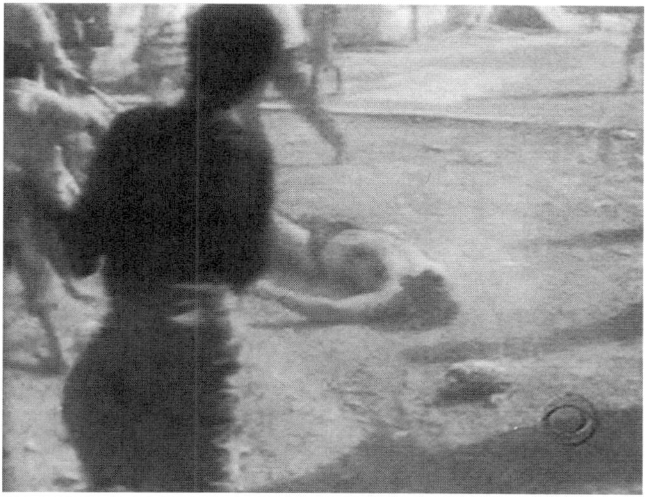

Figure 12.12 Mogadishu, Somalia, October 4, 1993; video still from CBS, October 4, 1993. *Source*: Reuters via CNN.

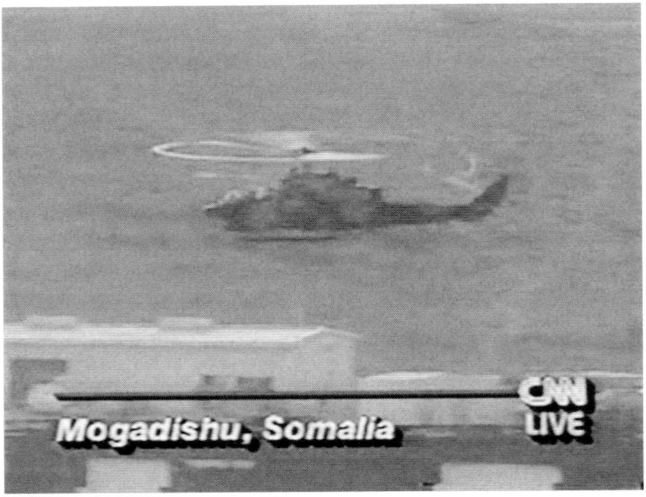

Figure 12.13 Mogadishu airport, Somalia, December 9, 1992; live video still from CNN. *Source*: CNN.

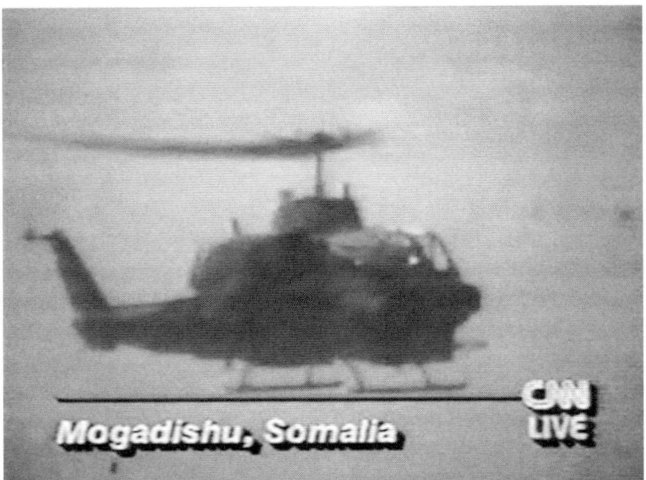

Figure 12.14 Mogadishu airport, Somalia, December 9, 1992; live video still from CNN. *Source*: CNN.

humanitarian intervention, seems to come true in a multiple and almost perverse way here: not simply that images—there, of starving children—could shame governments into action, but that armies will undertake humanitarian rescue missions for the publicity value alone, and that publicity could also bring the mission to an end.

What happened there? We are not finished understanding the complex of clan politics and paramilitary violence, the liquidation of the postcolonial and post–Cold War state, famine and even starvation, and the succession of interventions, humanitarian and armed ones, and then nation-building which followed them.[19] But the images (from the starving children to the gun-belted fighters, the brightly lit landing, and the camcorder pictures of a helicopter pilot held hostage and a dead soldier dragged in the street) and the phrases (Mad Max vehicles, warlords, the photo op invasion and the CNN effect, and the Mogadishu line) have already decisively shaped the interpretation and practice of humanitarian interventions in the decade since that fateful night in the lights.[20]

The lesson of those lights was already clear, the morning after the event, to the grand old man of American foreign policy, George Kennan, who awoke that morning in December 1992 to watch the soldiers landing in real time, surrounded by reporters and interviewed on the beach, and offered a harsh assessment of the damage. He told his diary, and then the opinion page of the *New York Times*, that he had finally seen enough:

> If American policy from here on out, particularly policy involving the use of our armed forces abroad, is to be controlled by popular emotional impulses, and particularly ones provoked by the commercial television industry, then there is no place—not only for myself, but for what have traditionally been regarded as the responsible deliberative organs of our government, in both executive and legislative branches.[21]

Kennan—the architect of the Cold War, the author of the doctrine of "containment," Mr. X himself—watches his era end on his television, not with the fall of the Berlin Wall and the reunification of Europe, nor with the great borderless coalition and its triumph in the Gulf War, but with chaos on an African beach, disaster breaking out of new world order with such energy and confusion that it threatens to tear apart the institutions of government and publicity themselves. What is threatened in Mogadishu, not by the clans but by the cameras and the soldiers who are drawn to them, is nothing less than the basic structures of ethics and American democracy—responsibility and deliberation. The rational consideration of information, with a view to grounding what one does in what one knows, now seems overtaken and displaced by "emotion," and responses are now somehow "controlled" or, better, remote-controlled by television images. What disappears beneath the image or behind the screen is the place of politics itself.

There is, Kennan confesses, not only no place for him but no place at all for a decision, for the organs that regulate the link between knowledge and action. Television—that virtual place—displaces the public place, substituting emotion for reason, immediacy for the delay proper to thought.

In somewhat more complex, but no more theoretical terms Virilio has suggested that this phenomenon, the displacement of the traditional rational–critical experience of the public sphere by what is nicknamed "emotion," characterizes in general contemporary televisual publicity:

> The space of politics in ancient societies was the public space (square, forum, *agora* . . .). Today, the public image has taken over public space. Television has become the forum for all emotions and all options. We vote while watching TV. [. . .] We are heading toward a cathodic democracy, but without rules. [. . .] There is no politics possible at the scale of the speed of light. Politics is the time of reflection. Today, we no longer have time to reflect; the things that we see have already taken place. And we must react immediately. [. . .] Is a real-time democracy possible? An authoritarian politics, yes. But what is proper to democracy is the sharing of power. When there is no longer time to share, what do we share? Emotions.[22]

This compelling immediacy of the media, the magnetic pull of the image and the microphone, has been testified to by the highest officials of our government and military.[23] Images, they certify, do make things happen and sometimes too quickly. We can and should dispute the contention that discussion or sharing disappear in the putative instantaneity of the live transmission (as if it does not have its own temporality, its own internal structure, its delays and frames and decisions) but there is no debating the claim that the image (and especially the image of catastrophe) has the power to circumvent or pressure political institutions, and not just in democracies.[24]

In Somalia events did seem dictated by this CNN effect, with the attendant displacement of deliberation by emotion and hence the short-circuiting of the public sphere, whether it was a matter of the starving children, the proud international forces, or the dead American soldier. But what then of Bosnia, where everything seemed to be visible as it happened, and yet, on the contrary, it is said, virtually nothing happened in response? As David Rieff wrote, "no slaughter was more scrupulously and ably covered" and "it [did] no good"—"we failed":

> [T]he hope of the Western press was that an informed citizenry back home would demand that their governments not allow the Bosnian Muslims to go on being massacred, raped, or forced from their homes. Instead, the sound bites and "visual bites" culled from the fighting bred

> casuistry and indifference far more regularly than [they] succeeded in mobilizing people to act or even to be indignant.
>
> (pp. 223, 222, 216).

If the lesson of Somalia was that cameras made things happen, and sometimes too quickly, Bosnia seemed to tell the opposite story: a brutal combination of overexposure and indifference. Somalia was hyperactivity; Bosnia inactivity, just watching. This was the clichéd meaning for which Sarajevo became the metonym. We are back to where we started: let me cite a few examples, from war correspondents themselves, of the trajectory that travels from a certain expectation about the putative power of images to despair at their failure and even to anger, from Mogadishu to Sarajevo.

Only nine months into the siege, in a dispatch that won her the first of many prizes for coverage of Sarajevo, CNN's Christiane Amanpour reported on a creeping despair with the televisual:

> Take any day in the life of this city. The sights are so familiar, perhaps they have lost their impact. [. . .] Around noon another mortar falls. More people are killed and injured. They are rushed to the hospital. The emergency ward is full. Surgeons labor to save lives. The operating theatre is awash in blood. Early on in the war the staff were patient with photographers, hoping perhaps their pictures would shock the world into doing something. The world has done nothing and the doctors have lost hope and patience.[25]

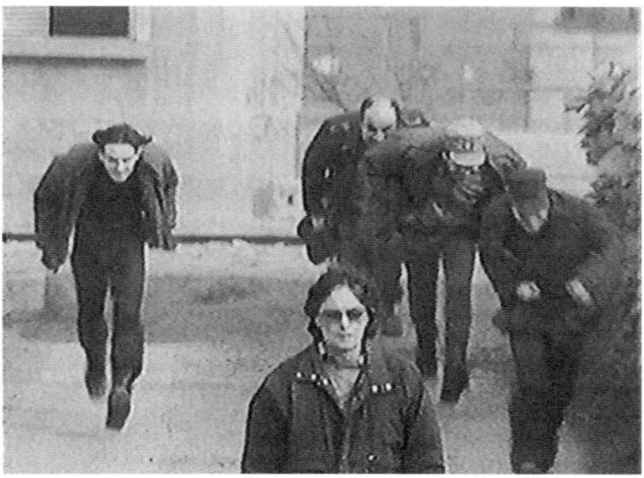

Figure 12.15 Sarajevo, Bosnia- Herzegovina, January 1993; video still from CNN, January 26, 1993. *Source*: CNN.

Years later Giles Rabine, reporting live for France 2 from Sarajevo on July 13, 1995, just after the fall of Srebrenica, commented simply that, after 39 months of televised siege, "the Sarajevans have had enough of being interviewed, being filmed, being photographed; they've had enough of us watching them die, live, without trying to do anything to save them. And who's to say they're wrong?"

They were not wrong. Roger Cohen of the *New York Times* took this as the premise for his searching front page report on "postmodern war" in the besieged city one Sunday in May 1995. Postmodern for many reasons, but mainly because it's a matter of images, of what the reporter finds to be a dangerously blurred boundary between event and representation, and of a certain paralysis, the apparent re- or dis-location of the field of knowledge and action to the screen of a monitor and the entry of those representations back into the field of the things and events they ought simply to represent. Here is his lead for an article headlined "In Sarajevo, Victims of a 'Postmodern' War":

> Faruk Sabanovic, a pale and gentle-featured youth, is a thoroughly modern victim of war. He lies in the main hospital here with a video of the moment when he was shot and became a paraplegic.
>
> There he is, outside the central Holiday Inn, walking briskly across the street, his hair ruffled by the wind. The crack of a shot echoes in Sarajevo's valley. He falls. He lies on his side. He is curled in an almost fetal position. A United Nations soldier looks on, motionless.
>
> A Sarajevan man arrives, screaming abuse at the soldier, who eventually moves his white United Nations armored personnel carrier. This slight movement is enough to cover the civilian as he rushes out to retrieve Mr. Sabanovic, whose lithe body has turned limp.
>
> "It's strange when I watch the video, I feel like it's somebody else," said Mr. Sabanovic, who is 20. "But I remember it so well. After I was hit, I felt my legs in my chest. Then I saw my feet. I tried to move them. But I could not. This United Nations soldier was looking at me. He did nothing. He just looked. For me, it was so long."

The scene is shocking, doubly so by virtue of the videotape. The civilian victim is not only crippled by a sniper but is also in possession of the images of his attempted murder. The reporter can thus not only interview the person but watch TV with him. And the image is somehow not just of Faruk Sabanovic or of what happened to him on the street in Sarajevo; it is for Cohen an allegory, an image of something else, more confusing, an image of the confusion and loss of orientation—in images—which have affected our sense of reality itself. Watching this tape, with its inert star next to him, Cohen seems paralyzed by the sight of people watching: "Faruk lies . . . with a video"; "a United Nations soldier looks on, motionless"; "'this United Nations soldier was looking at me—he did nothing—he just looked.'"

Thanks to images like these, we are all like that UN soldier, just looking, or like the cameraman, waiting. That is their rich allegorical meaning, their hermeneutic supplement: they mean the inaction that they demand of their producer and their viewer.

> The images capture more than the maiming of Mr. Sabanovic; they capture the increasingly surreal and sordid nature of the three-year-old Bosnian war. A civilian is shot on a city street; a television cameraman, waiting at a dangerous crossroads to see somebody killed or mutilated, films the shooting; a soldier sent by the United Nations as a "peacekeeper" to a city officially called a "safe area" watches, unsure what to do and paralyzed by fear. The elements of this troubling collage are also elements of what some military analysts are now calling "postmodern" or "future" war.

In the tape, in the hospital, Cohen sees an image from Sarajevo and in it the whole new troubling thing metonymized. As the space and time of what happens shifts onto the screen, even "there" in Sarajevo, all sorts of boundaries are collapsing with it. He enumerates the transformation or the decay that coincides with the emergence of the videotape: states are replaced by militias or other informal groupings; armies and peoples become indistinguishable; central authorities disappear; and "live images of suffering, distributed worldwide, sap whatever will or ability there may be to prosecute a devastating military campaign." Looking is not acting, in Sarajevo or in New York, and for Cohen the diffusion of images goes hand in hand with a more disturbing dispersion or evisceration of the conditions of action: lost are centrality, authority, borders and clear distinctions, principles, and all the rest.

The triumph of images figures this for Cohen: images sap the will in war, he says, and yet paradoxically it is a war of images, fought with images. Remember:

> Mr. Sabanovic got in the way at a particularly dangerous Sarajevo crossroads. That is why there was a cameraman there to film his near-death. Because the spot is treacherous, the chances are good that a few hours of patience by a cameraman will be rewarded with compelling images of a life being extinguished or incapacitated.

A "compelling image" is, of course, a weapon, and the cameramen sometimes seemed like the best gunners the Bosnian Government had, being deprived of almost all other military equipment. Certainly most of the journalists in Sarajevo understood this, and recognized that their work was not simply impartial. Didn't Somalia suggest, after all, that images could be compelling, that tele-guided public opinion could force action?

Cohen, in the late spring of 1995, had seen enough to withdraw that conclusion. There are no compelling images: "Thus, just as the world has long watched the crushing of Sarajevo—so endless as to become increasingly unreal—the people of Sarajevo may now watch from their hospital beds the moment they were crippled, so abruptly that comprehension is difficult."

The image sparks a crisis, not just in action but in comprehension, and the sentence that speaks of it also tells the story of a more profound disturbance. "People can watch the moment they were crippled, so abruptly that comprehension is difficult," he writes, but just what exactly happens so abruptly, the crippling or the watching? Surely Cohen means the suddenness of the rifle shot itself, caught on tape, but his dangling modifier betrays the ambiguity he is most alert to, the difficulty of discerning event and video repetition. In the face of this difficulty, Cohen proposes some reservations, or some objections, which although they are not formalized, and hesitant at best, do constitute something like a systematic critique of this "postmodern" condition—it troubles him and with him, "reality". Sarajevo becomes surreal, unreal, endless; there is too much watching, too much mediation, even there in Sarajevo, so that even the subject of the image is himself alienated from it, split from himself. "I feel like it's somebody else," says Sabanovic, now sharing the position of the immobilized one who just watches. The sniper and what Cohen calls "the twisted video" together reduce everyone to a paraplegic—inert, paralyzed by fear, just looking. And yet Cohen finds a moral for the story in the prone 20-year-old, "a strength and a conviction that rise far above the banal violence of his video with its succinct accounting of a directionless war in which civilians die live on camera." Without direction, pulling the very subject of the image apart from himself, the war of "live death" comes to mean for Cohen at once an excess of imagery and a failure of the promise of those images—no action, no comprehension, only difficulty and a certain indetermination. Faruk Sabanovic, for his part, thanks the camera: the United Nations, he says, is "just here to ease consciences. . . . And I know they brought me to the hospital in their ambulance only because the camera happened to be there. I have to say that I despise them."

So in the end the two viewers of the tape disagree about its effects while agreeing that it has one, and these opinions recapitulate what I think represents the crisis of a certain idea of publicity. The symmetrical opposition of the interpretations—Mogadishu and Sarajevo—confirms that what is in question is the theoretical status and the actual function of the public image. Sabanovic believes in the CNN effect: "they brought me to the hospital only because the camera happened to be there." Cohen fears that the camera and the watching cripples our responses, that "images sap the will."

The strong version of his hypothesis has also been articulated by Jean Baudrillard, who thus forms a symmetrical pair with Virilio.[26] Baudrillard suggests that "Bosnia exemplifies total weakness," "the West has to watch helplessly," in a "military masquerade where the virtual soldier . . . is paralyzed and immobilized" (p. 87). And "the Bosnians . . . end up finding the whole situation unreal, senseless, and beyond their understanding. It is hell, but a somewhat hyperreal hell, made even more so by their being harassed by the media and humanitarian agencies . . . thus they live amid a type of spectral war" (p. 81).

Some American commentators have drawn radical conclusions from this proposition, and although it is in a certain sense highly disputable, there is nevertheless something extremely important at stake here which this radicalization can help clarify. In a recent collection of essays, called *This Time We Knew* and edited by Thomas Cushman and Stjepan Mestrovic, the editors attempt to measure the significance of what seems an obvious failure: the last time around, we might have been able to say we didn't know what was happening, but throughout the second genocide in Europe in this half century, we have no such excuse. Because of television and the rest of the "daily barrage of information and images," it is not possible for "even the most disinterested viewer to ignore the grim reality of genocide." Their "Baudrillardian" hypothesis:

> Lack of action proceeds . . . from the fact that the mediated images of the world are mere representations that lend an air of unreality to the things they represent. . . . Media watchers lose touch with reality, . . . stand passively by or engage in self-serving forms of ineffective action, . . . [their] voyeurism and individualism feed[ing] on televised images of evil."
>
> (p. 79)

And that means that the crisis is not merely one of inaction. In fact, what is lost in Bosnia is nothing less than the Enlightenment, and with it the discovery of the public sphere as the site where knowledge and action are articulated. They feel obliged to ask, then, "whether there is any relationship between the degree or extent of public information and practical or moral engagement by those who receive it" (p. 7).

> The important point is that there is a sharp discrepancy between what we know and what we do, and this discrepancy has been neglected in most previous analyses. Yet this gap between knowledge and action is full of meaning for apprehending history as well as the present. In addition, this contrast causes us to rethink the success of the so-called Enlightenment project: the passive Western observation of genocide and other war crimes in the former Yugoslavia amounts to a toleration of the worst form of barbarity and gives us pause to wonder whether, behind the rhetoric of European progress and community, there is not some strong strain of irrationality that, if laid bare, would call into question the degree of enlightenment the civilized West has managed to attain at the century's end.
>
> (p. 8)

It is not clear just how far these two are willing to go in "calling into question" the Enlightenment axioms that the "gap between knowledge and action" in Bosnia provokes. The specter of "irrationality"—always opposed to a normative reason—and the progressivist hint in the word "attain" suggest that they remain committed to the project that has become questionable. But what happens if we seize on this insight—that the Enlightenment and its public sphere are in question—and try to move beyond the simple desire to recover it, to rescue it from its temporary loss. Suppose it were, precisely, the problem.

What failed in Bosnia? We often say that *we* failed, and we imply that *we* are just this well-known public of the "so-called Enlightenment project." But the more we rely on and retreat to the sense that the public sphere collapsed, the more we shore up just the notion whose apparent solidity may be implicated in the disaster. What if the belief in this public was part of the failure, if the faith in the obviousness, the evidence or self-evidence of the pictures and the automatic chain of reasoning they inspire, was not what failed but the very failure itself? What is at stake is the program which expects that, as David Rieff put it, "one more picture, or one more story, or one more correspondent's taped stand-up in front of a shelled, smoldering building would bring people around, would force them to stop shrugging their shoulders, or like the United Nations, blaming the victims". One more picture would force something to happen—what if just that expectation about information and illumination was part of the problem?

To draw out the most radical conclusion from Cushman and Mestrovic: what if it is some part of the "Enlightenment," and not its failure but rather

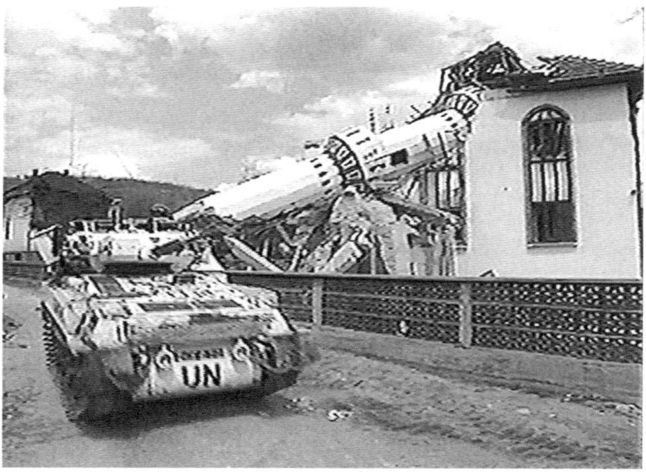

Figure 12.16 Warrior armored vehicle, Cheshire Regiment/UNPROFOR, Amhici, Bosnia-Herzegovina, April 22, 1993; video still from CNN, April 23, 1993.

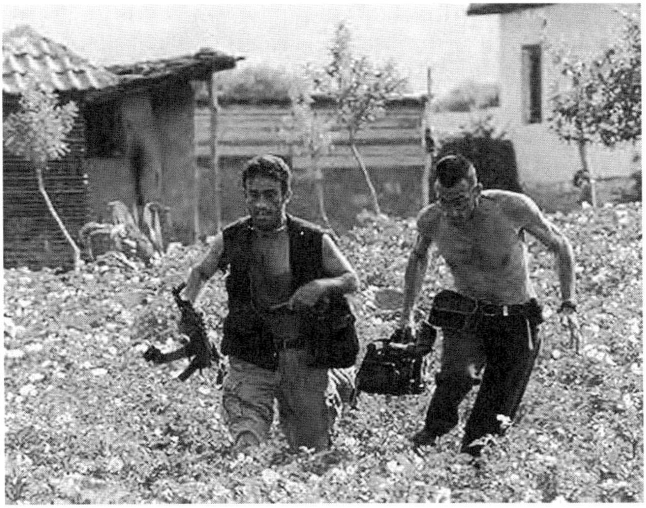

Figure 12.17 AP TV cameraman Miguel Gil Moreno de Mora (left, with camera), Kosovo, 1998; photograph.

the faith we put in the informative power of images, that didn't just fail to stop what happened but allowed it to go on? What if, because the cameramen and the images were there, and because they are supposed to make a difference simply by virtue of what they showed, the disaster continued?

Hypothesis: to the extent that we imagine or take for granted the articulation between knowledge and action, which seems to define the public sphere, it is bound to fail. But what can only be thought of as a failure in those terms is, in another sense, the success of a political strategy, and if we continue to think that images by virtue of their cognitive contents, or their proximity to reality, have the power to compel action, we miss just the opening of "new fields of action" (Benjamin) that they allow.

So what if we think about this understanding of publicity not as a failure or as the reemergence of irrationality, but as an alibi, a conceit, or a consolation? These are words I borrow again from David Rieff, in *Slaughterhouse*, who suggests that it is this that has failed in Bosnia: a naive hope, a consolation, and a conceit, the consolation of images and the dream of public information. Here's the collapsed public in a sentence: "People . . . console themselves with the thought that once they have the relevant information, they will act. It is an old conceit" (p. 41). "It was the conceit of journalists . . . [:] if people back home could only be told and shown what was actually happening in Sarajevo . . . then they would want their governments to do something" (p. 216).

The conceit or fantasy of this kind of public sphere must, after Bosnia if nowhere else, contend with what we could call the rule of silence—no image speaks for itself, let alone speaking directly to our capacity for reason. Images always demand interpretation, even, or especially, emotional images—there is nothing immediate about them. This implies a second rule, that of unintended consequences or misfiring—the story of Bosnia is that images which might have signified genocide or aggression or calculated political slaughter seemed for so long to signify only tragedy or disaster or human suffering and hence were available for inscription or montage in a humanitarian rather than a political response. So what failed in Bosnia is an idea or an interpretation—and a practice—of publicity, of the public sphere as the arena of self-evidence and reason, an idea which now must be challenged, not to put an end to the public sphere but to begin reconstituting it.

As it happened, the images were open enough to demand only that we "do something," and the problem concerns, in short, this something. The naive consolation is precisely that its content or meaning is self-evident, even analytically implied by the information itself, by "what is actually happening." And "do something" they did, in fact, something which amounted to, as Rieff puts it in the sharpest phrases of his book, "administering the Serb siege" (p. 147) and "becom[ing] accomplices to genocide" (p. 189). The combination of the traditional tasks of peacekeeping—which require military observers to be stationed on or between the front lines, and hence in the zone of any possible offensive military operations like air strikes—and the new humanitarian tasks of escorting convoys across lines of confrontation meant that the "humanitarian" operation was an active impediment to any other action. Not just the "fig leaf" which Rieff too lightly calls it (p. 189), but an affirmative choice: "the wish that there be no intervention" (p. 176). And this project was best accomplished by undertaking the other intervention: stationing peacekeepers close enough to Bosnian Serb forces that they would either be targets of Western air strikes or easy hostages for the Serbs, and escorting the convoys that always made it necessary not to, as the UN put it, "compromise the humanitarian mandate" by antagonizing the aggressor. "This convergence of interest between the UN and the Chetniks was not an exceptional situation," as Rieff says, but the very structure of the situation (p. 175).

And it happened thanks to the images, from which we expected something rather different. But images, information, or knowledge will never guarantee any outcome, force or drive any action. They are, in that sense, just like weapons or words, a condition, but not a sufficient one. Still, the only thing more unwise than attributing the power of causation or of paralysis to images is to ignore them altogether. If they can condition some action—and indeed, in Sarajevo and elsewhere, that's exactly what happened—then it is only at the risk of this very indirection, the unexpected outcome, we might say: here, the humanitarian one. We cannot, at least not without repeating what seems to me to be the basic strategic error here, not expect the unexpected—we cannot count on the obviousness of the image, fall for the conceit that information leads ineluctably to actions adequate to the compulsion of the image, precisely because images are so important. There is no compulsion, only interpretation and reinscription, and the image dictates nothing.

This fate of the image—left to wander and to drift from context to context, nothing but surface and frame—is what we can call, borrowing words from the reporter in Sarajevo, its "banal violence," the banality of a "succinct accounting" on video. The image has no guaranteed meaning, and remains only to testify, to demand, to induce a responsibility—even if, as Avital Ronell argues about the videotape of Rodney King being attacked by the LA police, "it is a responsibility that is neither alert, vigilant, particularly present, nor informed."[27] The responsibility of the viewer is co-extensive with the lack of self-evidence of the image: it dictates nothing, compels nothing. It can always be used, though, which is to say that it can and must always be interpreted, and the terrible failure of Bosnia was that a certain understanding of the public sphere—"the thought that once [people] have the relevant information, they will act"—allowed or even produced an interpretive complacency. "Surely one more picture, or one more story, or one more [. . .] stand-up [. . .] would bring people around, would force them to stop shrugging their shoulders"—nothing is less sure, less certain, precisely because we think that it is.

The question of surveillance teaches us, finally, that there is no 'finally' where its images are concerned. Images never speak for themselves, never make anything in particular happen, even if they seem often to make something happen and are now indispensable in war. In Bosnia, they opened a gap, issued a call, and in response came the humanitarian option, displacing all others.[28] This means that the accounting, however succinct, does not stop—the image remains, without guarantees, always available for reinterpretation and reuse, of necessity the focus of an endless vigil and a struggle for reinscription. The battle takes place in public, in fact the public sphere is constituted by the irreducibility of this battle, not the public as the last refuge of that dream or consolation of information properly acted upon, but another public, space and time, virtual and visual and nevertheless real enough, tenuous, uncertain, where everything is open to abuse and appropriation; shaky ground indeed.

* * *

Some years ago, Virilio warned in *Le Monde Diplomatique* that, far from merely offering new opportunities for exploration or relaxation, the media of which telegraph, telephone, radio, and film were the merest announcements have by now radically accelerated and generalized the transmission of event and signification, and indeed absolutized it to the point of instantaneity, such that places no longer matter. He wanted us to believe that, when surveillance is

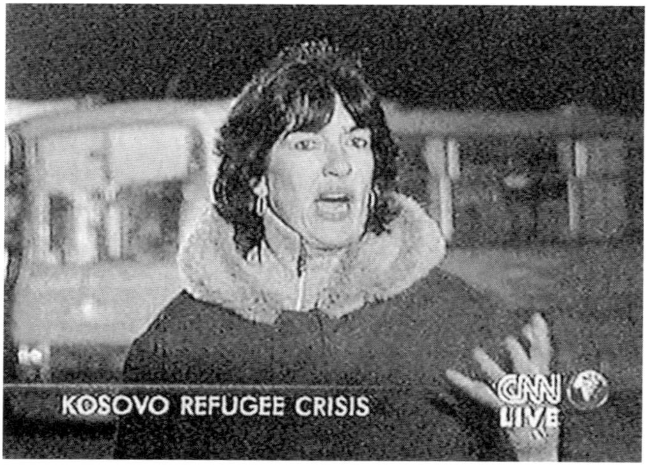

Figure 12.18 Christiane Amanpour, reporting from the Kosovo border, spring 1999, video still from CNN, date unknown.

ubiquitous and its output moves at the speed of light, new media now threaten to deprive us of places altogether, inducing not simply vertigo or disorientation but a more radical "de-situation." In an age of real-time communications, of "instantaneous, globalized information," Virilio saw nothing less than the disappearance of the world itself in a "tyranny of absolute speed."[29]

In other words, the vertigo is absolute and unceasing, depriving us of ourselves. Far from creating a new world citizenry, a virtual community of humanity freed of allegiances to anything other than other humans as such, blown by the technologies of "anywhere" out of the local particularities of place and identity, what disappears here is humanity, the relation to the other. When what happens there happens here too, in real-time, for Virilio what we lose is the fold of reflection, the gaps and delays that make decisions possible and debatable, that divide them into and across more than one instance.

"This goes beyond CNN. Actually, CNN is history," Virilio told a reporter. "And it has nothing to do with the current surveillance of parking lots and street corners by security cameras. [. . .] We're witnessing today the deployment of a new, global tele-surveillance system whose impact will be far more profound than that of the traditional television."[30]

But as surveillance goes global and speed crosses the barrier into instantaneity, do time and responsibility, and with them the possibility of democracy, disappear? Storage and montage happen in so-called "real-time," of course but excess time is built into the transmission, into the mediation that defines any technology of inscription. "No longer time to share?" Not quite.

In addition to the vertigo of acceleration, there is also a more subtle vertigo of deceleration, of slow motion. What Benjamin called "the dynamite of the tenth of a second" means that fast and slow cannot simply be opposed to one another. What is destabilized is the privilege of the present, the experience of the human subject and its self-present reflection (deliberation, reason, judgment), which would seem to regulate the transformations of speed. But only the most classical metaphysics of the subject and of presence can see in this the end of politics, the disappearance of the public sphere in the pure surface of the image. And only an excessive commitment to some ideology of the real imagines that time and decision evaporate in the light of the television screen. What speed teaches us is that this surface is itself folded, temporally or rhythmically complex and heterogeneous, that there is always an "interior" lag which divides the subject from itself. It is this division which makes possible, in fact, the sharing that defines democratic conflict. We cannot simply say, "warning! slow down!"—as if the distortions of speed could be undone and the self-identity of the present reinstated, and with them an anachronistic definition of the political, the public, and the instance of decision. We can say, though, that the vertigo of deceleration—the slow motion of even the fastest and most "compelling" image—tears us apart from our solid selves and opens the possibility of a decision, even of a properly political relation to others, in the question it poses. We are not quite out of time, but the image does not provide the answer for us either.

Notes

*The authors and editors are grateful to *PMLA* for allowing the chapter to be reprinted.

1. Paris: Galilee 1999; *Strategy of Deception*, trans. Chris Turner, New York: Verso, 2000 (trans. modified when necessary). "Hasty" because, as the example of Richard Falk shows, some left opponents of the war later became convinced of its necessity and justification. See "Kosovo Revisited," *The Nation*, April 10, 2000.
2. Miles Hudson and John Stainer, *War and the Media*, Phoenix Mill (UK): Sutton Publishing, 1999, 300.
3. Cardinal Jean-Marie Lustiger, Archbishop of Paris, "I am Ashamed as a Man . . .," interview by Zlatko Dizdarevic, *Oslobodjenje* (European edition), January 6, 1994; http://www.bosnet.org/archive/bosnet.w3archive/9405/msg00021.html. See also Lustiger quoted in Zlatko Dizdarevic, "What Kind of Peace Is This?," *New York Times Magazine*, April 10, 1994, 21: "Here in Sarajevo, hundreds of TV crews parade before our very eyes; dozens of foreign journalists, reporters, and writers. Everything is known here, right down to minutest details, and yet, nothing."
4. "While America Watched: The Bosnia Tragedy," Peter Jennings Reporting, ABC News, broadcast 30 March, 1994.
5. David Rieff, *Slaughterhouse: Bosnia and the Failure of the West*, New York: Touchstone, 1996, 140.
6. Think of the feature films already available about the conflict in the former Yugoslavia, even skipping Marcel Ophuls' war reporter epic-doc *The Troubles We've*

Seen, from the British *Welcome to Sarajevo* to the Spanish *Territorio Commanche* to the Serbian *Pretty Village, Pretty Flame*, the war apparently cannot be portrayed without putting a reporter and cameras at the center of the action, which is to say, the war cannot be presented and its story told without putting its immediate presentation and the tellers of its story, then and there, at the heart of the story itself. See "Media Art in the Balkans," a special issue of *Afterimage* 28, no. 4, January/February 2001, esp. the article by Dina Iordanova.

7. Mark Danner, "The US and the Yugoslav Catastrophe," *New York Review of Books* November 20, 1997: 56–64; "America and the Bosnia Genocide," *New York Review of Books* 4 Dec. 1997: 55–65.
8. And consider, if you can bear it, the former president of the United States, whose extraordinary (in so many ways) confession, in Kigali in March of 1998, of his and our inaction in the face of the Rwandan genocide put him squarely—and immobile—in front of the television: "Today the images of all that haunt us all: the dead choking the Kigara River, floating to Lake Victoria. [. . .] We did not act quickly enough after the killing began. [. . .] All over the world there were people like me, sitting in offices, day after day after day, who did not fully appreciate the depth and the speed with which you were being engulfed by this unimaginable terror" ("Remarks by the President to Genocide Survivors, Assistance Workers, and US and Rwandan Government Officials," March 25, 1998, Kigali Airport, http://clinton4.nara.gov/Africa/19980325-16872.html). For further information, see Samantha Power, "Bystanders to Genocide, " *The Atlantic Monthly*, September 2001: 84–108.
9. Michael Ignatieff, *The Warrior's Honor: Ethnic War and the Modern Conscience*. New York: Metropolitan/Holt, 1997.
10. Rony Brauman, "La pitié dangereuse," in Brauman and René Backmann, *Les médias et l'humanitaire: Ethique de l'information ou charité-spectacle*, Paris: CFPJ, 1996, 9–60.
11. Alex de Waal, "Becoming Shameless: The failure of human-rights organizations in Rwanda," *Times Literary Supplement* February 21, 1997, 3–4.
12. Bernard Kouchner, *Le malheur des autres*, Paris: Odile Jacob, 1991, 210.
13. The literature on the CNN effect is already significant, including books (all in the Bibliography) by Benthall, Neuman, Seib, Minear et al., Strobel, Gow et al., Girardet, Rotberg and Weiss, Ignatieff, Moeller, Gowing and Livingston.
14. Alex de Waal, *Famine Crimes: Politics and the Disaster Relief Industry in Africa*. Oxford: James Curry, and Bloomington: Indiana UP, 1997, 65.
15. Walter Benjamin, "The Work of Art in the Age of Mechanical Reproduction," in Harry Zohn trans., *Illuminations*, New York: Schocken, 1969, 217–51.
16. Roger Cohen, "In Sarajevo, Victims of a 'Postmodern' War," *New York Times*, May 21, 1995, 1. Cohen retells the story in *Hearts Grown Brutal*, 377–9.
17. Marshall McLuhan: "As electrically contracted, the globe is no more than a village. Electric speed in bringing all social and political functions together has heightened human awareness of responsibility to an intense degree. It is this implosive factor that alters the position of the Negro, the teenager, and some other groups. They can no longer be *contained*, in the political sense of limited association. They are now *involved* in our lives, as we in theirs, thanks to the electric media" (5).
18. McKenzie Wark, *Virtual Geography: Living with Global Media Events*, Bloomington: Indiana UP, 1994.
19. See, on the trade of images for food, McKenzie Wark, "Fresh Maimed Babies," *Transition* 5, No. 1, Spring 1995, 36–47.

20. I have addressed some of these questions in "Live from . . ./En direct de . . .,"in Elizabeth Diller and Ricardo Scofidio, *Back to the Front: Tourisms of War/Visite aux armées: Tourismes de guerre* (Caen: FRAC Basse-Normandie, 1994), 130–63.
21. George Kennan, "Somalia, Through a Glass Darkly," *New York Times*, September 30, 1993, A25.
22. Paul Virilio, *L'écran du désert: Chroniques de guerre*, Paris: Galilée, 1991, 71–2.
23. Larry Minear, Colin Scott, and Thomas G. Weiss, *The News Media, Civil War, and Humanitarian Action*, Boulder: Lynne Reinner, 1996, 46.
24. See Samuel Weber, "Television; set and screen," in *Mass Mediauras: Form Technics Media*, Stanford; Stanford University Press, 1996, 108–28.
25. Christiane Amanpour, Cable News Network, January 26, 1993. That was in January 1993, when she was just back from Somalia; in July of that year she again reported something similar:

    > Those of us who've stayed in Sarajevo—which is the majority of the press corps—have been welcomed throughout most of this year. People have looked at us as their conduit to the West, and perhaps have looked at us as being able to jog, perhaps, some conscience in the West. The attitude towards us has been changing—certainly, in the last month. People have seen that really nothing has changed, that promises have been made and broken—promises from the Western allies have been made and broken. And people look to us and sort of vent their frustrations on us. . . . So the tide is turning somewhat, but that's a measure of people's despair, I suppose.
    >
    > (CNN, *Larry King Live*, 30 July 1993).

26. Jean Baudrillard, "When the West Stands In for the Dead" and "No Pity for Sarajevo," trans. James Patterson, in Thomas Cushman and Stjepan G. Mestrovic (eds), *This Time We Knew*: *Western Responses to Genocide in Bosnia*, New York: New York University Press, 1996, 87–9 and 80–4.
27. Avital Ronell, *Finitude's Score: Essays for the End of the Millennium*, Lincoln: University of Nebraska Press, 1994, 305.
28. See David Campbell, "Atrocity, Memory, Photography: Imaging the Concentration Camps of Bosnia—The Case of ITN versus *Living Marxism*," *Journal of Human Rights* 1, 1 (2002): 1–33. "it is simplistic to assume that an image, in and of itself, can provoke a particular reaction the possibility of which did not exist prior to the production of that image."
29. Paul Virilio, "Alerte dans le cyberspace," *Le Monde Diplomatique*, August 1995; "Red alert in cyberspace," trans. Malcolm Imrie, *Radical Philosophy* 74, November–December 1995, 2–4.
30. Bruno Giussani, "For a Philosopher, the Net Is a Whole New Perspective," *New York Times* on *the Web*, December 9, 1997, http://www.nytimes.com/library/cyber/euro/120997euro.html.

Works Cited

Baudrillard, Jean. "When the West Stands In for the Dead," trans. James Patterson. Cushman and Mestrovic. *This Time We Knew: Western Responses to Genocide in Bosnia* eds. Thomas Cushman and Stjepan Mestrovic. New York: NYU Press (1996).

——. "No Pity for Sarajevo," trans. James Patterson. Cushman and Mestrovic. *This Time.*

Benjamin, Walter. "The Work of Art in the Age of Mechanical Reproduction," trans. Harry Zohn. *Illuminations*. New York: Schocken, 1969.

Benthall, Jonathan. *Disasters, Relief, and the Media*. London/New York: I. B. Tauris, 1993.

Brauman, Rony. "La pitié dangereuse," in Rony Brauman and René Backmann, *Les médias et l'humanitaire: Ethique de l'information ou charité-spectacle*. Paris: CFPJ, 1996.

Campbell, David. "Atrocity, Memory, Photography: Imaging the Concentration Camps of Bosnia—The Case of ITN versus *Living Marxism*," *Journal of Human Rights* 1 and 2, 2002.

Clinton, William Jefferson. "Remarks by the President to Genocide Survivors, Assistance Workers, and US and Rwandan Government Officials," The White House, Office of the Press Secretary (Kampala, Uganda). March 25, 1998. Kigali Airport, Kigali, Rwanda. http://clinton4.nara.gov/Africa/19980325-16872.html.

Cohen, Roger. "In Sarajevo, Victims of a 'Postmodern' War," *The New York Times*, May 21, 1995: 1, 12.

——. *Hearts Grown Brutal: Sagas of Sarajevo*. New York: Random House, 1998.

Cushman, Thomas and Stjepan G. Mestrovic (eds) *This Time We Knew: Western Responses to Genocide in Bosnia*. New York: New York UP, 1996.

Danner, Mark. "The US and the Yugoslav Catastrophe," *New York Review of Books*, November 20, 1997: 56–64.

——. "America and the Bosnia Genocide," *New York Review of Books*, December 4, 1997: 55–65.

De Waal, Alex. "Becoming Shameless: The Failure of Human-Rights Organizations in Rwanda," *Times Literary Supplement*, February 21, 1997: 3–4.

——. *Famine Crimes: Politics and the Disaster Relief Industry in Africa*. Oxford: James Curry, and Bloomington: Indiana UP, 1997.

Dizdarevic, Zlatko. "What Kind of Peace Is This? Sarajevo's 700 Days," trans. Ammiel Alcalay. *The New York Times Magazine* April 10, 1994: 36–41, 64, 68–69, 78.

Feher, Michel. *Powerless by Design: The Age of the International Community*. Durham: Duke University Press, 2000.

Girardet, Edward. (ed.) *Somalia, Rwanda, and Beyond: The Role of the International Media in Wars and Humanitarian Crises*. Crosslines Special Report 1. Dublin: Crosslines, 1995.

Gow, James, Richard Paterson, and Alison Preston (ed.) *Bosnia by Television*. London: British Film Institute, 1996.

Gowing, Nik. "Real-Time Television Coverage of Armed Conflicts and Diplomatic Crises: Does it Pressure or Distort Foreign Policy Decisions?," Working Paper 94-1, Joan Shorenstein Barone Center on the Press, Politics, and Public Policy, Harvard University, Cambridge, MA. June 1994.

Giussani, Bruno. "For a Philosopher, the Net Is a Whole New Perspective," *New York Times on the Web* December 9, 1997, http://www.nytimes.com/library/cyber/euro/120997euro.html.

Ignatieff, Michael. *The Warrior's Honor: Ethnic War and the Modern Conscience*. New York: Metropolitan/Holt, 1997.

——. *Virtual War: Kosovo and Beyond*. New York: Henry Holt, Metropolitan Books, 2000.

Keenan, Thomas. "Live from . . .," *Back to the Front: Tourisms of War/Visite aux armées: Tourismes de guerre*, edited by Elizabeth Diller and Ricardo Scofidio, 130–63. Caen: FRAC Basse-Normandie, 1994.

——. "Do something," *Alphabet City* 7: Social Insecurity, 2000: 44–6.

Kennan, George. "Somalia, Through a Glass Darkly," *The New York Times* September 30, 1993, A25.

Kouchner, Bernard. *Le malheur des autres*. Paris: Odile Jacob, 1991.

Livingstone, Steven. "Clarifying The CNN Effect: An Examination of Media Effects According to Type of Military Intervention." Research Paper R-18. Joan Shorenstein Center on the Press, Politics, and Public Policy, Harvard University, Cambridge, MA. June 1997.

McLuhan, Marshall. *Understanding Media: The Extensions of Man*. Cambridge: MIT Press, 1994 [orig. 1964].

Minear, Larry, Colin Scott, and Thomas G. Weiss. *The News Media, Civil War, and Humanitarian Action*. Boulder: Lynne Reinner, 1996.

Moeller, Susan. *Compassion Fatigue: How the Media Sell Disease, Famine, War and Death*. New York/London: Routledge, 1999.

Neuman, Joanna. *Lights, Camera, War: Is Media Technology Driving International Politics?* New York: St. Martin's, 1996.

Power, Samantha. "Bystanders to Genocide," *The Atlantic Monthly*, September 2001: 84–108.

Rieff, David. *Slaughterhouse: Bosnia and the Failure of the West*. New York: Touchstone, 1996.

Ronell, Avital. *Finitude's Score: Essays for the End of the Millennium*. Lincoln: University of Nebraska Press, 1994.

Rotberg, Robert I., and Thomas G. Weiss (ed.). *From Massacres to Genocide: The Media, Public Policy, and Humanitarian Crises*. Washington: Brookings Institution, and Cambridge, MA: World Peace Foundation, 1996.

Seib, Philip. *Headline Diplomacy: How News Coverage Affects Foreign Policy*. Westport: Praeger, 1997.

Strobel, Warren P. *Late-Breaking Foreign Policy: The News Media's Influence on Peace Operations*. Washington: US Institute of Peace, 1997.

Virilio, Paul. *L'écran du désert: Chroniques de guerre*. Paris: Galilée, 1991.

——. Paris: Galilée 1999; *Strategy of Deception*, trans. Chris Turner. London/New York: Verso, 2000. (Trans. modified when necessary.)

——. "Alerte dans le cyberspace." *Le Monde Diplomatique* August 1995; "Red alert in cyberspace," trans. Malcolm Imrie. *Radical Philosophy* 74, November–December 1995: 2–4.

Wark, McKenzie. *Virtual Geography: Living with Global Media Events*. Bloomington: Indiana UP, 1994.

——. "Fresh Maimed Babies." *Transition* 5:1, Spring 1995: 36–47.

Weber, Samuel. "Television; set and screen," *Mass Mediauras: Form Technics Media*. Stanford: Stanford University Press, 1996.

13
ELSI Priorities for Brain Imaging*

Judy Illes, Raymond De Vries, Mildred Cho, and Pam Schraedley-Desmond

How might neuroimaging disrupt the mysteries of human nature, spirituality, and personal identity? Rather than waiting for an ethical agenda to emerge from some unpredictable combination of the concerns of ethicists and researchers, the attention of journalists, or after controversy is sparked by research that cannot be retracted, we queried key figures in bioethics and the humanities, neuroscience, media, industry, and patient advocacy in focus groups and interviews. We identified ethical, legal and social issues (ELSI) that highlight researcher obligations and the nonclinical impact of the technology at this new frontier.

Modern techniques for imaging the brain have given medical scientists and clinicians powerful new tools to measure brain function and to demonstrate links between the brain and behavior (Illes et al. 2003). The ability to map how we reason through difficult moral decisions, make conscious and unconscious value judgments, and how we cooperate and compete with others presents a significant challenge to traditional ways of understanding human nature and agency and raises significant and controversial ethical questions. Much remains to be said about the epistemological challenges of using brain mapping data for understanding personal identity (Illes and Racine 2005), but our human curiosity about these topics is inescapable. The field is proceeding apace, but with comparably less focused attention on the ethical and practical issues that are following in its wake, a situation that may create unpredictable roadblocks to progress. In order to replace "reactive ethics"—where the public, ethicists, and regulatory agencies must grapple with ethical issues raised by science and its applications after research has already been completed—with "proactive ethics" (De Vries 2004; Cho et al. 1999), we organized several focus groups and interviews with stakeholders with interests in the neuroimaging sciences. We borrowed the term "ELSI"—ethical, legal and social implications—from the National Human Genome Research Institute's extramural funding program, and queried participants about the future of the field and about the roles they perceive for themselves, as professionals, as citizens, and as consumers of the products of science.

Methods

Focus groups and individual interviews

We assembled leaders and representatives from bioethics, neuroscience, law, and the humanities from major academic institutions, the National Institutes of Health and private sector sponsors of research, along with interested members of the media, patient advocate groups, and of the neuroimaging industry. They were selected on the basis of their publications, leadership in relevant professional organizations or academic institutions, and expressed interest in the imaging neurosciences, and were recruited by invitation of the first author (JI). Background reading was made available by direct e-mailing of key journal articles to participants and by access to others available at http://neuroethics.stanford.edu. With IRB approval, we held six focus groups around the country (with a total of 57 participants) and individual interviews (N=15). Because our goal was to learn how participants defined the ethical and policy issues presented by neuroimaging—using fMRI as a model—we asked them simply to describe their enthusiasm and concerns about neuroimaging. This allowed us to gain an understanding of the areas our participants found most acute and most promising.

We organized five focus groups by professional background, with a separate group for bioethicists, neuroscientists/neuroimagers, patient advocates/research sponsors, journalists, and religious scholars. Because a few participants represented more than one area there was some overlap between groups. Our sixth focus group comprised individuals from a variety of disciplines who serve on the Advisory Board for one of the grants funding this work. They were included to allow comparison with responses from the more homogeneous groups.

Focus groups were facilitated by a chair who instructed the group and by a moderator whose role was to ensure the flow of discussion and even

Table 13.1 Primary composition of focus groups and interviews

Primary Focus Group Composition	**Group (N)**	**Interview (N)**	**Total (N)**
Bioethics	11	1	12
Neuroscience Neuroimaging	12	3	15
Patient Advocacy Research Sponsorship	10	2	12
Print and Radio Media	2	7	9
Community and Academic Religious Scholars	9	0	9
Mixed Group	13	0	13
MR Equipment Vendors	N/A	2	2
Total	57	15	72

participation by all members. Focus groups lasted for approximately one hour and were audio taped. Each participant was given approximately two minutes for an opening comment. The remaining time was allocated to discussion.

As a supplement to our focus group data, we conducted interviews (in person or by telephone) with 15 individuals who indicated a desire to participate but who could not attend their respective focus groups. Each interview lasted approximately 20 minutes. Interviews were audio taped or recorded by written notes.

Analysis

We transcribed all focus group sessions *in extenso* and analyzed the transcripts using a strategy grounded in the inductive methods common to qualitative research (Boyd et al. 2003; DeVries et al. 2002; Glaser and Strauss 1967; Strauss and Corbin 1998; Schatzman and Strauss 1973; Stewart and Shamdasani 1998). Given our goal, we developed an analytic approach that mapped the focus group conversation and allowed us to measure both the quality and quantity of participant comments. All team members read the transcripts, after which we developed a list of themes and subthemes discussed in the focus groups. One member of the team (RDV) then divided the transcripts into coherent units based on the themes identified. Because the unit of analysis in focus group research is the *group*, not the *individuals* in the group, we did not attempt to correlate mention of a certain topic with characteristics of individual members (such as gender, age, ethnicity, highest degree). Instead we mapped themes of the conversation, dividing the transcripts into blocks with a consistent theme. In some cases these blocks would extend across the comments of several focus group members, in other cases, a single contribution from a focus group member included two or more separate blocks. Next, in order to insure interobserver agreement, two of us (JI and MC) coded the transcripts independently, identifying the themes and subthemes found in each block. Together we reviewed codes for matches and discrepancies and achieved consensus. This coding method allows us to compare the discussions about fMRI across different focus groups both qualitatively (looking at the content of the comments) and quantitatively (measuring the percentage of each discussion devoted to specific topics). Because the data from individual interviews were different in nature from the focus groups—they were conducted on a one-on-one basis and did not have the benefit of open discussion—they were not coded per se, but were used for cross-reference to our focus groups and to enrich our understanding of those data.

Results

Our analysis of the focus group transcripts yielded a total of 333 thematic blocks revolving around three major areas of emphasis: (1) researcher obligations, (2) the clinical and nonclinical impact of technology, and (3)

Table 13.2 Priority areas and emerging focus groups themes for the imaging neurosciences. Shown is the percentage of subthemes occurring in each focus group discussion per the total number of coded subthemes in that focus group. Totals may deviate from 100 percent due to rounding

	Bio-ethicists	Neuro scientists	Patient Advocates	Journalists	Religious Leaders	Mixed Group
I. Researcher Obligations						
Experimental Issues *(%)*						
Data acquisition	4	0	11	0	4	2
Data interpretation, New challenges	5	20	16	0	10	21
Research Ethics *(%)*						
Informed consent	8	2	24	0	0	2
Incidental findings	0	2	3	0	2	0
Conflict of interest	5	0	0	0	0	2
II. Technological Impact						
Non-clinical *(%)*						
Growth of industry	11	16	10	0	4	12
Media and public opinion	5	23	7	57	10	19
Uses, approaches, policy	33	11	5	38	2	12
Clinical *(%)*						
Diagnosis	3	4	3	0	0	2
Prediction	3	7	5	0	0	0
Treatment and intervention	3	4	7	0	4	15
Privacy and confidentiality	1	9	4	0	2	10
III. Theoretical & Philosophical Issues						
Precedent *(%)*						
Prior challenges	3	4	1	5	18	0
Personhood *(%)*						
Consciousness	18	0	4	0	45	4

philosophical and theoretical issues. These themes, and the subthemes that formed them, are mapped by focus group and shown in Table 13.2.

Note the varied areas of emphasis in each focus group: the bioethics group was most concerned with nonclinical applications of the technology, as was the neuroscientist group; the patient advocate group paid most attention to the obligations of researchers; journalists, not surprisingly, focused on the role of the media; the mixed focus group (advisory board) distributed their concerns broadly; and the religious scholars' primary concern was with issues theoretical and philosophical. In the following sections we use the rich detail of our transcripts to flesh out the nature of these issues and concerns.

Obligations of researchers

The first major theme under researcher obligations—experimental issues—included challenges associated with the acquisition and interpretation of neuroimaging data. The acquisition of images raised methodological issues related to accuracy, standards, and reproducibility; problems of interpretation include over- and misinterpretation of data, handling of artifacts, and understanding and translating results. We heard many comments about the need for more care in the translation of research for lay audiences. A participant in the neuroscience group noted:

> To the extent that the science isn't really ready, or is over-generalized, or over-interpreted, it has both a negative impact in terms of how it's used in society . . . and what that does in the long run in terms of our science.

Another neuroscientist called attention to how the lack of formal theory creates problems for the interpretation of complicated neuroimages:

> Why is it that neuroscience and psychology, which sit above physics and below economics and sociology and population biology, standout as the only scientific disciplines that don't have formal theory as part of their backbone? Now, there are a million excuses: you can say that it's too complicated. But the weather is pretty complicated, but we at least we have to try . . . [O]ther fields, I don't think are quite as susceptible [to these problems of interpretation] because they know that their theories and the intellectual tradition in the field grows out of an understanding of the complex.

A third member of the neuroscientist group commented on the pressures imposed by the contrast between the ease of collecting neuroimages and the difficulty of interpreting them:

> You know, it's easy to generate data, there's a hungry public for it, and [neuroimages] carry a lot of face validity and we are not being taught

> how to really do the science in a way that, you know, presents it in its full complexity that it actually reflects.

Problems of interpretation led one member of the neurosciences community who also provides leadership at a major funding agency, to invoke the metaphor of a *sausage maker* to describe the way imaging techniques are used:

> I think we need to pay more attention to the specific validation of whatever the imaging modality is with the actual, behavioral, observable, clinical context. I think that it's quite simplistic at the moment and that simplistic interpretation of the pretty pictures . . . is a potential danger. . . . I think that we haven't paid enough attention to the, and made known, to the consumers of these technologies what the advantages and limitations of the different imaging modalities are. So we have PET, we have SPECT, we have fMR, we have MR, we have all of these, but we [also] have the concomitant physiological techniques. In our research laboratories, we tend to throw everything at our subjects to see what comes out of the sausage maker. But it's not necessarily done in a paradigmatic way, it's not necessarily hypothesis driven.

With respect to issues of research ethics, the second theme under researcher obligations, focus group members raised concerns about informed consent. Like all researchers and ethicists, participants worried about how to get consent from someone with impaired capacity to consent, but they also had concerns about the special problems with consent inherent to neuroimaging. In particular, they pointed to the difficulty of properly informing a patient or a research subject about the risks and benefits of neuroimaging, given the great uncertainty about what the image portends for a person's health. As one research sponsor stated:

> You are talking about the situation where we have insufficient knowledge about the actual predictive validity of what we see. But even if it were, even if we had the predictive validity, there are the ethical issues about what do you do about it if the current state of therapy is such that there's nothing that would change in terms of how that patient is treated.

In an interview, a clinical investigator highlighted how problems with consent bias the recruitment of subjects and thus the direction of research and development:

> In autism, the spectrum is huge (low intellectual functioning to high)—[the] high functioning are easy to deal with. So technology is very much high functioning oriented—[the] low functioning are being neglected.

Technology impact—nonclinical

Growth of the industry in terms of increasing professional pressure and expanding commercial development, were emerging subthemes for technology impact and diffusion outside the medical arena. Of concern to a participant in the bioethics group, for example, was:

> [T]he contextual issues of the way this science and the way these new technologies will be developed given the current political economic environment of research funding in the US and . . . the breaking down of barriers between industry and the Academy.

Public opinion was a separate subtheme of this area, encompassing issues surrounding the public's interpretation, perception, and understanding of the limitations of neuroimaging results. Our participants were concerned about the shaping of public opinion by the media, especially given its selective coverage of research and the power of the brain map as a visual image. Closely related to discussions both about researcher obligations, as described above, and to issues of public opinion, were comments about the interaction of neuroscientists with the media (one interviewee defined members of the media as "*predators looking for stories*"). A neuroscientist noted an immediate need for "responsibility in terms of educating the public and making sure what is actually coming out is appropriate."

Echoing this call for responsibility, another neuroscientist, a leader of a major academic institution, called attention to the role that researchers have in the spread of misinformation:

> The issue is that it's not just the general public and the media, it's scientists who don't understand complexity [of neuroimages], just don't understand. They [scientists] think linearly, they can think one spot at a time, one gene at a time. And while . . . what we are seeing [is] a statistical hotspot that is part of a big network, people don't think that way. And so the press doesn't get it from nowhere.

Our participants frequently referred to the power of neuroimages: "There is something very seductive about brain imaging data." A journalist, who is also a neuroscientist, explained the nature of this seductive power:

> I think neuroimaging papers are intrinsically newsworthy because they obviously touch on things that are . . . of profound human importance and we can all identify with, you know, trust, love, moral reasoning, justice, racial prejudice. . . . I mean, you can hardly fail to attract a journalist's attention if you claim to have discovered something about how the brain controls those things.

The subtheme of uses, approaches, and policy includes jurisprudence (i.e., the use of neuroimages in the courts), prediction of behavior in custody and abuse cases, and other forensic uses of neuroimages. The potential of this technology to be used for surveillance and screening—a concern for all groups—was particularly worrisome for bioethicists. One bioethicist described his fears about the:

> potential for the use of neuroimaging in the forensic context and in the law, [the use of] diagnostic neuroimaging with psychiatric disorders, [and] the potential for new forms of surveillance. A scenario that comes to my mind, so to speak, is that we could identify areas of the brain that are associated with hostile or violent ideation and that we could screen everybody who comes into a sensitive public area, like an airport or train station, and footnote that, I am persuaded there is no science, registered science, that is published in the United States that isn't read by somebody in the CIA, or the NSA, or many, many contractors. There are tens of thousands of contractors. Many of them are very competent scientists who work for the agencies. So, I'm sure all of that is of great interest to those guys as well. And, it's warranted for people in bioethics to think about those things.

Participants in other focus groups also expressed concern about the use of neuroimaging to monitor and predict behavior in public areas, for lie detection, and for discrimination in employment or insurance. A representative from industry, in reply to an interview question about uses of fMRI outside the academic or medical setting, mentioned its possible use for national security:

> You can also think of the question in airports "Are you a terrorist?" and then you try to find out . . . and I'm a little bit skeptical of putting these things in schools or airports.

Recalling lessons learned from genetics, a participant in the group of religious scholars worried about the use of fMRI to reify biological explanations of differences in behavior and, by implication, the possibility of creating hierarchies among groups based on these differences:

> One of the dangers that I see in neuroimaging is as we learn more about patterns of activity in the brain, we will, I guarantee it, start looking comparatively between groups of people. Do men do this more than women do? Do women do this more than men do? What about different ethnic groups?

In the context of this subtheme, we also heard about the varied roles of different disciplines in bringing ethical thinking to neuroimaging, such as foreseeing problems, raising questions, and considering the future impact and consequences of frontier neurotechnology on society. For example, a participant in the patient advocacy and sponsor group, noting the American proclivity for seeking legal remedies, commented on the need for neuroscientists to call upon the expertise of lawyers:

> [In creating] any type of guidelines, or proactively setting up standards . . . the legal profession [should be included] because they will, in their job play a huge part. . . . [I]f something goes wrong, they can be your worst enemy or your biggest friend. And if you get them involved early on, you can prevent them from being your worst enemy. Because any ethical issues that arise, any resulting problems, culturally at least in this country, is immediately going to [the courts].

A member of the bioethics community, participating in that same mixed group, commented on the need for a bioethical presence in neuroimaging: "The role of bioethicists is to raise questions about where technology might lead us."

An equally prominent bioethicst describes another role for her discipline in striking a careful balance between benefit and risk: "We need to protect knowledge for individual privacy rights but still advance knowledge for society."

Commenting on the role of industry, a vendor foresaw the need for policy to the ethical challenges of neuroimaging in the future: "We have not had to face the issue of [the use of imaging for] job selection but will develop a public and political position when needed for both marketing and selling."

Technology impact—clinical

While issues of clinical impact did not emerge as prominently as the other themes, focus group members did identify a number of challenges. Here also we heard about the power of neuroimages, in this case about how diagnoses are made more forceful and more real when they are conveyed by pictures. We are also concerned about introduction of technology before it is ready. Commenting on technical readiness and introducing new technology as a complement or replacement to an existing gold standard, a neuroscientist commented: "It's a very different matter when somebody shows you a picture and says, this means you are depressed."

We also heard questions about the wisdom of using neuroimaging for diagnosis and predictive screening. One vendor stated: "A sensitive, delicate issue is who do you screen? Where do you draw the line? Will you have Alzheimer's later in life? Will you be susceptible?"

A clinical bioethicist, who is also a physician, expressed his concern about the prematurity of clinical applications of this technology:

> I guess the first worry I have is the 'not-ready-for-prime-time' worry about jumping in and announcement of results or even entering into

> clinical practice before the results are really validated in the same way that's been a real problem in genetics.

A clinical researcher agreed:

> Interpretation of research is extremely difficult to understand even for people within the field. How do you report back to parents? How do you use this technology clinically? This is especially difficult in fMRI—with low N's and complex technique. Very few physicians truly understand the technique. One must understand physics and mathematics—it takes years to understand just the analysis. What do we tell parents or people who are affected?

A science writer, commenting on health policy in her interview, warned about the temptation to use neuroimaging because of public misconceptions about the technology:

> The biggest danger in neuroimaging is not what we can do or what the technology allows, but what society believes it can do. Society has been promised results that are greater than the reality. If neuroimaging is used as diagnosis, accuracy doesn't matter. The technology is accepted and policy is shaped, even if it is inaccurate—that is terrifying.

Our participants also noted that the clinical use of neuroimaging is intimately related to the development of the technology. In an interview, one leading manufacturer of state-of-the-art MRI equipment stated, "technical development, together with clinical developments . . . that's basically setting the trend for neuroimaging development."

A second vendor would concur:

> What drives us in the functional neuro arena today is the priorities of a lot of the researchers that we collaborate with that really drive what we're doing. Now we're beginning to see a transition into more of a clinical arena . . . I would argue that functional neuroimaging has a role in personalizing medicine—based on genetic makeup; predisposition to disease.

The subtheme of treatment and intervention included comments about the use of neurotechnology for monitoring therapeutic response, neurocognitive enhancement, personalized medicine, understanding the effects of drugs, feedback to patients about neuroimages, and other issues including follow-up care. Focus group members raised questions about distributive justice—access, affordability, and other economic implications—with regard to the clinical use of neuroimaging.

Privacy and confidentiality as related to "real-world" clinical applications included commentary on labeling, stigma, and issues commensurate with privacy in genetic testing. In spite of these concerns, optimism about neuroimaging remains. A patient advocate in the mixed group remarked:

> I am not as concerned about stigma [referring to concerns about stigma raised by others in group]. . . . I think a lot of the clients I work with and a lot of the people in my organization will tell you that having an additional tool is something that is very positive.

Philosophical and theoretical issues

Participants raised philosophical and theoretical issues that have long attended the study of the brain. For example, a participant in the neuroscience group cautioned, "there is an area for this and an area for that and then suddenly we are modern day phrenologists running on a high budget."

While issues pertaining to philosophical notions of consciousness, spirituality, faith, and ideas about selfhood emerged in many of the focus groups and interviews, they were a particularly significant theme in the bioethicist and religious scholar focus groups. In this context we heard concerns about how our concepts of personal identity are challenged by the use of the brain as a shorthand way of talking about the person, the individual, and the self. Some participants discussed how these challenges are compounded by the demystification of human experience. Religious scholars were the most vocal on this subject. One religious scholar noted:

> [There is the danger of] reducing spirituality to merely a "thing," of reducing concepts of spirit, the concept of mind. You already have problems reducing the mind to the brain, [now] there are other issues of reducing spirituality to the mind.

Not all religious scholars shared this concern. One commented:

> The advantage of this research is that it reminds us of how embodied we are as individuals. . . . So on the one hand we want to be non-reductionist. We don't want to say just because we are bodies doesn't mean our religious experience isn't true. On the other hand there is an extraordinary value in learning that every dimension of the human experience has a physiological component—if not even just substructure to it.

A journalist who frequently covers brain imaging stories summarized the issues at play here:

> The world of science, neuroscience, is becoming more complicated and nuanced. The irony is that people need to understand better but don't.

> They need to better understand promises and risks. Spirit and neuroscience are closely united in how we self-identify.

In an intriguing bridge between themes that span basic science and philosophy, an academic leader in religious studies warned about using old ways of thinking to understand the new type of information provided by neuroimaging:

> There's a tendency to divide brain functions into cognitive functions, emotional functions, and so on. That's perhaps the standard that is used in western science, but it raises very important questions about whether those kinds of classificatory systems are themselves problematic. . . . We must be careful that we don't import the categories that are comfortable for us . . . because once the science is framed in certain categories, those categories tend to take over. There's a very important ethical issue right at the outset about the development of the science itself and the way it will categorize things.

Discussion

Responding to an image of bioethicists as naysayers, one of our participants asked: "When is ethics positive?" Our answer to this question is: Now. Rather than waiting for ethical, legal, and policy issues to emerge for neuroimaging, necessitating defensive responses, we have taken a positive step to identify those issues preemptively. From our sample of key members of the bioethics, humanities and neuroimaging communities, consumers and vendors, we heard how neuroimaging is perceived to have the potential to change the way we view ourselves and others, to give us the power to modify behavior, to place things that were previously thought of as unknowable into the realm of the knowable, and to create peculiar problems for researchers and clinicians (Illes and Racine 2004; Farah and Wolpe 2004; Moreno 2003). We conclude from the data that researcher obligations and the perceived non-clinical impact of the technology are areas of first priority and action.

Armed with these insights we can begin to respond to the ELSI challenges associated with neuroimaging. Our findings on the concerns and enthusiasm for neuroimaging are supported by observations about the publication of, use, and public reaction to application of these technologies. For example, questions about the possible misuse and commercialization of neuroimaging are already on the public agenda, brought to the fore by the recent controversy over industry-sponsored neuroimaging for studies of product preference (Editorial 2004). Backlash against this work arose from use of expensive neuroimaging resources for the purposes of improving marketing—arguably a potential form of human manipulation—and bold statements about what the technology can deliver. In a letter sent to the Senate Commerce Committee urging an immediate halt to this work, an organization called Commercial Alert wrote:

> What would happen in this country if corporate marketers and political consultants could literally peer inside our brains, and chart the neural activity that leads to our selections in the supermarket and the voting booth? What if they then could trigger this neural activity by various means, so as to modify our behavior to serve their own ends? . . . "Orwellian" is not too strong a term for this prospect.
>
> (www.commercialalert.org 2004)

The issue of when neuroimaging technologies are ready for clinical application, identified by our groups is particularly salient, given the tens of thousands of fMRI citations in PubMed, and 761 fMRI abstracts at the 2004 annual meeting of the *Society for Neuroscience* alone (Raichle 2004). Such volume of activity makes it difficult to justify to the public that the technology is not ready for prime time. These problems are made even more acute by the emergence of a sense of "neuro-realism" in the press (Racine et al. 2005)—an expression of how the study of a phenomenon with neurotechnology such as fMRI uncritically reifies that phenomenon. This is the case even if results from related studies are identified as preliminary or inconclusive. Public impatience with the pace of science is already visible: single photon emission computed tomography (SPECT) scans are now directly marketed to adult consumers worried about Alzheimer's disease and to parents of underachieving school-age children, and neuroimaging data from EEG recordings and PET scans are being used in our courtrooms (see e.g., *State of Iowa vs. Terry Harrington* [284 N.W.2d 244; 1979 Iowa Sup.]).

It is imperative for researchers to be attuned to, and to take ownership of these issues (Greely 2004), especially if the widening scope of neurotechnology in the public domain comes to lead, rather than trail, our well-intentioned but still academic neuroethics debates. Our findings suggest that taking ownership means being clear about the implications and limitations of research results, including limits of accuracy and validity, how the data may or may not be interpreted, and being careful not to prevent inferences that drift beyond the data. It also means being aware of how the public or media might misperceive research results or misuse neuroimaging technologies in and outside of the medical environment, and actively working toward policies that will discourage and limit this.

This study is not without shortcomings. Our data represent a relatively small sample of thinkers. The diversity and degree of expertise in neuroimaging of the participants are both a potential strength and a weakness. Our results are limited by the time constraints and dynamics of focus group studies. Nonetheless the data defined some first priority issues for the continued ethical consideration of neuroimaging as one prototype for neuroscience today; more studies are needed to validate these results and identify new priority areas as they emerge. With increasing concern about advances in neurotechnology (Fukuyama 2002; PBC 2003), it is imperative that we actively search for solutions that promote and protect good science.

Note

*The authors and editors are grateful to Taylor and Francis for allowing the chapter to be reprinted from the *American Journal of Bioethics*.

References

Boyd, E. A., M. K. Cho, and L. A. Bero. "Financial Conflict of Interest Policies in Clinical Research: Issues for Clinical Investigators," *Academic Medicine* 78 (2003): 769–74.
Brammer, M. "Brain scam?" *Nature Neuroscience* vol. 7 no. 10 (2004): 1015.
Cho, M. K., D. Magnus, A. L Caplan, D. McGee, and the Ethics of Genomics Group. "Ethical Considerations in Synthesizing a Minimal Genome," *Science* 286 (1999): 2087–90.
De Vries, R. "How can we Help? From 'Sociology in' Bioethics to 'Sociology of' Bioethics," *Journal of Law, Medicine and Ethics* 32 (2004): 279–92.
De Vries, R., B. C. Martinson, and M. S. Anderson. "Beyond FFP (Falsification, Fabrication and Plagiarism): Scientist's Perceptions of the Ethical Context of Research," presented at the Office of Research Integrity Meeting, Washington DC (November 2002).
Editorial. "Brain Scam?" *Nature Reviews Neuroscience*, vol. 7, no. 7 (2004): 683.
Farah, M. and P. R. Wolpe. "Monitoring and Manipulating Brain Function: New Neuroscience Technologies and their Ethical Implications," *The Hastings Center Report* (May/June, 2004): 35–46.
Fukuyama, F. *Our Posthuman Future: Consequences of the Biotechnology Revolution*. New York: Picador, 2002.
Glaser, B. and A. Strauss. *Discovery of Grounded Theory: Strategies for Qualitative Research*. Chicago: Aldine Publishers, 1967.
Greely, H. "Ethical Dimensions in Neuroscience," *Society for Neuroscience*, San Diego, CA (October, 2004): 23–7.
Illes, J. and E. Racine. "Imaging or Imagining? A Neuroethics Challenge Informed by Genetics," *American Journal of Bioethics*, vol. 5, no. 2 (2005): 1–14.
Illes, J., M. Kirschen, and J. D. Gabrieli. "From Neuroimaging to Neuroethics," Nature Neuroscience 6 (2003): 250.
Moreno, J. D. "Neuroethics: An Agenda for Neuroscience and Society," *Nature* 4 (2003): 149–53.
President's Council on Bioethics. *Beyond Therapy*. Washington DC: US Government Printing Office, 2003.
Racine, E., O. Bar-Ilan, and J. Illes. "FMRI in the Public Eye," *Nature Reviews Neuroscience*, 6 (2) (2005): 159–64.
Raichle, M. *Ethical Dimensions in Neuroscience*. Society for Neuroscience, San Diego, CA, October 23–7, 2004.

Schatzman, L. and A. Strauss. *Field Research Strategies for a Natural Society*. New Jersey: Prentice-Hall, 1973.

Stewart, D. and P. Shamdasani. "Focus Group Research: Exploration and Discovery," in *Handbook of Applied Social Research Methods* (eds) L. Bickham and D. Rog. Thousand Oaks, CA: Sage, 1998.

Strauss, A. and J. Corbin. *Basics of Qualitative Research: Techniques and Procedures for Developing Grounded Theory.* Thousand Oaks, CA: Sage, 1998. http://www.commercialalert.org/index.php/category_id/1/subcategory_id/82/article_id/259. Accessed September 17, 2004.

14
The End of the World Picture? Lyotard, Technology and the Human

Simon Malpas

The collection of works of art, cultural practices and critical theories that tend to be identified as postmodern have, as two of their defining characteristics, an ambiguous relation to contemporary developments in technology and a fascination with questions of image and presentation. Again and again, postmodern theory and culture examines the consequences of recent technological revolutions in order to explore their effects on our experiences of and potentials to act within today's world. For many thinkers, writers and artists, the postmodern is marked quintessentially by an ontological transformation of the status of the image. Recent developments in digital technology have, it is often claimed, changed fundamentally the relation between image and reality, signification and referent, and such changes have provided postmodern theorists with key points of focus from which can be unfolded more widely ranging analyses of contemporary thought and culture.

In his essay 'Something Like: "Communication ... without Communication"', Jean-François Lyotard, one of the most influential thinkers of the postmodern, presents the impact of new technologies on art and experience as precipitating a crisis for the human sciences that leads the philosopher to ask a series of key questions:

> Is it the case that in this crisis . . . there is simply the loss of something (donation or presentation) without there being some gain? We are losing the earth (Husserl), which is to say the here-and-now, but are we gaining something and how are we gaining it? Can the uprooting which is linked to the new technology promise us an emancipation?[1]

Those familiar with many of the recent commentaries on Lyotard's work, might well expect his answers to such questions to be entirely negative: new technologies offer no possibility of gain, the loss they impose is absolute, and the very idea of emancipation has vanished. It is probably not overstating the case too strongly to say that Lyotard has frequently tended to be read as a late twentieth-century incarnation of the machine-wrecking Luddites of the industrial revolution. Let me, at this point, cite just a couple of examples of such commentary. Stuart Sim presents Lyotard as calling 'for a campaign against techno-science and all its works', and claims that he 'rails against techno-science'.[2] Similarly, Simon Cooper argues that Lyotard, albeit eventually unsuccessfully, seeks to 'mount a radical critique of technology through developing a strategy committed to undermining modern frameworks legitimated by reason'.[3] Such claims suggest that there is little in Lyotard, other than outright rejection, that is likely to be relevant to the analysis of contemporary imaging technology. This, however, is not the case. Although there are undoubtedly passages of scepticism towards many of the emancipatory claims made on behalf of the new technologies in his writings and, to a certain extent, these two critics do capture elements of his work accurately, I want to argue that Lyotard's response to technology, and particularly those imaging technologies associated with contemporary artworks, is considerably more nuanced than such categorical assertions allow.

The aim of this chapter, then, is to explore the ways in which Lyotard analyses the threats and promises of new forms of imaging technology in terms of the ways in which they inform and are informed by phenomenology. In order to do this, I want to explore his critique of the humanist idea that technologies and images are tools to be put to work by the autonomous rational subject. I hope to show that the impacts of technological thought and phenomenological philosophy on his work allows Lyotard to develop a more complex and nuanced account of what is at stake in our relation to the ontological crisis he sees as precipitated by contemporary developments in imaging technology.

The end of the world

> While we talk, the sun is getting older. It will explode in 4.5 billion years. It's just a little beyond the halfway point in its expected lifetime. It's like a man in his early forties with a life expectancy of eighty. With the sun's death your insoluble questions will be done with too. . . . You explain: it's impossible to think an end, pure and simple, of anything at all, since the end's a limit and to think it you have to be on both sides of that limit. . . . Now this is true of limits belonging to thought. But after the sun's death there won't be a thought to know that its death took place. That, in my view, is the sole serious question to face humanity today.[4]

This challenge, posed by 'HE' to the philosopher 'SHE', opens Lyotard's essay 'Can Thought go on without a Body?' With the demise of the sun, the material basis for all thought and experience will be annihilated, marking an end, as HE puts it, of all philosophising. The limit this event imposes will be absolute: the complete extinction that is threatened, by putting an end to thought, will remain irrecoverable by the limit-transcending movement of dialectical sublation. The task of contemporary thought, the 'sole serious question to face humanity today', HE claims, is to find a means to carry thought beyond this disaster, to hold out the possibility of continuing the dialectic by producing a vehicle capable of allowing thought to survive the destruction of the solar system and the material ground it provides.

This might seem a somewhat alarmist opening to a philosophical essay. An event that is unlikely to occur for four and a half billion years, however devastating one might imagine it to be, is hardly the most pressing concern for contemporary culture and is unlikely to cause anyone but the most radical of futurologists sleepless nights. But this, of course, is not the point. What Lyotard is using this 'fable' to do is develop a series of questions about the relation between thought, technology and sense, and to explore the question of thinking in relation to transformations in technology that, even today, are fundamentally altering the ontological status of experience and the image.[5] His presentation of the end of the world makes explicit a series of crucial tensions between phenomenological philosophy and contemporary technoscience in terms of their respective analyses of the human capacity to engage cognitively with sensations of the world. What will emerge from Lyotard's thinking of this disaster is a relation between phenomenology and technology that disrupts familiar accounts of both, calling into question both their coherence and ground, and transforms the status of the image.

The two voices in Lyotard's essay each develop a response to this event. First, HE cites the demise of the sun as presenting a technological imperative:

> The hardware [the body] will be consumed in the solar explosion taking philosophical thought with it (along with all other thought) as it goes up in flames. So the problem of the technological sciences can be stated as: how to provide this software [thought] with a hardware that is independent of the conditions of life on earth. That is: how to make thought without a body possible.[6]

For the technologist, then, the question posed by the explosion of the sun must focus on creating a technological system, a new 'hardware', capable of maintaining thought in the absence of the nourishment required for the survival of the body. Despite the recognition that such hardware must work outside of the 'binary mode' of contemporary computing and deal not just with 'units of information (bits), but with intuitive, hypothetical configurations', HE still appears to present experience in terms of a binary model: the task of technology is to secure the hardware that will allow thought-programmes to 'run'.[7] In other words, HE retains a notion of mind–body dualism in which the biological can be replaced by the technological, a process which is rapidly becoming more and more possible with today's advances in medical science, in order to project a stage of development at which the body can, once and for all, be dispensed with. This mode of analysis of the relation between technology and the human is probably already familiar, not only from contemporary science fiction with its depictions of androids, world-dominating mega-computers and totally immersive virtual realities, but also from those postmodern theoretical analyses of cyborg life that one finds in the influential work of such thinkers as Donna Haraway and Sadie Plant.[8]

In contrast to this apparent dualism, SHE presents a significantly different account of the relation between experience, thought and body; one that raises explicitly the problem of imaging. Such a thinking machine, SHE argues, must be capable of reproducing the full complexity of human perception:

> A direct, focussed vision is always surrounded by a curved area where visibility is held in reserve yet isn't absent. . . . Perceptual 'recognition' never satisfies the logical demand for complete description. In any serious discussion of analogy [between human and machine] it's this experience that is meant, this blur, this uncertainty, this faith in the inexhaustibility of the perceivable, and not just a mode of transfer of data onto an inscription-surface not originally its own. . . . Real 'analogy' requires a thinking or representing machine to be *in* its data *just as* the eye is in the visual field. . . . It's a matter (to use the attractively appropriate locution) of 'giving body' to the artificial thought of which they are capable.[9]

This notion of the 'reserve' in perception marks a crucial distinction from the dualist analysis of imaging presented by HE: the perceivable is 'inexhaustible' precisely because the inclusion of the perceptual mechanism (the 'eye' or 'representing machine') in the perceptual field transforms that field in every act of perceiving. The act of perception, in producing an image for thought, necessarily generates the reserve that makes such a presentation 'absolutely unique every time' and constitutively open to subsequent perceptual modifications.[10]

Already recognisable in this passage are the categories of phenomenology. Although Lyotard doesn't refer explicitly to Edmund Husserl at this point, it seems likely that the model of embodied perception being alluded to here is that which the latter introduces in *Ideas I* and *II*, the notion of a hyletic substrate, '*a human being's total consciousness is in a certain sense, by means of a hyletic substrate, bound to the Body*, though, to be sure, the lived experiences are *no longer* directly and properly *localised*; they no longer form a stratum of the Body'.[11] In other words, according to Husserl, immanent within each intentional act of perception are two components: first, the intentional dimension that generates meaning, which Husserl refers to as its noetic content, and, second, a non-intentional and non-localisable sensuous content referred to as *hyle* (which Husserl locates in the 'what it is like' of such sensations as pleasure, pain, nausea, and so on). The hyletic substrate, which has 'nothing of intentionality' in them, is what allows perceptual content to be intended by withdrawing from and within that intending to produce the noematic object.[12] While this hyletic component is capable of becoming the noetic content of a fresh perception, this second act would bring with it a further hyletic substrate, and so on indefinitely.[13] The consequence of this is that perception is irreducible to a calculable, programmable or reproducible intentionality as it always also contains a supplementary sensuous, in this instance bodily, substrate that is generated by the presence of the perceiver in the perceptual field and makes that field inexhaustible.

SHE, by focussing on this aspect of the phenomenology of perception, presents thought as a 'suffering': 'This soliciting of emptiness, this evacuation – very much the opposite of overweening, selective identificatory activity – doesn't take place without some suffering.'[14] This suffering, which Lyotard elsewhere refers to as passibility and links to Kantian aesthetics, is the withdrawal that occurs within presentation to disrupt programmatic identification thereby making the image singular and untotalisable.[15] SHE links this passibility to gender difference, and argues that the gendering of thought opens an irreducible aspect of human perception: 'To that which without gendered difference would only be a neutral experience of the space-time perceptions . . . gendered difference adds the suffering of abandonment because it brings to neutrality what no field of vision or thought can include. . . . And that force is desire.'[16] It is desire that differentiates human perception from technological scanning: in contrast to the electronic mice that scan and process their surroundings to find their way out of mazes, human beings engage with images in terms not just of knowledge, but of suffering, passibility and desire. For SHE, this is the aspect of experience that technoscience's dualistic analysis of thought fails to explore, and yet is necessary if anything like 'thought' might be able to go on without a body.

Are we to take this simply as an attack on technoscience? That seems to me overhasty. HE doesn't get a chance to respond to the comments of SHE. Because of its form, the direct speech of the two voices is unaccompanied by any narratorial commentary, 'Can Thought go on Without a Body?' makes no attempt to reconcile their perspectives, and deliberately leaves the titular question unanswered. In fact, one might suppose that the point of the essay is to present the two positions as being subject to what Lyotard calls a 'differend': 'a case of conflict, between (at least) two parties, that cannot be resolved for lack of a rule of judgement applicable to both of the arguments'.[17] Technological and phenomenological descriptions of thought both circulate around the complexity of bodily perception, speaking the problem in different, apparently irreconcilable, languages, and, because of this, are incapable of providing a final answer to the question.

The arguments of SHE are a response to, and would make little sense outside of the context of, those of HE. The essay then, contra the comments of the critics cited at the beginning of this chapter, does not mark a rejection of technology, but rather a problematisation of what might be at stake in the analysis of contemporary imaging technologies.

In what follows, my aim is not to attempt to provide the resolution that Lyotard's essay refuses, but rather to explore the implications of the differend between phenomenology and technology that he sets up here, and that occurs throughout his discussions of the postmodern. Before moving on to a more detailed analysis of Lyotard's postmodern analyses of imaging technologies, I want briefly to introduce the phenomenological engagement with technology, ground and the image that is perhaps the most influential precursor to postmodern theory, and certainly to Lyotard's arguments, which occurs in Martin Heidegger's explorations of technology and the 'world picture'.

Image and ontology: The age of the world picture

According to Heidegger, modernity is, essentially, the 'age of the world picture', and this becoming picture of the world produces, and is produced by, technology. In a series of important essays and lectures written between 1938 and 1962, he presents technology as an inescapable horizon for contemporary existence that is at once 'the supreme danger' and, yet, also the site at which the 'coming to presence of truth' might remain possible.[18] There is no space here to trace out the full range of questions this work poses, so, in lieu of such a discussion, I want to focus on one aspect: Heidegger's analysis of technological thinking as, essentially, a mode of production of images that frame human experience and being.

According to Heidegger, the most fundamental challenge facing modern philosophy is produced not just by the new technologies themselves but by the processes of what he calls technological thinking. This essence of technology is, he argues, 'by no means anything technological', and so long as 'we represent technology as an instrument, we remain held fast in the will to master it', a will to mastery which lies at the heart of the technological thought that treats the world entirely in terms of means and ends.[19] Accordingly, existence and experience have become essentially technological: our day-to-day being in the world has come to be 'Enframed' (*Gestellen*) by technology, and this 'coming to presence of Enframing is the danger'.[20] What he means by 'Enframing' is 'the way in which the real reveals itself as standing-reserve'; or, in other words, the attitude towards the world that encounters the object as material to be used as a means for an end, as an energy or resource whose value lies in its potential to be exploited.[21] The world of objects comes to be experienced as a stockpile of resources that humankind puts to work in order to shape its destiny. For Heidegger, this approach to the world generates a contradiction, which, he argues, gives rise to the 'supreme danger':

> As soon as what is unconcealed [the world] no longer concerns man even as object, but does so, rather, exclusively as standing-reserve, and man in the midst of objectlessness is nothing but the orderer of the standing-reserve, then he comes to the very brink of a precipitous fall; that is, he comes to the point where he himself will have to be taken as standing-reserve. Meanwhile man, precisely as the one so threatened, exalts himself to the posture of lord of the earth. In this way the impression comes to prevail that everything man encounters exists only insofar as it is his own construct.[22]

In thinking the world as a standing-reserve for technological innovation and presenting itself as the master of this process, humankind precipitates itself towards a situation in which it too is in danger of becoming nothing more than the material to be put to work in the name of progress. Thought in terms of standing-reserve, the object of experience becomes entirely programmable, calculable, and the suffering and passibility of experience is excluded. The subject, in turn, becomes the sort of programmatic thinking machine that SHE sets out to criticise in Lyotard's essay, and, as a consequence of this, a cog in the wheels of technologically driven industrial or political machination.

For Heidegger modern technology gains its determinative power only on the basis of a prior ontological decision that gives rise to what he calls the 'world picture'. This ontological decision occurs in the emergence of the modern subject and humanism, which finds its clearest and most influential formulation in Descartes:

> When man becomes the primary and only real *subiectum*, that means: Man becomes that being upon which all that is, is grounded as regards the manner of its Being and its truth. Man becomes the relational centre of that which is. . . . Where the world becomes picture, what is, in its entirety, is juxtaposed as that for which man is prepared and which, correspondingly, he therefore intends to bring before himself and have before himself, and consequently intends in a decisive sense to set in place before himself. . . . Wherever we have the world picture, an essential decision takes place regarding what is, in its entirety. The Being of whatever is, is sought and found in the representedness of the latter.[23]

What Heidegger is presenting here is the argument that, in the generation of a subject–object ontology, experience and knowledge become matters of the representation or picturing of an object for and by the central subject. What is necessarily excluded by such an ontology is, quite precisely, the hyletic substrate that Husserl and Lyotard insist accompanies all perception: the intentionality Heidegger refers to in this passage is a purified noesis that 'sets in place' the object as a means or model to be programmed and processed towards some end. Suffering and passibility are relegated by this technological rationality to the level of mere and contingent affects. As Lyotard argues, 'passibility as . . . belonging to an immediate community is repressed nowadays in the general problematic of communication, and is even taken as shameful'.[24] And the effect of this exclusion of passibility is to generate a conception of knowledge as a calculating representation produced by and for the knowing subject:

> Knowing . . . calls whatever is to account with regard to the way in which and the extent to which it lets itself be put at the disposal of representation. Research has disposal over anything that is when it can either calculate it in its future course in advance or verify a calculation about it as past. Nature, in being calculated in advance, and history, in being historiographically verified as past, become, as it were, 'set in place' [*gestellt*]. Nature and history become objects of a representing that explains. Only that which becomes an object in this way *is* – is considered to be in being. . . . This objectifying of whatever is, is accomplished in a setting before, a representing, that aims at bringing each particular being before it in such a way that man who calculates can be sure, and that means be certain, of that being.[25]

Calculation paints the world picture: in Heidegger's characterisations of scientific and historiographical methodology the spaces of nature and the past are laid out for a rational knowledge which is able to verify and objectify all of their determinations in order to project a future that is entirely predictable. The objectivity of experience is founded on its transformation of the world into a collection of present but spatially and/or temporally distant objects whose relations, meanings and effects can be categorised, classified and organised by the scientist in the here-and-now as standing-reserve for productive processes. And, crucially, as Heidegger asserts, '[o]nly that which becomes an object in this way *is* – is considered to be in being'.

The parallels with Lyotard's analysis in 'Can Thought go on without a Body?' are clear: the threat of extinction that Lyotard fables as the death of the sun is, for Heidegger, immanent in modern experience as technological thinking excludes all that resists reduction to objectivity. Humanism, according to Heidegger's analysis, produces an inherently self-contradictory account of the image: its attempt to present the human being as 'the relational centre of the universe' capable of manipulating images and objects for its own ends has its relation to the world inverted by the technology it treats as a mere tool to become itself an object of technological thinking, and, what is more, an object that is reduced to the status of a standing-reserve to be shaped and transformed in the processes of production. In the light of this thinking of technology and the world picture, which generate the crisis that Lyotard evokes in the quotation with which this chapter opened, the key question is how thought can preserve passibility in the face of the rapidly increasing power, scope and impact of contemporary technologies.

The end of the world picture? Phenomenology, technoscience and the postmodern image

As I said at the beginning, postmodern theory and culture respond ambiguously to technological development. For many postmodern thinkers, and perhaps this is what makes them postmodern, the technologisation of thought and experience that Heidegger depicts as the impending 'supreme danger' has come to fruition in contemporary culture, and is now inescapable. In order to differentiate Lyotard's thought from the more resigned responses to this state of affairs, I want to cite just one version of postmodernism, albeit possibly the most notorious, that seems to embrace Heidegger's danger as 'just the way things are'. Jean Baudrillard asserts that

> [a]bstraction today is no longer that of the map, the double, the mirror or the concept. Simulation is no longer that of a territory, a referential being or a substance. It is the generation by models of a real without origin or reality: a hyperreal. The territory no longer precedes the map, nor survives it. Henceforth, it is the map that precedes the territory – *precession of simulacra* – it is the map that engenders the territory. . . . But it is no longer a question of either maps or territories. Something has disappeared: the sovereign difference between them that was the abstraction's charm.[26]

What Baudrillard is indicating in this, now canonical, postmodern assertion is that the common-sense notion of representation in which signs stand in for a pre-given reality, image it, and can be judged on the basis of their relation with it, has come to be replaced by a structure of simulation in which such 'reality' is no longer the focus, the basis or the controlling instance of signification. The immediacy of simulation makes abstraction, and therefore, representational imaging impossible. For Baudrillard, the meaning and value of a sign is produced and determined entirely in terms of its relation with other signs, and 'reality' is thus 'produced from miniaturised units, from matrices, memory banks and command models. . . . It is a hyperreal: the product of an irradiating synthesis of combinatory models in a hyperspace without atmosphere.'[27] There is no outside of simulation: the Cartesian decision that generates the world picture for the subject has itself been caught up in the picture, and the humanist subject has succumbed to

Heidegger's 'supreme danger' by being turned entirely into standing-reserve for the 'combinatory models' of postmodern hyperreality.

The key consequence of this move into hyperreality is a transformation of experience, epistemology and the image from a model based on representational correspondence or authenticity to one premised on performance: in the world of the simulacra, the true or the real is simply what works most efficiently. The result of this is that Baudrillard's analysis of postmodern simulation presents contemporary culture as a perverse, hyperreal literalisation of Hegel's assertion about the dialectical relation between reality and rationality: 'What is rational is actual; and what is actual is rational.'[28] For Baudrillard,

> [t]here is no longer any critical and speculative distance between the real and the rational. There is no longer really even any projection of models in the real . . . but an in-the-field, here-and-now transfiguration of the real into model. A fantastic short-circuit: the real is hyperrealised. Neither realised nor idealised: but hyperrealised. The hyperreal is the abolition of the real not by violent distinction, but by its assumption, elevation to the strength of the model.[29]

The postmodern, according to Baudrillard's analysis, marks the point at which the real has been entirely replaced and 'transfigured' by rationalised technological models and, as a result of this, the possibility of a 'critical and speculative distance' between them has collapsed. If, as Hegel argues, the progress of knowledge is marked by the development of a rational grasp of reality through the continual sublation of the limits between them, Baudrillard claims that this process is completed in postmodern hyperreality. The effect of this, however, is very different from the Hegelian ideal of the realisation of absolute knowing and universal freedom. Instead, in a grotesque parody of the speculative dialectic, a 'fantastic short-circuit', it marks an absolute loss of critical distance through an inability to distinguish between the two poles of the opposition: we can no longer use reason to map the real because, due to the precision and complexity of our models and technologies, the two have become identical as rationality 'assumes' the real. What Baudrillard means by this is that with the advent of new sciences and technologies, the models that can be produced to understand the world have become more real, more sophisticated and more accurate than reality itself. Humanity has become lost in a realm of hyperreality that refuses us the distance to stand back from our experiences and question them; refuses us, in other words, the possibility of anything other than Heidegger's technological objectivity. There is thus no longer any world picture for Baudrillard, as even the minimal critical distance that would bring such a picture into focus has been absorbed into the immediacy of hyperreal simulation.

This seems to me to present a pretty hopeless scenario. Baudrillard's writing veers between celebration of postmodern simulation and a passive longing for some event of such extreme proportions that simulation is itself short-circuited by the shock re-emergence of the real.[30] What is less clear is whether it allows room for any form of critical questioning or response.

Like Baudrillard, Lyotard frequently appears to present the postmodern as a site in which the imminent threats of technological thinking that Heidegger predicts have come to some sort of fruition. He famously describes the 'postmodern condition' as an 'incredulity toward metanarratives' in which the systems of epistemological or ethical organisation that have guided thought and experience from the beginnings of humanist philosophy no longer hold sway, and have come to be replaced by a single form of ordering based on the aim of 'optimising the systems performance', which, he claims, 'entails a certain level of terror, whether soft or hard: be operational . . . or disappear'.[31] The threat issued by the operational imperative of contemporary knowledge production is, for Lyotard, its capacity to present all aspects of experience and knowledge as, precisely, standing-reserve to be put to work to maximise the performance of the system:

> The system seems to be a vanguard machine dragging humanity after it, dehumanising it in order to rehumanise it at a different level of normative capacity. The technocrats declare that they cannot trust what society designates as its needs; they 'know' that society cannot know its own needs since they are not variables independent of the new technologies.[32]

This appears to share Baudrillard's position. However, in contrast to an entirely hermetic notion of simulation, Lyotard's writing seeks to question the consequences of the performative criteria of technological development and its impact upon social existence. His work presents a much more troubled analysis of the effects of technological thought by exploring in detail the status of imaging, and is continually concerned to find spaces for resistance. Such resistance cannot, for Lyotard, be based on notions of autonomous subjectivity or humanism: one cannot simply step outside of the inhumanity of the operational imperative of late capitalist technoscience. Rather, resistance must be produced immanently from within this system, and critique must aim to interrupt its operation.

In contrast to this technological inhuman of the operational imperative, Lyotard points towards another form of the inhuman: the potential for being taken hold of by surprising and uncanny transformative possibilities that cannot be predicted, explained or mastered by technologically based systems of reason. He locates this sense of the inhuman in what he calls the 'anguish of a mind haunted by a familiar and unknown guest which is agitating it, sending it delirious but also making it think'.[33] This, quite precisely, is the suffering and passibility generated in the perception of the image that SHE insists upon in 'Can Thought go on without a Body?'

For Lyotard, then, experience is the product of a conflict between two inhumans: the inhuman systems of capitalist development and its technological–operational thinking threaten to extinguish anything that is not relevant to the optimisation of the system, and yet within the experiences produced by this system lies the uncanny strangeness of another inhuman that is a potential site of resistance:

> [T]he question I am raising here is simply this: what else remains as 'politics' except resistance to this inhuman [system]? And what else is left to resist with but the debt to which each soul has contracted with the miserable and admirable indetermination from which it was born and does not cease to be born? – which is to say, with the other inhuman?[34]

Without the inhuman indetermination at its heart the human ceases to be able to resist the other form of the inhuman, that of the developmental system. The differend between them, like the differend between technoscientific discourse and phenomenology, is irreducible and necessary. For Lyotard it is not simply a question of somehow rejecting technology and all its works; rather, his response to the new imaging technologies is based upon a continued phenomenological analysis of the intentional suffering generated by the images they produce that will serve to disrupt the operational imperative of technological thinking by introducing the irreducible and infinite differend of passibility. The question of whether we have or will come to the end of the world picture cannot definitively be answered, but must continue to be held open if anything like the human is to survive.

Notes

1. Jean-François Lyotard, 'Something Like: "Communication ... Without Communication"', in Lyotard, *The Inhuman: Reflections on Time*, trans. Geoffrey Bennington and Rachel Bowlby, Cambridge: Polity, 1991, p. 116.
2. Stuart Sim, *Lyotard and the Inhuman*, Cambridge: Icon Books, 2001, pp. 5 and 9.
3. Simon Cooper, *Technoculture and Critical Theory: In the Service of the Machine?*, London: Routledge, 2002, p. 90.
4. Lyotard, 'Can Thought go on without a Body?' in Lyotard, *The Inhuman*, 1991, p. 8.
5. This thought experiment is taken up again, explicitly as a 'fable', in Lyotard's *Postmodern Fables*, trans. Georges Van Den Abeele, Minneapolis: University of Minnesota Press, 1997.
6. Lyotard, p. 13.
7. Lyotard, p. 15.
8. See, for example, Donna Haraway's *Simians, Cyborgs, and Women: The Reinvention of Nature*, London: Routledge, 1991, or Sadie Plant's *Zeroes + Ones: Digital Women + The New Technoculture*, London: Fourth Estate, 1997. Although both

of these writers present technology as offering the potential to challenge the straightforwardly dualistic analyses of sexual difference, they tend to do so while leaving an, albeit modified, account of mind–body dualism in place.

9. Lyotard, pp. 16–17.
10. Lyotard, p. 17.
11. Edmund Husserl, *Ideas Pertaining to a Pure Phenomenology and to a Phenomenological Philosophy. Second Book: Studies in the Phenomenology of Constitution*, trans. Richard Rojcewicz and André Schuwer, London: Kluwer, 1980, pp. 160–1.
12. Edmund Husserl, *Ideas Pertaining to a Pure Phenomenology and to a Phenomenological Philosophy. Second Book: General Introduction to a Pure Phenomenology*, trans. F. Kersten, London: Kluwer, 1982, p. 172.
13. In his essay '"Genesis and Structure" and Phenomenology', Jacques Derrida provides a detailed analysis of this irreducibly problematic relation between *hyle* and *morphe* as a means of calling into question 'the unity of the world from which transcendental freedom releases itself'. See Derrida, *Writing and Difference*, trans. Alan Bass, London: Routledge, 2001, pp. 193–211. I want to argue that Lyotard's exploration of the impact of new technologies, when read in relation to Heidegger's analysis of the 'world picture', opens up related questions about freedom and the transcendental.
14. Lyotard, 'Can Thought go on without a Body?' p. 18.
15. In 'Something Like: "Communication ... Without Communication"', Lyotard offers the following definition of passibility: 'If we are in a state of passibility, it's that something is happening to us, and when this passibility has a fundamental status, the donation itself is something fundamental, originary. What happens to us is not at all something we have first controlled, programmed, grasped by a concept. ... Passibility, in contrast, has to do with an immediate community of feeling demanded across the singular aesthetic feeling, and what is lost is more than simple capacity, it is propriety', pp. 111 and 117.
16. Lyotard, 'Can Thought go on without a Body?' pp. 21–2.
17. Jean-François Lyotard, *The Differend: Phrases in Dispute*, trans. Georges Van Den Abeele, Manchester: Manchester University Press, 1988, p. xi.
18. Martin Heidegger, 'The Question Concerning Technology', in Heidegger, *The Question Concerning Technology and Other Essays*, trans. William Lovitt, New York and London: Harper & Row, 1977, pp. 26 and 30.
19. Heidegger, 'The Question Concerning Technology', pp. 4 and 32. This notion is developed in much more detail in his discussions of Leibniz's 'principle of reason' in Martin Heidegger, *The Principle of Reason*, trans, Reginald Lilly, Bloomington and Indianapolis: Indiana University Press, 1996.
20. Martin Heidegger, 'The Turning', in Heidegger, *The Question Concerning Technology*, p. 41.
21. Heidegger, p. 23.
22. Heidegger, pp. 26–7.
23. Heidegger, pp. 128–30.
24. Lyotard, 'Something Like: "Communication ... Without Communication"', p. 117.
25. Heidegger, p. 127.
26. Jean Baudrillard, *Simulations*, trans. Paul Foss, Paul Patton and Philip Beitchman, New York: Semiotext(e), 1983, p. 3.
27. Ibid., p. 3.
28. G. W. F. Hegel, *Elements of the Philosophy of Right*, trans. H. B. Nisbett, Cambridge: Cambridge University Press, 1991, p. 20.

29. Jean Baudrillard, *In the Shadow of the Silent Majorities: Or, the End of the Social and Other Essays*, trans. Paul Foss, Paul Patton and John Johnston, New York: Semiotext(e), 1983, pp. 83–4.
30. This characterisation of Baudrillard is one that he himself appears to endorse in his, not entirely persuasive, attempt to read what took place on 11 September 2001 as just such an event in *The Spirit of Terrorism*, London: Verso, 2003.
31. Jean-François Lyotard, *The Postmodern Condition: A Report on Knowledge*, trans. Geoff Bennington and Brian Massumi, Manchester: Manchester University Press, 1984, p. xxiv. This notion of 'incredulity toward metanarratives' is frequently read by critics as Lyotard's celebration of postmodern freedom. It is not. I discuss what it might entail in more detail in Simon Malpas, *Jean-François Lyotard*, London: Routledge, 2003.
32. Ibid., p. 63.
33. Jean-François Lyotard, 'Introduction: About the Human', in Lyotard, *The Inhuman*, p. 2.
34. Ibid., p. 7.

15
Word and Image in Online Education*

Andrew Feenberg

My adventures in distance learning

Once the stepchild of the academy, distance learning is finally taken seriously. But not in precisely the way early innovators like myself had hoped. It is not faculty who are in the forefront of the movement to network education. Instead politicians, university administrations, and computer and telecommunications companies have taken the lead. But proposals for a radical "retooling" of the university emanating from these sources are guaranteed to provoke instant faculty hostility.

This is a story about my role in the recent transformation of distance learning from pariah to standard. The drama unfolds in the late 1990s when I found myself in the paradoxical position of defending my own understanding of distance learning against both its foes on the faculty and its advocates in the administration.

In 1981 I worked on the design team that created the first online educational program. This was the School of Management and Strategic Studies at the Western Behavioral Sciences Institute in La Jolla, California (Feenberg, 1993; Hamilton and Feenberg, 2005). The school offered courses taught by humanistic social scientists addressing major issues, such as globalization, environmentalism, urban planning, philosophy of technology, and so on. For nearly ten years, I helped with the operation of the school, trained teachers, and myself taught courses in it.

At the time online education was essentially untried. The equipment was expensive and primitive. We used Apple IIE's with 48K of memory and 300 baud modems. (Multiply by 1000 and 100 respectively to get 2000 averages.) The complexity of basic computer operations in the 1980s was such that it took a full page of printed instructions just to connect. A variant of e-mail called computer conferencing was the only available electronic mediation.

Computer conferencing was suited to our application since it facilitated the sort of many-to-many communication that goes on in the classroom, but no one knew how to use it for education. None of us had ever been a student in an online class or seen one in operation, and we did not know the answers to the most elementary pedagogical questions, such as how to start a class, how long or short messages should be, and how often the teacher should sign on and respond to the students.

We soon discovered that computer conferencing was not very useful for delivering lectures, and of course it could not support any graphical contents, even the simple drawings teachers like to scribble on the blackboard. But these limitations led us to explore a Socratic pedagogy based on virtual classroom discussion that proved quite successful.

The school grew to include over 150 students in 26 countries around the world. It pioneered many of the features of online education taken for granted today. These include typical teacher and student roles and relationships, techniques for organizing discussion in a virtual classroom, ways of combining aspects of technical moderating and educational leadership, the use of informal chatting and "café" conferences, specialized client-server software, and so on.

Other experiments soon benefited from our example and added their own contributions. Among the earliest were online classes at the New Jersey Institute of Technology, The New School for Social Research, The University of Arizona, Tucson, The Ontario Institute for the Study of Education, and the Open University in England.

These experiments were all championed by enthusiastic professors who involved their students in an adventure on the frontiers of technology. At first growth was slow, but in the early years of this century online education has become a standard feature of the modern university. The widespread acceptance of online education dates from the late 1990s, when it was taken up enthusiastically by university administrators.

In 1998 I heard rumors that something called online education was coming to the university at which I was then teaching, San Diego State University, under the sponsorship of Microsoft, Hughes Aircraft, Fujitsu, and MCI. This initiative, called CETI, was supposed to build a $300 million information infrastructure to support virtual learning on our multicampus system. Our classrooms and dorms were to be hardwired to the Internet; we were to have video conferencing, various computer based teaching aids, electronic distance learning, and production facilities for marketable prepackaged courses to be sold by the CETI consortium for a profit.

CETI was opposed by most faculty and students. There were two main objections. First, both teachers and students doubted the educational value of networking, and second, some faculty members were upset by the commercial goal of CETI, the delivery of higher education through the market outside the context of a university community. What was once a daring faculty innovation had come to be perceived as a big business takeover of the campus (Noble, 1997).

I am no more enthusiastic about trading an academic job for one at Microsoft than the next faculty member, but this unqualified rejection of online education contradicts our experience at the Western Behavioral Sciences Institute. There the virtual classroom was a place of intense intellectual and human interaction.

Literally hundreds of highly intelligent comments were contributed to our computer conferences each month by both students and teachers. The quality of these online discussions surpasses anything I have been able to stimulate in my face-to-face classroom. As CETI became a common topic of discussion on my campus, I wondered why my colleagues did not share my interest in this innovative medium.

My puzzlement was soon to end. Our new system-wide chancellor, Dr. Charles Reed, was due for a get-acquainted visit. As he was leaving I finally had an opportunity to ask him the question that most bothered me: What is the pedagogical model that has guided CETI? The chancellor looked at me as though I had laid an egg, and said, "We've got the engineering plan. It's up to you faculty to figure out what to do with it." And off he went: subject closed!

Would you build a house this way or design a new kind of car or refrigerator? Surely it is important to find out how the thing is going to be used before committing a lot of resources to a specific plan or design. Yet this was not at all the order in which our chancellor understood the process. Why not? I would guess it is because he did not conceive of the technology of online education as a system, including novel pedagogical challenges, but as an infrastructure, an "information superhighway," down which we faculty were invited to drive. And just as drivers are not consulted about how to build the roads, so faculty were not much involved in designing the educational superhighway.

But this overworked metaphor is altogether inappropriate. In the case of educational computing, the choice of infrastructure will largely determine the applications. If corporations rather than faculty are consulted about this choice, the outcome will be entirely different from the ideal of educational community to which faculty is attached by their culture and traditions. The ambition of CETI to make and market computer and video based courses illustrated that difference.

The CETI story has a significant ending. Public outcry against it grew gradually as faculty and students protested on campus, in the newspapers, and before legislative committees. Legal and financial questions were raised about mixing public and private assets, and finally the companies pulled out one by one. The initiative collapsed and was replaced by a more modest plan paid for out of public monies, as is proper. The faculty shed no tears over having to wait a bit longer for their first ride on the electronic *infobahn*.

The fate of CETI is emblematic of many similar initiatives. After an initial burst of enthusiasm, administrators encountered problems and for the most part gave up on the commercial ambitions that had at first inspired their investment in online education. But the investment remained. No university president was prepared to say he had made a mistake in supporting the purchase of the equipment required by his online education initiative. The slippery concept of "online education" gradually returned to something more like the original meaning we gave it in the early 1980s. In this form universities remained committed to expanding network opportunities despite the lack of savings.

Rather than replacing the existing faculty, online education was presented to them as an enhancement of the classroom. Faculty was encouraged to use the available equipment. The automated alternative promised by computer companies and futurologists disappeared from the radar screens. Text-based "learning management systems" such as Web CT and Blackboard were acquired and teachers used them to supplement ordinary classroom teaching. These universities were little changed by the widespread introduction of the new technology. Meanwhile, adult educational institutions such as the University of Phoenix developed very large online education programs in which text based online classes played a central role.

Education and economics

CETI teaches an important lesson about the different ways in which most administrators and faculty understood distance learning and its technology during the dot com boom, of which this story is an episode. I will try to sketch what I take to be these different perspectives. Of course generalizations such as those I am about to formulate do not apply universally, but it is a fact that the distance learning debate polarized, and to some extent still polarizes, around two hostile positions that usually correspond to the different concerns of administration and faculty.

For too many administrators the big issues were not educational. The fiscal implications of electronic distance learning were what interested them. Administrators hoped to use new technology to finesse the crisis in higher education spending, and to accommodate exploding enrollments of young people and returning students. Innovations like video conferencing and automated online education were supposed to improve quality through the use of "star" professors while cutting costs of delivery. Students in virtual classrooms would need no new parking structures. What is more, courses could be packaged and marketed, generating a continuous revenue stream without further investment.

But how could new technology accomplish the existing educational mission for larger numbers at a discount? The failure of the projects of the 1990s has not discouraged advocates of the agenda elaborated then. They continue to propose two main solutions.

Video conferencing allows a professor to address a large number of students in remote locations. Live interaction can be supported by a two-way video feed.

The physical presence of teachers and students in the classroom can be reproduced electronically at some cost, but more students can be served without expanding existing campuses. This approach has some successful applications in remote geographical areas and with adult learners.

Automation offers a more radical solution with large start up costs but promising far greater savings in the long run. In an automated system, the teacher's physical presence in the classroom is reproduced on CD-ROM or made available over the Internet. Exciting computer based graphical materials can replace dull textbooks. Research on the Internet can replace hours spent in libraries. Testing and grading can be done online. Even essay tests can be graded by powerful programs for textual analysis (Foltz, 1996).

The key to automation is to separate out informational "content" from "process." A small number of well paid "content experts" will work as "star" performers, while the delivery process is deskilled so that inexpensive tutors can handle interaction with students. In a really low cost solution, discussion can be replaced by automated exercises. Eventually it will be possible to dispense with campuses altogether. Students will pick out courses at an educational equivalent of Blockbuster and "do" college at home without ever meeting a faculty member or fellow student (Agre, 1998).

Is this for real? Unfortunately many people think it is. Coopers and Lybrand published a white paper in which they claimed that 25 packaged courses can take care of half of community college and 35 percent of four-year college enrollments (Coopers & Lybrand, 1997). They are convinced that students will learn just as much if not more, and they will be free to study at their own pace. In educational terms, nothing fundamental will change except cost and convenience, those two favorite selling points marketers like to emphasize.

It's quite a vision, but few faculty buy it. Most faculty cannot imagine simply reproducing the learning experience of a face-to-face classroom online (Farber, 1998). Distance learning, like it or not, is a paradigm change, a change, many faculty fear, for the worse. Faculty skepticism is of course due in part to resistance to innovation and fear of change, as administrators charge. But they are, after all, the professionals and know something about the difficulties and opportunities of conventional classroom teaching. They have reasons to doubt that an item by item electronic replacement of their classroom is possible.

Faculty consistently anticipate specific losses with respect to face-to-face teaching in an electronic classroom. How, they ask, can one duplicate the learning experience of a highly interactive classroom on an electronic network, and how can one reproduce the wealth of informal human contacts that add so much to education on a campus? How can the intense moments of human interaction which mark our memories and our lives ever occur in a sterile electronic environment experienced in the isolation of the home? Students confirm what faculty suspect, that they are poor TV performers, that it is boring to watch them on the little screen. And both faculty and students complain that computer programs that are supposed to replace specific teaching tasks, such as guiding students through exercises, are often difficult to use or even incomprehensible.

On the other hand, faculty detect continuity in administration enthusiasm for cost-cutting at the expense of traditional educational roles and values. Between 1970 and 1995, the number of full-time faculty increased by about half, while over the same period part-time faculty grew by two and one half times. By 2000, part-timers have overtaken full-timers on college campuses. This worrying trend parallels the growth of the nontraditional or returning student population, which now constitutes the majority of students in higher education.

These students require different course schedules than the traditional ones to which faculty are attached. Largely because of this, adult education has developed outside the standard academic departments and procedures under direct administrative control. As a result, a vast parallel system of higher education has emerged in which faculty have low status and little power. Since it serves adult learners, precisely the students most likely to be open to distance learning, this parallel system has a free hand to experiment even if traditional universities resist.

These trends set a precedent for administration strategies in the late 1990s when the issue of distance learning was on the agenda at prestigious universities in a big way. A straight route down the information superhighway led from the deprofessionalization to the deskilling of higher education. The replacement of full-time by part-time faculty was to be merely the opening act in the plan to replace the faculty as such by CD-ROMs. A new economic model of education was sold under the guise of a new technological model. This is the route to what David Noble calls "digital diploma mills." Understandably, this is not a route many faculty wish to travel.

The question of distance learning technology

I believe there are two closely linked problems here. First, the source of innovation shifted from faculty to administration; and second, the nature of the innovation shifted as well, from text to video based communication. In what follows I will attempt to explain this linkage between actors and their preferred technological designs.

When faculty were lonely champions of the new distance learning technologies, their primary goal was pedagogical success. They had few resources and relied on inexpensive technologies such as e-mail. They were engaged through their vocation as teachers, their commitment to finding new and exciting ways to transmit knowledge and culture. Their principal allies were students interested in playing with computers, and occasionally companies willing to donate equipment. This was a world of tentative experiments in which the stakes were small and near-term expectations low.

The later administration dominated phase of the development of distance learning is very different. Now it's all about efficiency and, ultimately, money. And there is plenty of it for high tech approaches to education, if not to staff the French department. Contrary to the popular impression that the academic world is poor, American universities in fact spend about $200 billion a year, many times the revenue of the movie business (Oberg, 1998). Administrators command these resources and corporations know it.

Huge sums are involved in the purchase of elaborate networks. Corporations are major players and find a ready audience for their most expensive technologies among administrators. Big investments in technology today are supposed to pay off in savings on facilities and salaries tomorrow, although the details remain fuzzy. Pedagogical objectives take the back seat to prestige and budgetary ones. Faculty and students are not allies but obstacles to be swept along by the inevitable momentum of progress.

The shift from faculty to administration centered innovation is more than a shift in actors and their allies. It is also a shift in what might be called spontaneous philosophies of technology. By this I mean that administrators typically have a different vision of technology and what it can accomplish than faculty.

Perhaps this is due to the influence of corporations. Salesmen seem often to have the ear of administrators in a way faculty do not, and they use their access to sell not just devices but also the content/process distinction which gives plausibility to their claim to be able to revolutionize something called educational "delivery" without much attention to faculty insights into teaching. The faculty possess the "content" and supposedly the technology is there to "deliver" the existing classroom experience online. From the standpoint of this dubious doctrine, it seems natural to suggest that new tools be used simply to reproduce the classroom experience or better still, to automate its elements and deliver it as a package.

The aim of reproducing or automating the classroom feeds directly into a preference for video, which seems to offer the closest equivalent to the classroom experience. If administrators believe that, they may buy these expensive tools in the expectation that faculty will be able simply to pick them up and use them. This is naive: in the business world training employees to use new equipment is often more expensive than the hardware itself. But in fact universities do not seem anxious to make the enormous expenditures on adaptation and training that typically accompany the acquisition of complex new computer systems in the business world. A perverse fascination with capital investment seems to be involved.

Faculty, when they actually engage with the new teaching technology, sense immediately that it is not mature, that electronics is not "there" yet as a ready tool. In the actual experience of online education, technology is not a predefined thing at all, but an environment, an empty space faculty must inhabit and enliven. They have a craft relation to the technology rather than a development strategy. They try to get the feel of it and figure out how to animate it.

This difference is reflected in different technological emphases. While it would be nice to be a "star" professor in an automated virtual class, most faculty do not aspire to that exalted status. Video, with its complicated and intimidating apparatus, holds little attraction for either teachers or students. Talking heads just are not very entertaining. Of course this may change as high-speed access over the Internet becomes commonplace, but we are many years away from achieving this in campus settings much less in the home. And multicasting still poses technical problems faculty and students should not have to deal with. To the extent that they are interested at all today, most faculty appreciate the graphical capabilities of computers in a different connection, as aids to presenting information and for exercises in computer labs. But these applications are better compared to textbooks than to classrooms; they are supplements to, rather than replacements for, classroom teaching.

Although neither video conferencing nor automated learning have caught on with faculty, there is a long history of enthusiasm among at least a small group of them for interactive text based applications such as computer conferencing. These experiences go back to a time when there were no more elaborate alternatives; it is widely assumed that the introduction of image and sound renders earlier approaches obsolete. But that is a mistake. The latest equipment is not always the best for the task. Could it be that our earliest experiences with computer conferencing were not merely constrained by the primitive equipment then available, but also revealed something important about electronically mediated education? I believe this to be the case. Even after all these years the exciting online pedagogical experiences still involve human interactions and for the most part these continue to be text based.

But here is the rub: interactive text based applications lack the pizzazz of video alternatives and cannot promise automation, nor can they be packaged and sold. On the contrary, they are labor intensive and will probably not cut costs very much. Hence the lack of interest from corporations and administrators, and the eclipse of these technological options by far more expensive ones at the end of the 1990s. But unlike the fancy alternatives, interactive text based systems actually accomplish legitimate pedagogical objectives faculty can recognize and respect. There are good reasons for this.

Considered as an environment, the world of online interaction has properties that determine its appropriate use. Just as a concert hall is a space appropriate for different activities than a living room, so the electronically mediated spaces of computer networks are also suited to specific activities. It would of course be possible to conduct a class in a restaurant, or dine on a basketball court, but the results would likely be disappointing. Similar abuse of the online environment will also yield disappointing outcomes. But this is precisely what happens when attempts are made to reproduce a face-to-face classroom online.

The basic fact about computer networks is scarcity of bandwidth. Even with all the recent advances, we are far from being able to reproduce the actual experience of human proximity in space. Indeed, it is hard to imagine in what that would consist. What kind of network would make it possible to bump into someone on the way into class and make a new friend, to carry on a heated discussion after the end of the hour, to catch the professor's eye and exchange an instantaneous glance in which boredom or alertness is tacitly expressed?

On the other hand, we have a well-established method for communicating in a narrow bandwidth. It's called writing. And we have a rich experience of using writing to overcome the limitations of bandwidth. Writing is thus not a poor substitute for physical presence and speech, but another fundamental medium of expression with its own properties and powers. It is not impersonal, as is sometimes supposed. We know how to present ourselves as persons through writing; this is what correspondence is all about. And e-mail and blogging have introduced this technique of communication to a whole new generation. Nor is it harder to write about ideas than to talk about them; most people can formulate difficult ideas more easily in written form than in speech in front of an audience.

These considerations on writing hold the key to online education. The online environment is essentially a space for written interaction. This is its limitation and also its potential. Electronic networks should be appropriated by educational institutions with this in mind, and not turned into poor copies of the face-to-face classroom which they cannot adequately reproduce.

While interactive writing is the basic medium of expression on networks, in recent years we have learned to enhance the network experience with sound and image, and that is fine. We can expect these enhancements to develop gradually and perhaps someday to change the nature of online education. But for many years to come, writing will continue to be the basic medium of online expression, the skeleton around which other technologies and experiences must be organized to build a viable learning environment.

In online education as in the classroom, we must be careful to distinguish the basic medium from the enhancements and not to confuse their roles. Speech is the basic medium in the classroom, and we supplement it with labs, movies, slides, textbooks, computer demonstrations, and so on. Similar enhancements to the written medium are possible on networks. But confusing the medium with the supplementary enhancements leads to the pedagogical absurdity of teacherless education.

To replace online written interaction with the enhancements makes no more sense than to replace the teacher in the face-to-face classroom with labs, movies, slides, textbooks, and computer demonstrations. That was tried with educational television and computer-aided instruction long ago with no success.

What does this say about the ambition to replace campuses with virtual universities? Large markets for distance learning have emerged, and this is a blessing for many students who cannot attend college classes. But if we cut higher education loose from the traditional university and its values, the blessing will turn into a disaster. The best way to maintain the connection is through insuring that distance learning is "delivered" not just by computers but by living teachers, fully qualified to teach and interested in doing so online.

Then prepackaged materials will be seen to replace not the teacher as a mentor and guide but the lecture and the textbook. Interaction with the professor will continue to be the centerpiece of education, no matter what the medium. And of course for most people that interaction will continue to take place on campus if they have the means and the mobility to attend a college.

Conclusions

Let me summarize now the conclusions I drawn from these reflections.

First, administrators and businessmen should forget the idea that distance learning systems based on videoconferencing or star professors will replace face-to-face classroom education. The dream of automating the educational process has failed so often in the past that there is little reason to take it seriously on this, the nth round.

Second, politicians need to be realistic about the future costs of higher education. Distance learning is not going to be a cheap replacement for campuses. Some other solution to the parking problem will have to be found. The campus experience will remain in demand for the foreseeable future.

Third, the overselling of foolish ideas about technology should not be allowed to discredit the whole field of online education. We as faculty need to get beyond defensive contempt for this significant educational innovation and look at specific designs with legitimate pedagogical objectives in mind.

Fourth, the educational technologists themselves need to continue to work creatively with faculty and students to devise truly viable applications that fulfill real needs (Wilson, 1999). There are good reasons for sticking with interactive text based systems and supplementing them with visual and other online resources, rather than attempting to duplicate face-to-face education online. The design challenge of improving the original text based systems is well worth pursuing.

Fifth, we must give serious thought to the implications of student diversity. The influx of returning students over the past 25 years has had major benefits for many people who missed the opportunity to finish their schooling in adolescence. New educational formats have been developed that are more appropriate for working adults than the traditional residential college teaching schedule. But these innovations have gone along with a devastating deprofessionalization that has gutted the occupation of university professor of security and respect for approximately half of all current faculty. The idea that distance learning can now deskill the already half deprofessionalized profession is deeply offensive to faculty and out of touch with the best current thinking about how to employ advanced technology (Feenberg, 1999).

However, negativism is not enough. The faculty's failure to demand the right and privilege of teaching returning students, to innovate new formats appropriate to their needs, and to exercise control of their education has led to the current situation. The systematic rejection of online education by some faculty has had no effect on the deprofessionalization of higher education. The dream of automation under cover of which this process goes forward deserves criticism, of course, but that should not become an alibi for ignoring real dangers and opportunities. The faculty must accept the responsibility for shaping distance learning, and in the process, it should also attempt to reclaim ground lost in the development of programs for returning students.

Notes

*An earlier version was published in the winter 1999 issue of *Crosstalk*. The author and editors are grateful for being given permission to reprint it here.

References

Agre, Philip. "The Distances of Education: Defining the Role of Information Technology in the University," *Academe*, September vol. 85, no. 5 (1998): 37–41.

Coopers & Lybrand. "The Transformation of Higher Education in the Digital Age," report based on the Learning Partnership Roundtable, Aspen Institute, Maryland, July 1997.

Farber, Jerry. "The Third Circle: On Education and Distance Learning," *Sociological Perspectives*, vol. 41, no. 4 (1998): 797–814.

Feenberg, Andrew. "Building a Global Network: The WBSI Experience," in L. Harasim (ed.), *Global Networks: Computerizing the International Community*. Cambridge, MA: MIT Press, 1993.

Feenberg, Andrew. "Wither Educational Technology?" *Peer Review*, September, 1999: 4–7.

Foltz, P. W. "Latent Semantic Analysis for Text-Based Research," *Behavior Research Methods, Instruments and Computers*, vol. 28, no. 2 (1996): 197–202.

Hamilton, Edward and Andrew Feenberg. "The Technical Codes of Online Education," *Techné, Journal of the Society for Philosophy and Technology*, vol. 9, no. 1(2005): 94–123.

Noble, David (1997). "Digital Diploma Mills: The Automation of Higher Education," http://classweb.moorhead.msus.edu/teach/noble.htm.
Oberg, Christopher. "The Price of Information Technology in the Future of Higher Education," Digital Diploma Mills Conference, Harvey Mudd College, 1998.
Wilson, Brent "Adoption of Learning Technologies: Toward New Frameworks for Understanding the Link Between Design and Use," *Educational Technology*, vol. 39, no. 1 (1999): 12–16.

Index

Note: Page numbers in *italics* indicate figures; those with a *t* indicate tables